Introduction to VLSI Technology

Introduction to VLSI Technology

Dr T. E. Price

Prentice Hall

New York London Toronto Sydney Tokyo Singapore

First published 1994 by
Prentice Hall International (UK) Ltd
Campus 400, Maylands Avenue
Hemel Hempstead
Hertfordshire, HP2 7EZ
A division of
Simon & Schuster International Group

Typeset in 10/12 pt Times
by Vision Typesetting, Manchester

Printed and bound in Great Britain by
Redwood Books, Trowbridge, Wiltshire

Library of Congress Cataloging-in-Publication Data

Price, T. E.
Introduction to VLSI Technology / T.E. Price.
p. cm.
Includes index.
ISBN 0-13-500422-5
1. Integrated circuits—Very large scale integration—Design and construction.
I. Title.
TK7874.P736 1994 93-29795
621.39′5—dc20 CIP

British Library Cataloguing in Publication Data

A catalogue record for this book is available from the British Library

ISBN 0-13-500422-5

---

1 2 3 4 5 98 97 96 95 94

Contents

Preface

There is an ever increasing demand for large and complex integrated circuits. Each new generation of microprocessor results in more functions, more memory, higher speed and usually no increase in cost. Memory chips of 1 Mbyte and 4 Mbyte are now in routine production and 16 Mbyte chips are being developed with 64 Mbyte and 256 Mbyte chips being planned. For these developments to continue there is an increasing need for electronic engineers and scientists to be familiar with the design and manufacture of integrated circuits. This will ensure that they have the opportunity to take part in this exciting technology.

This book is concerned with both the technology and the design of integrated circuits, but at a level that can be understood by undergraduates. It has been developed from the author's notes prepared for a course on microelectronics in the final year of a degree course. A basic understanding of solid-state physics and electronics is assumed, as would be expected for the majority of degree courses in physical electronics or electronic engineering. There are three themes, as follows:

1. technology
2. devices
3. circuits

The author has adopted the philosophy that designers of integrated circuits should also have an understanding of the manufacturing technology and the basic physics of semiconductor devices. The opposite is also true: process engineers need to understand the physics of the devices and the operation of the integrated circuits. The text is not intended as a reference book, although practising engineers in either design or production would find it useful in providing an overview of the complete process from fabrication to design. For more detailed information the author makes extensive use of *VLSI Technology* by S. M. Sze and *Physics of Semiconductor Devices*, also by Sze.

Computer Aided Design is now very advanced and is widely available for the simulation of analog and digital circuits, with SPICE (or its derivatives) being almost universally accepted for the simulation of individual circuit elements. For process simulation there is SUPREM and for device simulation there is MINIMOS and

PISCES, plus many other programs. Mention is made of the use of these programs, with particular reference to the use of SUPREM and SPICE, some versions of which are in the public domain and can be obtained for a nominal fee from Stanford University. The author has made extensive use of both programs (SUPREM II, SUPREM III, SUPREM IV, HSPICE, PSpice) to increase student motivation and their understanding of the process technology and device design and operation. There are also many programs for the generation of the geometrical layout of integrated circuits, particularly MOS circuits, and the author has made extensive use of PHASE for the design and simulation of small MOS-based circuits (other programs are provided by Mentor, Cadence and ES2, to mention only a few).

A number of problems are provided for many chapters in order to give some practice in the numerical values involved, and to allow students to work on their own, numerical solutions are provided, together with a Solution Guide at the end of the book. An Appendix of the more important equations and graphs is also provided. The author has included a simplified version of this Appendix with the examination questions that are used to assess this subject at the end of the semester or year. The questions can then be directed at the student's understanding of the content of the course rather than his/her knowledge of a large number of formulae. A number of design exercises have also been used based on process and circuit simulators to assess both skill and understanding.

The author would like to thank the many students who have unknowingly, through their responses and actions, shaped the contents of this book. Thanks are also due to the many industrialist friends who over the years have allowed the author to visit their plants and to discuss with them their varied projects in microelectronics. The author would also like to acknowledge Technology Modeling Associates and Silvaco for allowing him to use their process and device simulators.

T. E. Price, 1993

Introduction

This book is concerned with the technology of the integrated circuit, from the processes which are used to manufacture the circuits, to the operation of the individual devices which form the circuits, through to the basic concepts of designing integrated circuits. There has been a tendency to separate the three activities, but with the increasing complexity of integrated circuits the three are becoming much more interdependent. The process and device engineers need to know something about circuits in order to understand the needs of the circuit designer, and, equally, the circuit designers need to know something about the limitations of the manufacturing processes and the devices.

The technology of semiconductor devices and integrated circuits has progressed rapidly from the discrete point-contact and alloy transistors of the 1950s, to the small-scale integrated circuits of the 1960s and to the large-scale integrated circuits of the 1970s, and the very large-scale circuits of the present. The early solid-state devices used germanium because it was relatively simple to produce the single-crystal material which was required for the operation of the diodes and transistors. Silicon had many more desirable properties but was much more difficult to manufacture because of its higher melting point. However, once the problems had been overcome, the rate of development of silicon devices and silicon integrated circuits was rapid.

Electronics, and particularly microelectronics, has become all-pervading and appears in industry and commerce, in our homes, hospitals, offices, banks and even children's toys. Many of the integrated circuits are now regarded as simple components similar to discrete transistors, resistors and capacitors, and are mass produced by many different companies. The skill of the circuit designer lies in assembling these components into marketable products. However, the semiconductor industry has now matured to the extent that the electronic circuit requirements of a complete product can be incorporated into a single very large-scale integrated circuit. The implications of this trend are that circuit design of these very complex circuits must devolve to equipment manufacturers because the semiconductor manufacturers could not possibly employ enough circuit designers to cater for the ever growing demand for new products. For this to happen, circuit designers of the future need to be aware of the manufacturing processes if they are to take full advantage of the opportunities made possible by the developments in the technology. Equally important is the need for the process engineers to have a basic understanding of the operation of circuits. Microelectronics requires the circuit designer to be familiar with the manufacturing process, and the technologist to be familiar with circuits.

This text is offered as an introduction to the technology of very large-scale integration. It does not attempt to report on the very latest developments but, rather, to provide a good overview of those manufacturing processes which are most widely used. It also provides information on the operation of bipolar and field effect devices and it deals with the design and operation of a number of simple circuits which form the basis of the more complex very large integrated circuits.

The text is written primarily for undergraduates, but it would also be of value for short specialist industrial courses for engineers working in the semiconductor industry or for equipment manufacturers.

Chapters 1–9 deal with the manufacturing process and the quality and reliability aspects of VLSI manufacture. They cover the basic properties of semiconductors, oxidation, diffusion, ion-beam implantation, photolithography and metallization, and also failure mechanisms and quality control. Chapter 10 provides an overview of VLSI devices and circuits. Chapters 11, 12 and 13 deal with bipolar devices and simple bipolar circuits for both digital and analog applications. Chapters 14–17 deal with the metal-oxide–semiconductor devices, non-volatile memories and simple digital circuits in both the n-channel and complementary MOS technologies.

The author has not attempted to cover the detailed mathematics associated with, for example, diffusion or diffusion combined with oxidation, or the detailed analysis of circuits, preferring instead to use computer simulators wherever possible. There is a chapter on process simulation using SUPREM and early versions of this software are available for a nominal fee from Stanford University. Similarly, the author makes extensive use of SPICE for circuit simulation of both analog and digital circuits. (In the case of digital circuits the simulations are restricted to single gates.) Device simulators are also now becoming available with MINIMOS for MOS devices and PISCES for both bipolar and MOS devices. Other software packages can be used to produce layouts for MOS circuits which can then be simulated with the built-in simulator. Many universities and colleges now have such packages.

If simulators are to be used, and it is hoped that they would be, then it is essential to provide adequate assistance in the form of clear instructions and simple demonstrations which can then be developed into actual exercises. The problem with many CAD programs is the long learning curve for the different sets of instructions which are required to operate the programs. This time can be greatly reduced with carefully prepared demonstration programs, or data files, which can be added to and modified by the student as part of an assignment.

The author has presented much of this material in a one-year (three-term) course on Microelectronics. It could also be presented as a one-semester course. It assumes a basic understanding of circuit theory and semiconductor device physics and, therefore, should be placed towards the end of a degree programme. The author's impression is that the students have enjoyed the course and it has been pleasing to note that a number are now employed in the semiconductor industry.

Fabrication overview

Introduction

The fabrication of integrated circuits is now a well-established process, and even though developments are still taking place, the basic processes are not changing greatly. The main semiconductor is silicon and, while gallium arsenide is being used increasingly for very high-frequency applications, the advances that are continuously taking place in silicon technology mean that the switching speed of silicon devices continues to improve and to challenge the supremacy of gallium arsenide.

The production of silicon is usually separate from the manufacture of integrated circuits, and the majority of IC manufacturers purchase the silicon in the form of wafers some 100–200 mm in diameter. These wafers are approximately 1 mm thick, are highly polished and are ready for processing.

The manufacturing process consists of creating layers either in or on the silicon wafers. These layers are patterned into simple rectangular shapes by means of photographic processes. The shapes form the transistors, diodes, resistors and capacitors that form the integrated circuit. A complete integrated circuit may measure 10 mm × 10 mm, but within its boundaries there may be a million or more transistors.

This introductory chapter provides a simple overview of the manufacturing processes which are described in greater detail in the following chapters.

Silicon wafers

The IC manufacturer purchases silicon wafers according to the type and amount of impurity that they contain. Thus a wafer may be n- or p-type depending on whether the silicon has been doped with n- or p-type impurities during the production process. The most commonly used impurities for silicon are boron for p-type wafers, and phosphorus or arsenic for n-type. The impurity is distributed uniformly throughout the thickness of the wafer and the amount determines the resistivity of the silicon. The more impurity the lower the resistivity, and vice-versa. Typical values may be in the range 0.1–10.0 ohm cm.

For some applications it may be desirable to produce a wafer that has a thin layer of silicon on one surface with a different resistivity value, and possibly a different impurity type, from that of the substrate, as shown in Figure 1.

The substrate may be 0.5 mm or 500 μm (1μm = 10^{-6} m) thick, while the layer may be 1–20 μm thick. The resistivity may be higher or lower than that of the substrate and the impurity type may also be different. Thus an n-type layer may be formed on an n^+ substrate where the n^+ represents a low resistivity, or high impurity concentration; or an n-type layer may be formed on a p-type substrate.

The layers are produced by a process of epitaxy which takes place at a high temperature (900–1100 °C) in an enclosed chamber. A gas containing silicon is passed over the heated wafers and some of the atoms from the gas attach themselves to the surface atoms of the wafer, align themselves with the existing crystal structure, and grow into a layer that has the same physical structure as the substrate. The appropriate impurity is introduced into the gas and a small number of the impurity atoms also attach themselves to the silicon to control the resistivity of the layer.

The epitaxial process may be carried out by the manufacturer of the wafers, or alternatively, because it is an important step in the manufacture of integrated circuits, it may also form part of the IC manufacturer's production process.

Dielectric layers

Integrated circuits are formed from layers of n- and p-type silicon, together with layers of dielectric material, and metallic layers for electrical interconnections. A particular advantage of silicon is that its oxide (silicon dioxide) is a very good dieletric with a low dieletric constant (~4), and high breakdown voltage. It is also impervious to the ingress of moisture and it acts as a barrier to the diffusion of impurities.

Silicon dioxide is formed by heating the silicon wafer in an atmosphere that contains either oxygen or water vapour. This takes place in a furnace inside a quartz tube at temperatures of between 1100 °C and 1200 °C and may take a few minutes to

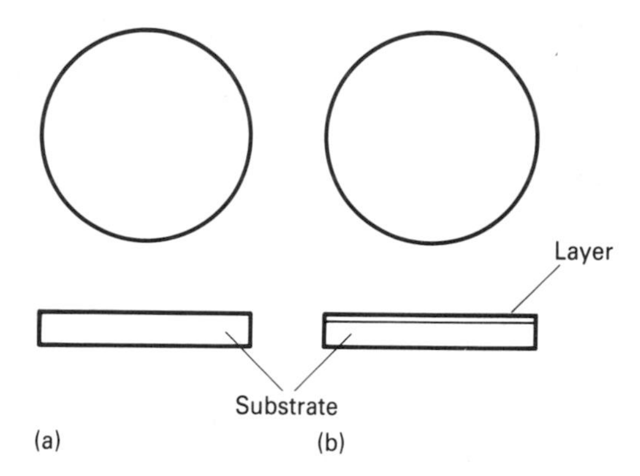

Figure 1 Silicon wafer (a) without a layer and (b) with an epitaxial layer.

many hours depending on the thickness of oxide that is required. Oxide thickness may range from 50 nm (1 nm = 10^{-9} m), for the gate oxides for MOS transistors, to 1.5 μm for oxide isolation regions where the increased thickness reduces interelectrode capacitance between the metal interconnections and the silicon substrate, which forms a ground plane.

An important feature of silicon oxide (silicon dioxide – SiO_2 – is usually referred to as silicon oxide) is the ability of the oxide to act as a mask to the introduction or diffusion of impurities into the silicon. A silicon wafer is first covered with a layer of oxide. Photographic methods are then used to cut (etch) holes in the oxide. The impurity is then introduced and wherever oxide covers the silicon the impurity is prevented from entering the silicon. This is shown diagrammatically in Figure 2.

The silicon wafer is first covered with a layer of oxide, as shown in Figure 2(a), and holes are etched in the oxide, as shown in Figure 2(b). The impurity is introduced through the openings in the oxide as a shallow layer, as shown in Figure 2(c), which is then driven into the wafer at a high temperature. An important feature of the drive-in is that the oxide in the opening is regrown, so that after drive-in, the surface of the wafer is again fully covered by oxide. In addition to driving the impurity deeper into the wafer, the impurity also moves laterally so that the junction between the original material in

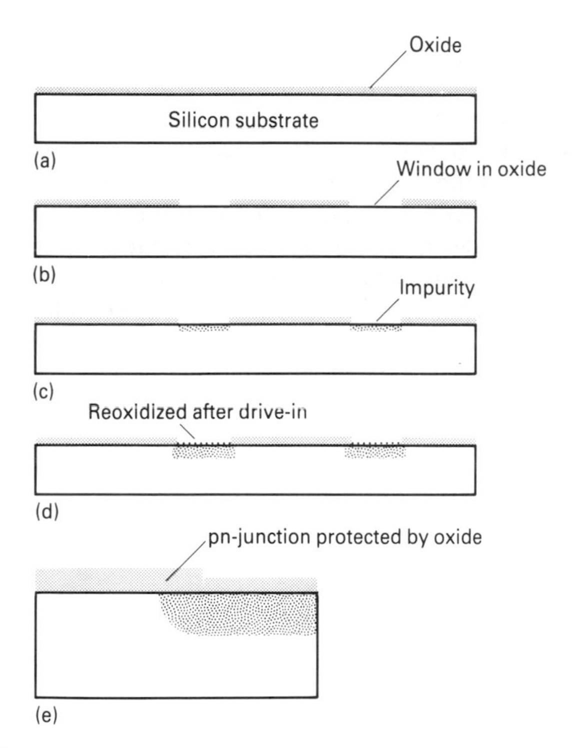

Figure 2 Formation of impurity regions in a silicon wafer with the aid of an oxide mask.

the substrate and the impurity introduced into the oxide window is protected by the original oxide. This is important for ensuring very low leakage currents when pn-junctions are reverse-biased.

The process of growing oxide, opening windows in the oxide and adding impurities is repeated many times to create the complex multi-layer structures that are required for an integrated circuit.

There are other types of dielectric layers, for example silicon oxide can be deposited rather than grown. In this form it can be used to cover metal tracks that are used to interconnect the different components in the circuit. The oxide coating protects the tracks from impurities that may be present in the plastic packaging materials, and as a result prolongs the life of the integrated circuit.

Another form of dielectric is silicon nitride. This material is also deposited rather than grown; it has the property of being very inert and does not react with the underlying silicon. It is used to form a mask on the wafer against oxidation in an oxidizing environment. Silicon nitride is used to produce integrated circuits that use oxide isolation to provide electrical isolation between the different components.

Semiconducting layers

Different resistivity layers can be formed in silicon by diffusion. The important point is that the added impurity must be present in a greater concentration than the original impurity. Thus a uniformly doped p-type wafer may have an impurity concentration of 1×10^{16} atoms per cm^3, to which may be added 1×10^{16} atoms per cm^3 of n-type impurity. This impurity is allowed to diffuse into the wafer at a high temperature and as it moves into the wafer its concentration reduces until it reaches 1×10^{16} atoms per cm^3. At this point the wafer reverts back to its original p-type form. The profile of impurity concentration versus distance into the wafer is shown in Figure 3.

With a diffused layer the thickness of the wafer is not affected. However, if an epitaxial layer is grown on the surface of the wafer the thickness is increased. The difference with an epitaxial layer is that the impurity concentration of the layer can be

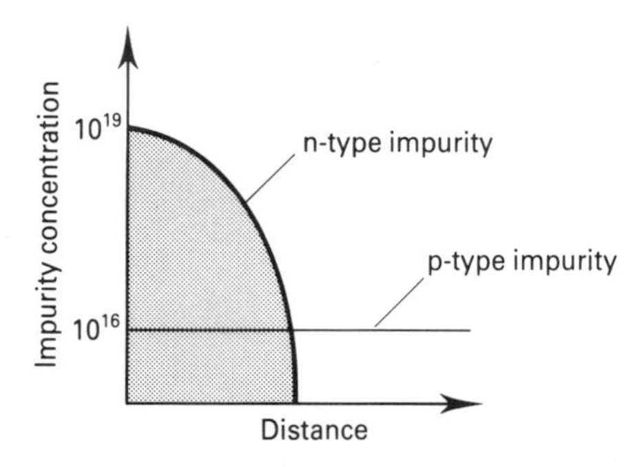

Figure 3 Variation of a diffused layer in a silicon wafer.

less than that of the original substrate, and an n-type layer with a concentration of 1×10^{16} atoms per cm^3 can be grown on a substrate with a carrier concentration of 1×10^{18}atoms per cm^3. The junction between the two layers is very abrupt, as shown in Figure 4.

Another form of silicon layer is polycrystalline silicon. This occurs when silicon is grown on an insulating substrate, for example a silicon wafer covered with silicon oxide. The silicon atoms from the gaseous silicon compound which is passed over the heated wafer are unable to align with other silicon atoms and as a result form random clusters of single-crystal silicon, or polycrystalline silicon. When the layers of polysilicon are doped with impurity to lower the resistivity, they can be used as conductors. The vast majority of integrated circuits that use MOS transistors use a combination of polysilicon and metal for the conductors.

Silicon can also be grown on other types of single-crystal substrate, such as sapphire (or aluminium oxide). Sapphire is an insulator, and when integrated circuits are formed in the silicon films that are grown on it, the electrical isolation between the different components is much greater than that in conventional circuits. The frequency performance of silicon-on-insulator circuits is considerably better than that of conventional circuits. These circuits are also more able to withstand the effects of radiation that might occur in outer space in satellite communication systems.

Diffusion

Mention has already been made of the process of diffusion as a means of introducing impurities into a silicon wafer. It is a natural physical process that applies to the flow of heat along a metal bar, the dispersal of smoke from a fire to the mixing of two perfectly still, coloured liquids in a container. Diffusion occurs when there is a high concentration of particles in one region and a much lower concentration elsewhere. The natural random motion of the particles will result in more of them moving away from the region of high concentration. After sufficient time has elapsed, the concentration is uniform, but if the process is frozen at some point in time then the

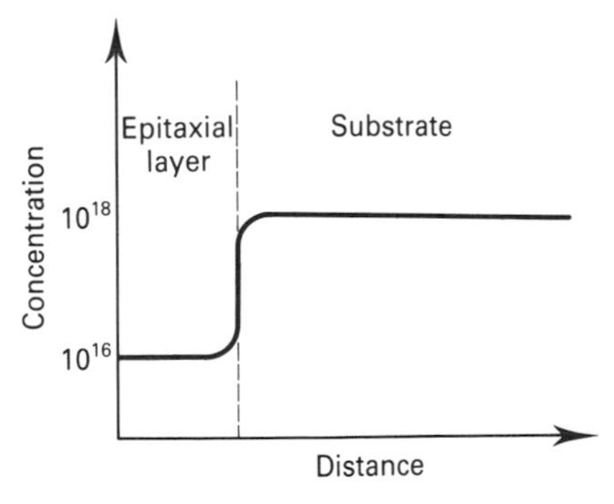

Figure 4 High-resistivity epitaxial on a low-resistivity substrate.

concentration profile that exists at that instant is retained. This is what happens in silicon.

The most commonly used impurities for silicon are arsenic and phosphorus for n-type impurities and boron for p-type impurities. These impurities can be introduced into silicon in a variety of ways and are then diffused into the silicon at high temperature (1000–1200 °C). After a certain time the wafers are removed from the source of heat and allowed to cool to room temperature, when the impurities are fixed into the crystal lattice. A typical diffusion profile is shown in Figure 3. The distance that the impurity travels into the silicon depends on the temperature and the time. The higher the temperature the easier the impurities can move through the crystal lattice, and, similarly, the longer the time then the further they travel.

The relationship between temperature, time and distance can be solved from simple physical laws that govern diffusion. Certain boundary conditions are applied depending on whether there is an unlimited supply of the impurity at the surface of the wafer or whether the supply is strictly limited. The unlimited supply results in an error-function profile, while the limited supply results in a Gaussian profile. These profiles are described in greater detail in Chapter 4.

The process becomes much more complicated when an oxide is growing at the same time as the impurities are diffusing. During the growth of the oxide silicon is consumed at the interface between the oxide and the silicon. The impurities contained in the surface layer of silicon that is being converted to oxide behave differently depending on the impurity. For boron the impurity atoms can move out of the silicon and into the oxide. When this happens the impurity concentration in the silicon changes and this change modifies the rate of diffusion into the silicon. Also, the fact that silicon is being consumed affects the impurity profile. In the case of phosphorus the impurity atoms are rejected by the oxide and as the silicon is consumed the excess phosphorus piles up in the silicon, increasing the concentration at the interface between the oxide and the silicon. These and other effects mean that simple hand calculations of impurity profiles are not very accurate except under very simplistic situations of diffusion in a non-oxidizing ambient. The effects have been modelled mathematically for computer simulations, and programs are available that can accurately model the process of diffusion under many different conditions.

To determine whether the profile is correct it is necessary to be able to measure the profile or some property of the diffused region that is affected by the profile. The simplest measurement is the resistance of the diffused layer. This is measured in a variety of ways, including the formation of test structures on the wafer that allow the resistance to be measured for a well-defined geometry, to the measurement of the resistance of an infinite layer of the diffused region. An infinite layer involves a test wafer where the impurity is diffused into the whole surface of the wafer. Metal test probes are then placed on the surface of the diffused layer and the resistance is measured. Using some mathematics it is possible to extract a value for the sheet resistance of the diffused layer.

The depth of the diffused layer usually involves sectioning the wafer, applying a stain to distinguish between the n-type and p-type regions and measuring the thickness of the layer under a high-powered microscope.

Neither of the above two measurements produces a profile, but simply a depth and a value of resistance that is dependent on the depth and the number of impurities in the diffused region. In order to measure the profile it is necessary to use more sophisticated methods. One of these requires a small piece of the wafer containing the diffused region to be bevelled at a shallow angle. Two metal probes are then placed on the silicon in the bevelled region and the current is measured. The probes are mechanically stepped down the bevel a few micrometers at a time and the current is measured at each position. A graph of current versus position is obtained which, with appropriate mathematical manipulation, is converted to a profile of impurity concentration versus depth. The second method that is commonly used involves a scanning electron microscope modified to produce a beam of ions which are directed at the sample of silicon that contains the impurity profile. The ions have to be sufficiently energetic to strike off silicon and impurity atoms from the sample. These secondary ions are then gathered and measured. The signal from the secondary ion mass spectrometer is proportional to the concentration of the impurities. A knowledge of the rate at which silicon is removed from the sample by the incident ion beam is used to create a profile.

Ion implantation

Ion implantation is increasingly being used to introduce impurities into silicon. The wafers are placed at the end of an acceleration tube in a high vacuum and ions are accelerated and directed at the wafers. Electrostatic scanning is used to deflect the beam of ions in a raster pattern in much the same manner that the beam of electrons is scanned in a TV cathode ray tube. The high velocity of the ions causes them to penetrate the surface of the silicon and come to rest a distance below the surface. A typical profile is shown in Figure 5.

The profile is Gaussian in shape and the peak occurs at a depth R_p, which is a function of the energy of the incident beam of ions. The width of the distribution dR_p is also a function of the energy. The peak value of the profile is dependent on the number

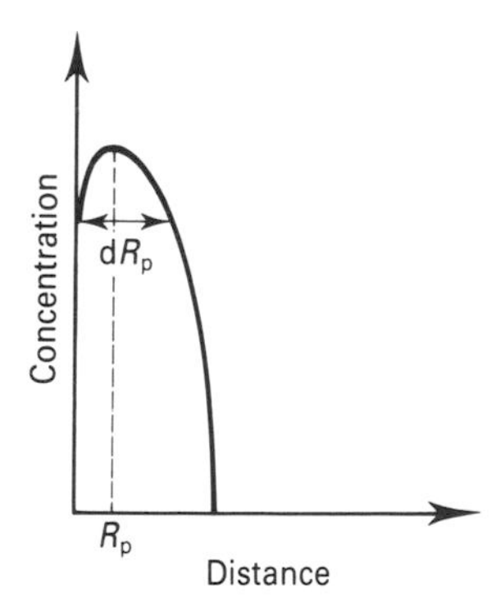

Figure 5 Typical profile formed by ion implantation.

of ions or the dose. The distance that the ions enter the silicon is usually quite small and diffusion must be used to distribute the ions to a greater depth.

When ions enter the silicon they cause considerable damage by forcing the silicon atoms apart. If the dose is sufficient then the surface layer becomes amorphous. In this state the usual electrical properties of pn-junctions are destroyed and the material behaves as a high-resistance conductor. The damage can be removed by heating the wafer for a few minutes at temperatures of 600 °C or greater. If the impurity is to be redistributed by diffusion then this process takes place automatically at the same time as the diffusion takes place. Whenever ion implantation is used to introduce impurities into silicon, then it must be annealed to remove the damage.

The advantage of ion implantation is that the number of impurities that are introduced (the dose) is controlled by simply measuring the electric current generated by the beam of ions: the greater the current, the greater the number of impurities. The total dose is obtained by integrating the current over time.

Oxides can be used to act as masks to confine the beam to certain locations where junctions are to be formed, but because the process takes place at room temperature it is also possible to use photoresists to mask the beam. The use of a photoresist mask greatly increased the flexibility of the ion implantation process to create junctions and also modify surface regions to assist control of the electrical properties of transistors, particularly the threshold voltage of MOS transistors.

Process simulators

The processes of oxidation, diffusion and ion implantation can all be described mathematically and the equations can be used to produce a computer program that can simulate the processes. One of these programs is SUPREM, which was developed at Stanford University. Version II is readily available and can be made to run on workstations such as SUN and HP/Apollo which have a Fortran compiler. Although version II has now been replaced by version III it is still perfectly adequate for evaluating the effects of temperature, time and concentration on the growth of oxides, epitaxy, ion implantation and/or the diffusion of impurities into silicon.

The data is provided in the form of a simple text file that describes the different steps of the process in the form of temperature, time, impurity type and its concentration. For an ion implantation the dose, energy and impurity are specified; for epitaxy the impurity, concentration, temperature, time and rate of growth are specified. The program uses finite difference methods to obtain the solution and to produce a graphical output showing the impurity profile or the oxide thickness.

SUPREM II and III provide one-dimensional solutions, but SUPREM IV is a two-dimensional simulator showing both transverse and lateral effects.

Access to SUPREM II or SUPREM III greatly enhances the interest and the understanding of the fabrication processes and provides the opportunity to investigate complex sequences of oxidation, epitaxy, ion implantation and diffusion as may be required to produce an npn bipolar transistor with a buried n^+ collector region.

Photolithography

The intricate patterns required to produce the different layers in the silicon are produced photographically in photosensitive resists. The original pattern is created as a two-dimensional drawing on a computer screen using a computer aided design drawing package. A different drawing is required for each layer and for a complex integrated circuit there may be ten to twenty layers. The computer data for each layer is used to control a photoplotter that reproduces the drawing for the layer on a glass photographic plate. For very-large-scale integrated circuits a block of four to eight identical images is produced, as illustrated in Figure 6.

The images in the reticle shown in Figure 6 may be × 10 larger than the actual integrated circuit. Within each image there is a complex pattern of squares, rectangles and tracks representing the pattern of shapes required to produce the pn junctions, contact windows or interconnections required for the circuit. The reticle is placed in a projection printer in much the same way as a 35 mm slide is placed in a slide projector and the image cast on to a screen. In the case of the projection printer the image is projected on to the silicon wafer, which is coated with a photosensitive resist. The resist is exposed and the wafer moved to a new position to expose another image. In this way the complete set of images is stepped and repeated on to the wafer. After exposure, the resist is developed to remove the exposed resist. In this description of the process it is assumed that a positive of the image is created on the wafer; the resist is known as a positive resist. It is also possible to use negative resists to create a negative image on the wafer. For some stages of the process a negative resist offers advantages and for others the positive resist is better.

The resist acts as a mask to a process that etches the oxide. The oxide may be etched with suitable chemicals containing hydrofluoric acid, or alternatively it may be etched in a partial vacuum with a plasma of ionized gas. Both processes result in the removal of the oxide and the exposure of the underlying silicon. After etching the oxide the photoresist is removed and the wafer, after careful cleaning, is ready for the next step of the process. The steps are summarized in Figure 7.

The photolithographic process is used many times during the manufacturing sequence to build the complex pattern of shapes for the components of the integrated circuit. Because the patterns are formed from a series of steps following one after the

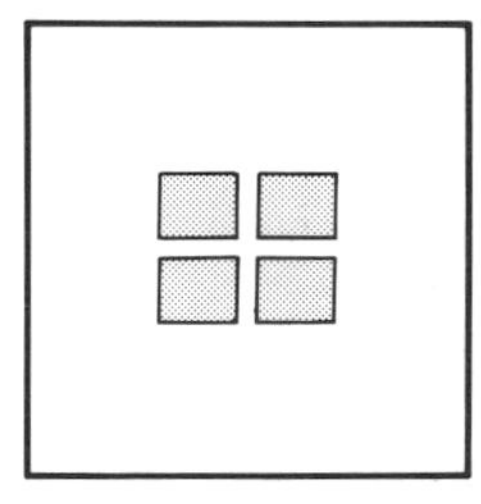

Figure 6 Illustration of a reticle for a projection printer.

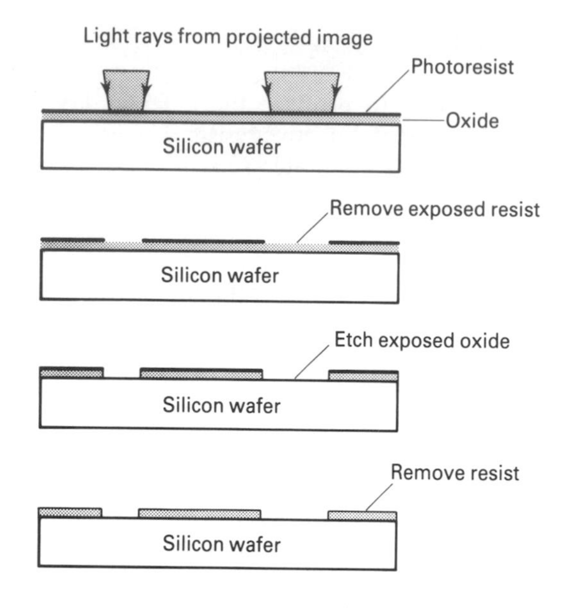

Figure 7 Stages in photolithography.

other it is important that the patterns of the individual steps align with each other. In the past this has been done by operators observing the patterns projected on to the wafer during the photolithographic process using a high-powered microscope, and then manipulating the wafer to position the patterns already on the wafer with the new images being projected on to the surface from the reticle. In more recent projection equipment, this operation is performed automatically by means of video cameras and optical recognition software that aligns fiducial marks on the wafer with similar marks in the projected image.

Ultra-violet light is used to expose the resist in order to improve the resolution and, therefore the ability to reproduce fine lines. To date, ultra-violet has been used to reproduce line widths of between 0.5 and 1.0 μm, but it becomes increasingly difficult to reproduce lines of smaller dimensions. Eximer lasers have been used and also electron beams, and these forms of exposure may extend the minimum dimension down to 0.2 μm. Below this it is likely that X-rays could be used, but this technology is still being developed.

Metallization

One of the final steps in the manufacturing process is the formation of the metal interconnections. When all the high-temperature steps have been completed and when all the silicon is once again fully covered with oxide, a photolithographic step is used to create windows in the oxide to form contacts. The complete surface of the wafer is then covered with a thin layer of metal (~1 μm) and photolithography is again used to define the metal tracks and to allow the surplus metal to be removed. For very complex circuits it may be necessary to use more than one layer of metal to form the interconnections. In this case a layer of low-temperature deposited oxide is formed, new contact windows are opened and a second layer of metal is deposited.

The metal is usually aluminium because it is very easy to deposit by a variety of means; it has high conductivity, forms good low-resistance contacts with silicon, is easy to etch and easy to make contact to via wires to the pins of the external package. The usual means of depositing aluminium is by some form of high-vacuum process involving evaporation or sputter processes. The main factors affecting the quality of the films are adhesion to the silicon wafer and uniform coverage, which must include the steep-sided steps that result from the many different oxidation steps and possible layers of polysilicon. The topography of the surface of the wafer at the metallization stage is like a mountain range with peaks and valleys that may differ in height by a factor of 10. It is important that the deposition process is able to cover the tops, bottoms and sides of these irregularities uniformly.

As the dimensions continue to shrink, the contact resistance becomes critical and the simple aluminium–silicon contact is no longer suitable. Other metals are deposited first to form a low-resistance contact with the silicon, and then the aluminium forms a contact with the second metal rather than directly with the silicon. These metals are usually the noble metals such as tungsten, molibdenum, platinum and palladium, which form alloys with the silicon, known as silicides. The silicides form very good contacts to the silicon and also provide a low-resistance contact for the aluminium. Other metal combinations are also being explored to cater for ever smaller contact areas.

CHAPTER 1

Some electrical properties of silicon

1.1 Introduction

The most important semiconductor material for VLSI (Very Large Scale Integrated) circuit manufacture is silicon. The technology is well established and while other materials are used, notably gallium arsenide, silicon predominates for the mass market of linear and digital integrated circuits. Before considering how the integrated circuits are manufactured it is worth while considering some of the basic properties of silicon. Materials which are used in the semiconductor industry are loosely classified according to their conductivity into semiconductors, insulators and metals. Silicon is a semiconductor, silicon oxide is an insulator and aluminium, which is extensively used in silicon integrated circuits, is a metal. All of these materials are important for the manufacture of integrated circuits.

1.2 Band model

The atoms of all elements comprise a central nucleus, which is positively charged, surrounded by negatively charged electrons. The electrons occupy clearly defined energy levels which are grouped to form energy bands, and it is the nature of the outermost bands which provides a means for classifying materials into conductors, insulators and semiconductors. The electrons furthest away from the nucleus, in the outermost band, interact with similar electrons in adjacent atoms to bind the atoms together.

In a metal only a small number of outer electrons are required to bind the atoms together and the remainder are free to move between the atoms and to take part in the conduction of electrical current. For an insulator, all of the electrons in the outermost band are used to bind the atoms together and at room temperature none are available for conduction. At higher temperatures, or if the electrons can be given sufficient energy from an external source, then some of the electrons break free and small currents can flow in the insulator.

In the case of silicon there are four electrons in the outermost band and under ideal conditions all four are required to bind with the electrons of four adjacent silicon atoms. However, at any temperature above absolute zero some of the electrostatic bonds which bind the atoms together break and a small number of electrons are released and are available for conduction.

The effects described above can be illustrated by means of a simple diagram showing the position of the outermost energy bands, as in Figure 1.1, where the energy bands comprise closely spaced energy levels. The outermost band occupied by electrons associated with an atom is known as the valence band. Moving further away from the nucleus the next band is the conduction band. In order that electrons may take part in the conduction process, electrons must reach the conduction band. For a metal, the gap between valence and conduction bands is non-existent and electrons can move freely between the two. However, for insulators and semiconductors there is a well-defined gap between the top of the valence band edge and the bottom of the conduction band edge. There are no energy levels for electrons to occupy in this gap as there are in both the valence and conduction bands. For insulators, the gap is sufficiently large ($>4.0\,\text{eV}$) at room temperature to prevent electrons at the top of the valence band transferring to the energy level at the bottom of the conduction band and, therefore, conduction does not occur. For the semiconductor, on the other hand, the gap is not so great ($1.12\,\text{eV}$ for silicon) and at temperatures above absolute zero a sufficient number of electrons cross the gap to allow conduction to take place.

This simple model of energy bands will be used again in later chapters to describe the operation of semiconductor devices and it is worth while considering the model in a little more detail. For an ideal semiconductor at room temperature a certain number of electrons are assumed to be in the conduction band. Since these electrons originate in the valence band there are an equal number of vacancies in this band. These vacancies are referred to as holes. Since a hole represents a lack of an electron and since an electron is negatively charged, the hole is assumed to have a positive charge of the same value as that of the electron. Holes in the valence band can take part in conduction in the same way as can electrons in the conduction band. A simplified band diagram for a pure semiconductor is shown in Figure 1.2. Here the bands for the valence and conduction regions are replaced by lines which represent the band edges – the upper

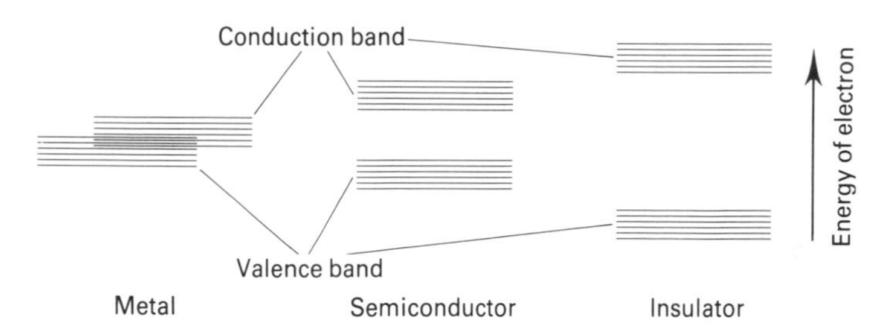

Figure 1.1 Simple band diagram model for metal, semiconductor and insulator.

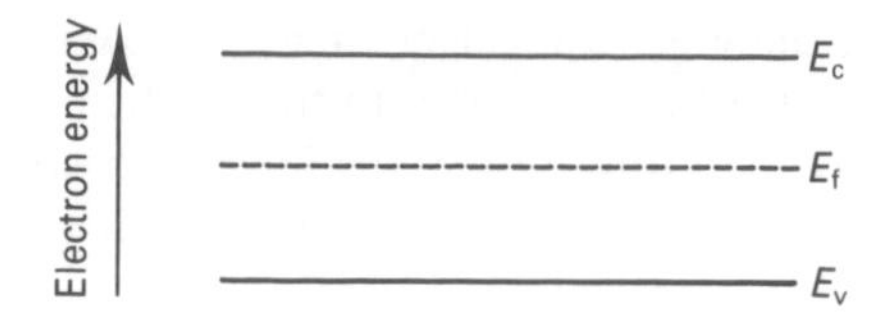

Figure 1.2 Band diagram for an intrinsic semiconductor.

edge of the valence band and the lower edge of the conduction band. The energy level for the valence band is denoted by E_v, which represents the upper energy level of this band, while for the conduction band there is an energy level E_c, which represents the lower energy level for this band. The difference between these two levels represents the forbidden energy gap E_g, where neither holes nor electrons can exist. At absolute zero the valence band is full of electrons, there are no holes and the conduction band is empty. Under these circumstances conduction cannot take place. For conduction to occur it is necessary for some of the electrons in the valence band to acquire sufficient energy to overcome the energy gap E_g in order to reach the conduction band. For each electron that leaves the valence band to transfer to the conduction band a hole is created in the valence band.

1.3 Intrinsic carrier concentration

For silicon, the energy gap is approximately 1.1 electron volts (eV) and at room temperature a sufficient number of electrons may acquire this amount of energy as a result of their random thermal motion to overcome this energy gap. As a result a relatively large number of electrons (n) will reach the conduction band with a corresponding number of holes (p) being produced in the valence band. For pure or intrinsic silicon:

$$n = p = n_i$$

where n_i is a constant at a particular temperature. At 300 K it is:

$$n_i = 1.45 \times 10^{10} \text{ per cm}^3$$

In Figure 1.2 an additional energy level E_i is shown mid-way between E_c and E_v. The position of E_i can be explained in terms of probability theory but for the simple models assumed in this text it is adequate to regard E_i as the mid-band energy level.

1.4 Extrinsic carrier concentration

An important feature of a semiconductor which makes it different from a metal is that it is very easy to change the carrier concentration by the addition of suitable dopants. For silicon, phosphorus and boron are the most commonly used dopants for this purpose.

The atomic dimensions of these elements are similar to those of silicon and they are easily incorporated into the silicon crystal structure without causing any undue mechanical stress. In the case of phosphorus there are five electrons in the valence band and when phosphorus is added to silicon, four of these electrons form bonds with adjacent silicon atoms, leaving the fifth electron very loosely coupled to the phosphorus atom. At room temperature this electron transfers to the conduction band where it can take part in the conduction process. When a dopant such as phosphorus is added to silicon it is referred to as a donor impurity and donates a carrier to the silicon. The number of electrons in the conduction band is thus increased from the intrinsic level of the pure silicon.

For a dopant such as boron there are only three electrons in the valence band and the bonding with silicon is achieved by removing an electron from the valence band of the silicon and leaving a hole. Therefore, the number of holes is increased above the intrinsic level. Boron is known as an acceptor impurity because it accepts an electron from the silicon.

Unlike the original intrinsic silicon, for which $n = p$, when a donor or acceptor is added such that n no longer equals p, the silicon is known as an extrinsic semiconductor. The number of extra electrons or holes is directly proportional to the number of impurity atoms since each impurity atom is assumed to either donate or accept one carrier (this is not true at high carrier concentrations when the number of electrically free carriers may be less than the chemical concentration of dopants). These additional carriers are known as majority carriers to distinguish them from the minority carriers associated with the intrinsic semiconductor. Under steady-state conditions a balance exists between the number of majority and minority carriers given by:

$$n_n \times p_n = (n_i)^2 = 2.1 \times 10^{20} \text{ per cm}^6 \text{ at } 300\,\text{K}$$

where n_n represents the number of majority carriers in n-type material and p_n represents the number of minority holes in n-type material. A similar relationship exists for p-type material.

For silicon doped with phosphorus there is an excess of electrons and the material is said to be n-type, while for silicon doped with boron there is an excess of holes and the material is said to be p-type. This distinction can be illustrated in the band diagram with the addition of a further energy level, as shown in Figure 1.3. The new energy level (E_f) is known as the Fermi level. For the definition of the Fermi level it is

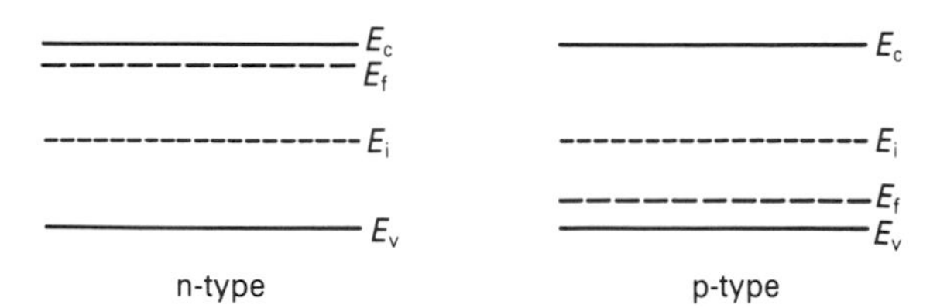

Figure 1.3 Band diagram for extrinsic semiconductors.

necessary to delve into quantum mechanics, but suffice it to say that the Fermi level represents an energy level at which the probability of finding an electron is $\frac{1}{2}$. For intrinsic silicon with an equal number of electrons and holes in the conduction and valence bands, respectively, the Fermi level is located at the centre of the bandgap. For an n-type semiconductor with a greater number of electrons in the conduction band the Fermi level is closer to the conduction band. Conversely, for p-type semiconductors the Fermi level is closer to the valence band. The position of the Fermi level is determined by the concentration of the impurities and is given by:

$$E_f - E_i = kT \ln (n_n/n_i) \text{ for n-type}$$

or

$$E_i - E_f = kT \ln (p_p/n_i) \text{ for p-type}$$

In practice, n_n and p_p are equal to N_d and N_a, respectively, where N_d is the donor dopant concentration and N_a the acceptor dopant concentration. The exact position of the Fermi level depends on the number of electrons or holes. If there are a large number of electrons then the Fermi level is near the conduction band, and, conversely, if there are a large number of holes it is near the valence band. Under certain circumstances the Fermi level is either above the conduction band or below the valence band. The material is then said to be degenerate. This condition is often required for contacts to minimize contact resistance.

The Fermi level acts as a reference level when considering the band diagram for pn junctions in diodes and transistors or the surface potential in the gate region of MOS transistors. The effect of externally applied voltages can be shown in the band diagram as a difference between the Fermi level for regions in the semiconductor with different resistivity or conductivity types.

1.5 Carrier transport

For a piece of n-type silicon the electrons are in a continuous state of motion due to thermal energy, as illustrated in Figure 1.4(a). The electrons collide with atoms and

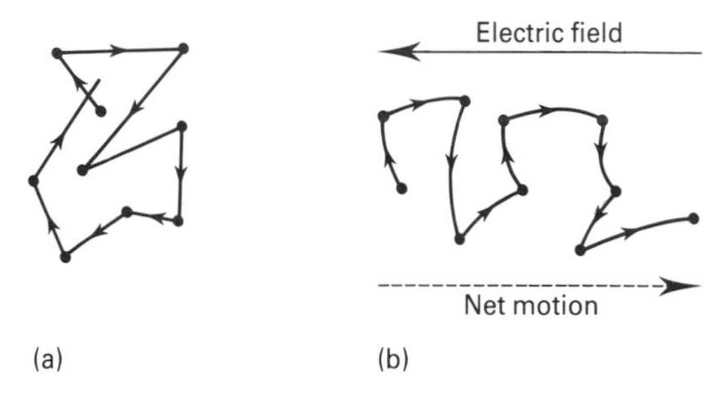

Figure 1.4 Representation of the random motion of an electron (a) with no applied electric field, (b) with an applied field.

make frequent changes of direction. This motion does not result in any overall displacement of the electrons and, therefore, there is no net transport of current. The random thermal velocity of the electrons is approximately 9.6×10^6 cm/s.

If an electric field is applied to the sample, then an additional force acts on the electrons and there is movement in a direction opposite to that of the electric field, as shown in Figure 1.4(b). In addition to the thermal velocity the electrons now have a drift velocity which is proportional to the electric field. The drift velocity is generally orders of magnitude lower than the thermal velocity.

The constant of proportionality between drift velocity and electric field is called the carrier mobility, i.e.

$$v = \mu \mathbf{E}$$

where μ has the units $\text{cm}^2\,\text{V}^{-1}\,\text{s}^{-1}$ and $\mathbf{E}$ is the electric field in V/cm.

The mobility provides a measure of the ease with which carriers move through a material. They are accelerated by the electric field but they are randomly scattered and lose the energy gained when they collide with silicon atoms (lattice scattering) and impurity atoms (impurity scattering).

Lattice scattering increases with temperature, while impurity scattering decreases with impurity concentration. The variation of mobility with total impurity concentration[1] for silicon at 300 K is shown in Figure 1.5.

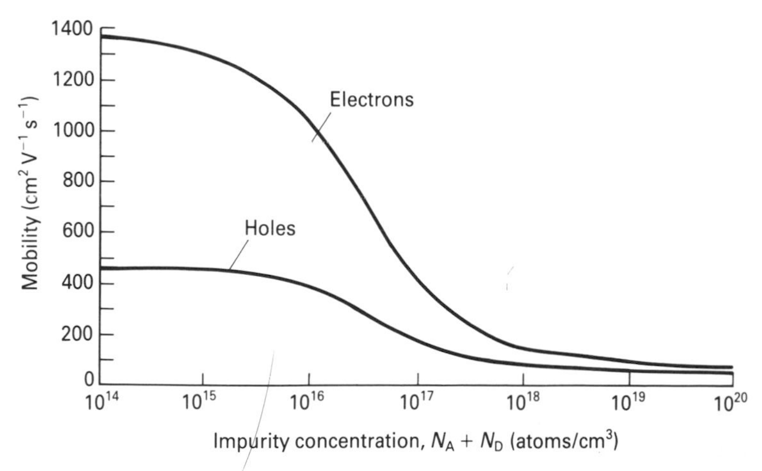

Figure 1.5 Electron and hole mobility in silicon at 300 K. [After Masetti[1] © 1983, *IEEE*.]

1.6 Resistivity

When carriers flow through a material the current density (**J**) is equal to the rate of flow of charge, i.e.

$$\mathbf{J} = nqv$$

where n is the number of carriers, q is the electronic charge and v is the drift velocity. Thus

$$\mathbf{J} = nq\mu\mathbf{E}$$

This equation can be written as:

$$\mathbf{J} = \sigma\mathbf{E}$$

where $\sigma(= nq\mu)$ is the conductivity in (ohm cm)$^{-1}$

The conductivity is the reciprocal of the resitivity ($\rho = 1/\sigma$ ohm cm). In general, when both electrons and holes are present the resitivity is given by:

$$\rho = 1/(qn\mu_n + qp\mu_p)$$

where μ_n is the electron mobility, μ_p is the hole mobility and n,p are the electron and hole concentrations.

For an extrinsic semiconductor one type of carrier predominates and the equation reduces to:

$$\rho = 1/(qn\mu_n) \text{ for n-type}$$

or

$$\rho = 1/(qp\mu_p) \text{ for p-type}$$

The variation of the resistivity with carrier concentration[2] is shown in Figure 1.6. These curves are for silicon with one impurity type. For silicon which contains both impurity types, then where the donor concentration (N_d) is greater than the acceptor concentration (N_a) the material is n-type, and the value of resistivity can be determined from the above equation by replacing n by $(N_d - N_a)$. The value for the mobility, however, depends on the total carrier concentration $(N_d + N_a)$, and can be obtained from Figure 1.5. When $N_a > N_d$ the material is p-type and $(N_a - N_d)$ is used instead of p for working out the resistivity.

1.7 Silicon wafer preparation

Silicon is used extensively in metal manufacturing industries, particularly of steel and aluminium where small quantities of silicon can greatly improve the quality of these metals. The amount of silicon used for the manufacture of semiconductors is very small in comparison. However, the silicon required for the semiconductor industry is much more refined and must be very pure, and in addition it must be in the form of a single crystal.

Single-crystal silicon is necessary for the successful operation of many semicon-

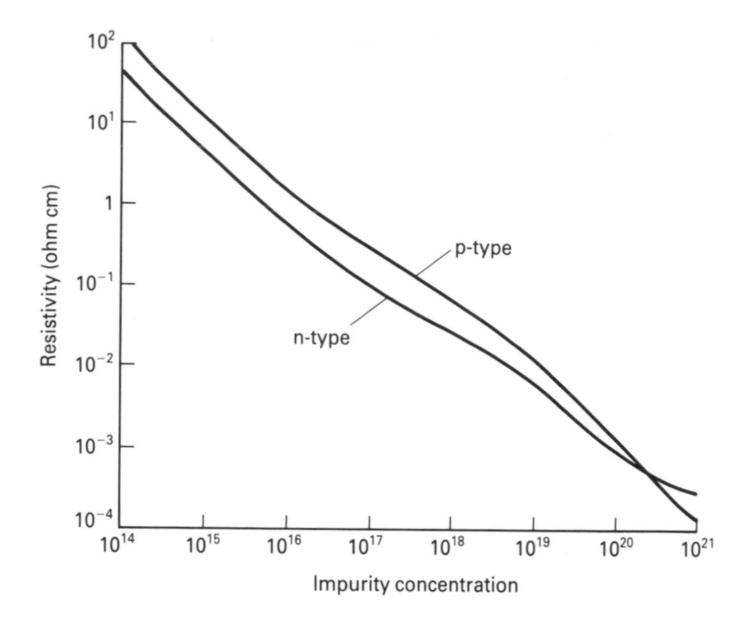

Figure 1.6 Variation of the dopant density with resistivity. [Data derived from J. C. Irvin[2] and used by permission © 1962, A T & T.]

ductor devices. Any irregularity in the arrangement of the atoms can reduce both the mobility and a property of minority carriers known as the lifetime. A regular atomic structure is achieved in single-crystal material.

It is usual to first convert the raw metallurgical silicon which is used by the metal manufacturing industries to a gas such as trichlorosilane ($SiHCl_3$), which can be purified by distillation. High-grade silicon is then obtained by high-temperature reduction of the trichlorosilane. At this stage the silicon is polycrystalline, that is, a regular atomic structure only exists over relatively small distances.

Single-crystal silicon is obtained by melting the polycrystalline material in a high-purity fused quartz crucible by either resistance or radio frequency heating. A 'seed' crystal of the required orientation is dipped into the melt (1400 °C) and slowly withdrawn (50–100 cm per hour) while at the same time the seed is rotated about its vertical axis. The rotation ensures mixing and results in more uniform growth. As the seed is withdrawn, silicon from the melt crystallizes out to form one continuous crystal having the same orientation as the seed. The diameter of the crystal is controlled by the temperature of the melt and the rate of withdrawal. The end result is a single crystal some 100–200 mm in diameter and 50–100 cm long. Impurities are added to the melt to produce silicon of the required resistivity.

The crystal is ground to the correct diameter to satisfy automatic wafer handling equipment and flats are ground along the length of the crystal corresponding to the crystal orientation and resistivity type, as shown in Figure 1.7. The crystal is then cut into wafers 0.5 mm thick for 100 mm diameter wafers, 0.7 mm for 150 mm diameter and 0.9 mm for 200 mm diameter.

The crystal orientation is determined by taking the direction of a plane through the single crystal. Because of the complex structure of the atoms within the single crystal the number of atoms in the plane of the surface through the crystal will depend on the direction taken by the plane. For semiconductors the most commonly used orientations are the ⟨111⟩ and the ⟨100⟩.

A photograph of 100 mm diameter wafers is shown in Figure 1.8. A single flat can be seen on the wafer lying on the bench which indicates a ⟨111⟩ orientation. Notice also the highly reflecting nature of the surface of the wafer which results from the combination of mechanical and chemical polishing which is used to remove surface damage caused when the wafers are sawn from the ingot. The wafers are usually transported in the fluroplastic carriers shown in order to protect the carefully cleaned and polished surfaces.

A distinguishing feature of the crystal orientation is the appearance of the surface after it has been chemically etched. For ⟨111⟩ material, microscopic etch pits are formed which are triangular in shape. The triangles are all precisely aligned with each other, with the sides being aligned with the underlying atoms. For ⟨100⟩ material the etch pits are square. In practice, the crystal orientation is determined by X-ray diffraction.

Surface damage is removed from the front surface by mechanical lapping and chemical etching to produce a flat and a highly reflective surface. The rear surface is ground to remove saw damage, but is not chemically polished. It is found that the damage left by mechanical grinding can act as a sink for unwanted impurities, particularly the heavy metals. This 'gettering' action of the rear surface can be

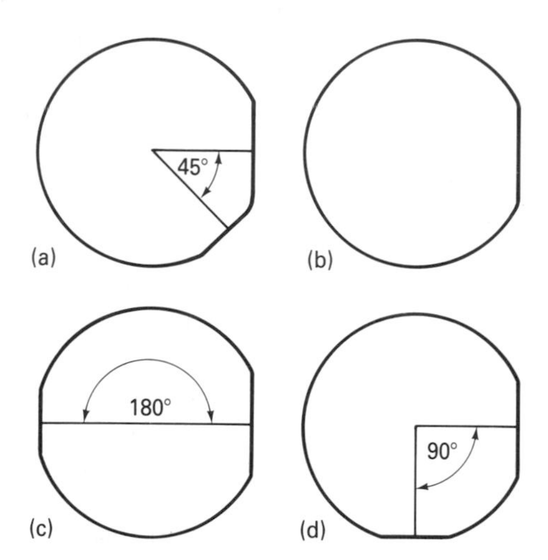

Figure 1.7 Wafer identification flats for (a) 111 n-type, (b) 111 p-type, (c) 100 n-type and (d) 100 p-type.

enhanced by special treatments such as ion implantation and rapid laser heating of the rear surface which promote gettering.

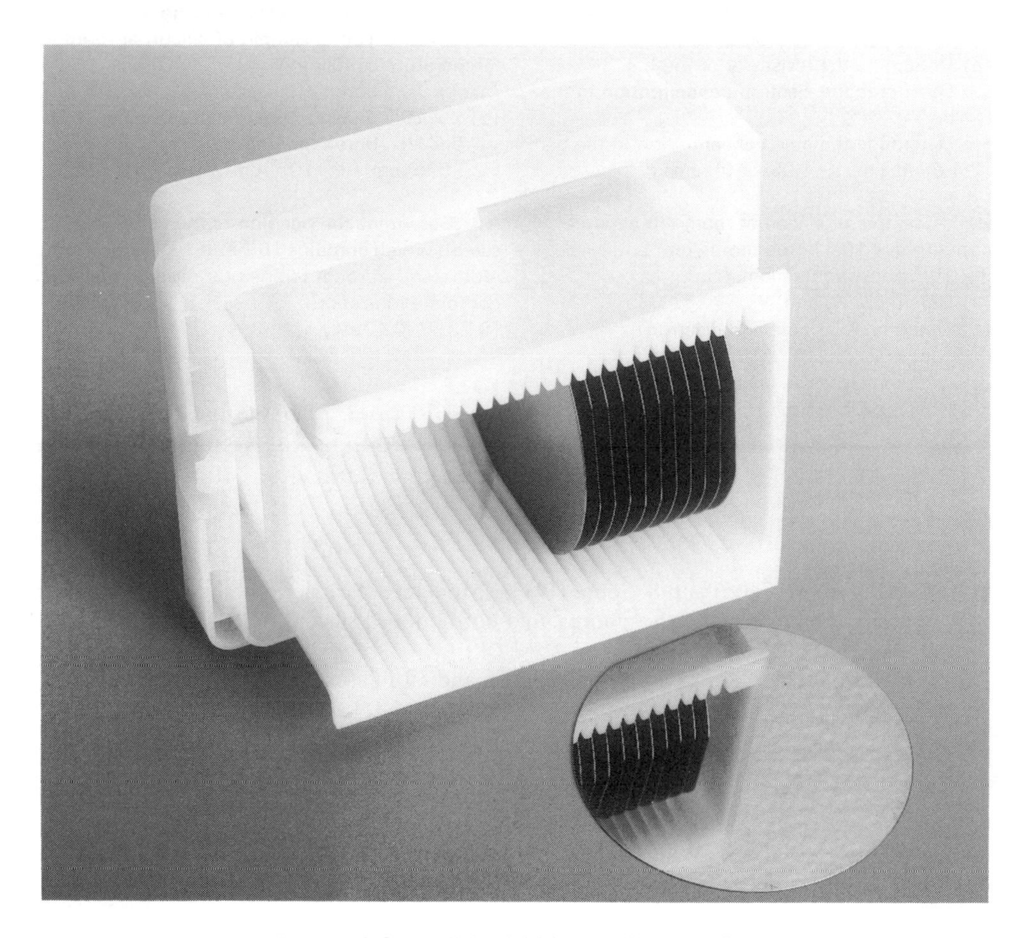

Figure 1.8 Polished 100 mm silicon wafers.

Problems

1. A bar of silicon at room temperature is doped with 2×10^{15} arsenic atoms/cm^3.

(a) Determine the resistivity of the bar.

(b) Determine the electron concentration in the bar.

(c) Determine the hole concentration in the bar.

[2.4 ohms cm, . . . , 1.05×10^5/cm^3.]

2. A sample of silicon at room temperature contains 1×10^{18} boron atoms/cm^3 and 3×10^{18} phosphorus atoms/cm^3.

(a) Determine N_d and N_a.

(b) Determine the hole and electron concentrations.

(c) Determine the resistivity of the sample.

[. . . , . . . , 2×10^{18}/cm^3, 10^5/cm^3, 0.02 ohm cm.]

3. Determine the electron and hole concentrations in a sample of silicon at room temperature containing

(a) 4×10^{17} boron atoms/cm^3;

(b) 2×10^{15} phosphorus atoms/cm^3 plus 5×10^{15} boron atoms/cm^3.

[. . . , 525/cm^3, 2×10^{15}/cm^3, 1.05×10^5/cm^3.]

4. Determine the position of the Fermi level in silicon which contains 10^{14} and 10^{18} phosphorus atoms/cm^3 at room temperature assuming complete ionization.

[0.23 eV, 0.47 eV.]

References

[1] G. Masetti, M. Severi and S. Solmi (1983), 'Modeling of carrier mobility against carrier concentration in arsenic, phosphorus, and boron doped silicon', *IEEE Trans. Electron. Devices*, **ED-30**, 7, 764.

[2] J. C. Irvin (1962), 'Resistivity of bulk silicon and of diffused layers in silicon', *Bell Syst. Tech. J.*, **41**, 387.

CHAPTER 2

Dielectric layers

2.1 Introduction

An important process during the manufacture of integrated circuits is the production of a stable and inert protective layer on the surface of the semiconductor wafer. The purpose of the layer is to protect the pn junctions from the ambient, to act as an electrical insulator for multi-layer metallization and to act as an impermeable mask for the selective introduction of dopants. An important advantage of silicon is that this layer can be formed by heating the silicon wafer in an oxidizing ambient such as oxygen or water vapour. The material which is produced is silicon dioxide. It is transparent, it has a dielectric constant of approximately 4, it is a very good insulator and it is readily etched into complex patterns which are subsequently used to define the n- and p-type regions which form the circuit elements of VLSI circuits.

In addition to thermal oxidation it is also possible to produce dielectric layers by thermal decomposition of suitable chemicals on to heated silicon wafers. The chemicals are passed over the wafers as a vapour and decompose to produce a dielectric layer. These layers can also be used on other semiconductors such as gallium arsenide, which does not have a stable thermal oxide.

2.2 Thermal oxidation

Thermal oxidation of silicon proceeds by one of two processes, 'dry' oxidation involving dry oxygen and 'wet' oxidation involving steam as follows:

$$\mathrm{Si} + \mathrm{O_2} \rightarrow \mathrm{SiO_2} \quad \text{dry}$$

$$\mathrm{Si} + 2\mathrm{H_2O} \rightarrow \mathrm{SiO_2} + 2\mathrm{H_2} \quad \text{wet}$$

Both processes take place at temperatures in excess of 900 °C and both require the oxidizing species to diffuse through the oxide to react with the silicon at the silicon–oxide interface. During the process silicon is consumed and the thickness of silicon consumed is related to the oxide thickness approximately by:

$$t_{si} = 0.45t_{ox}$$

This relationship is independent of the crystal orientation.

A process model of oxidation assumes that the silicon is covered by a layer of oxide with the oxide surface in contact with the gas phase consisting of either dry oxygen or water vapour. At temperatures in the range 900–1300 °C and at atmospheric pressure the oxygen or water vapour diffuses through the oxide to react with the silicon at the silicon–oxide interface. Initially, when the oxide is thin the oxide thickness is proportional to time as:

$$d_{ox} = B/A \cdot t$$

where B/A is a linear rate constant in μm/hr and t is the time (hr).

As the thickness increases, the rate of growth slows and the relationship between thickness and time becomes parabolic:

$$d_{ox}^2 = B \cdot t$$

where B is a parabolic rate constant in μm^2/hr.

Values of B and B/A can be determined experimentally and can be used in analytical models for the oxidation process for use in computer simulators. In practice, the above equations are only first-order approximations and for computer simulations much more complex relationships have been derived. For initial estimates of oxide thickness it is sufficient to use graphs of thickness versus time for a variety of temperatures and for dry and wet conditions, as shown in Figures 2.1 and 2.2[1,2,3].

2.2.1 Crystal orientation

It can be seen from Figures 2.1 and 2.2 that there are in fact two sets of curves, one for $\langle 111 \rangle$ silicon and one for $\langle 100 \rangle$ silicon. This results from the fact that the rate of growth is a function of surface potential which varies with crystal orientation. The two most commonly used crystal orientations for silicon are $\langle 111 \rangle$ and $\langle 100 \rangle$ and the surface energy associated with $\langle 111 \rangle$ silicon is greater than that associated with $\langle 100 \rangle$. The result is that the rate of oxidation for $\langle 111 \rangle$ silicon is greater than for $\langle 100 \rangle$ silicon. The difference becomes less significant at higher temperatures so that at 1200 °C the rate of growth is independent of crystal orientation.

2.2.2 Effect of pressure

Pressure is an added factor that affects oxidation, with increased pressure producing a significant increase in the oxidation rate. It can be shown that the product of time and pressure is a constant. This is illustrated in Figure 2.3, which shows thickness versus time for pressures of 1, 5, 10 and 20 atmospheres for steam at a temperature of 900 °C[4]. As an example 0.8 μm is produced in 10 hr at 1 atmosphere, but is produced in 2 hr at 5 atmospheres.

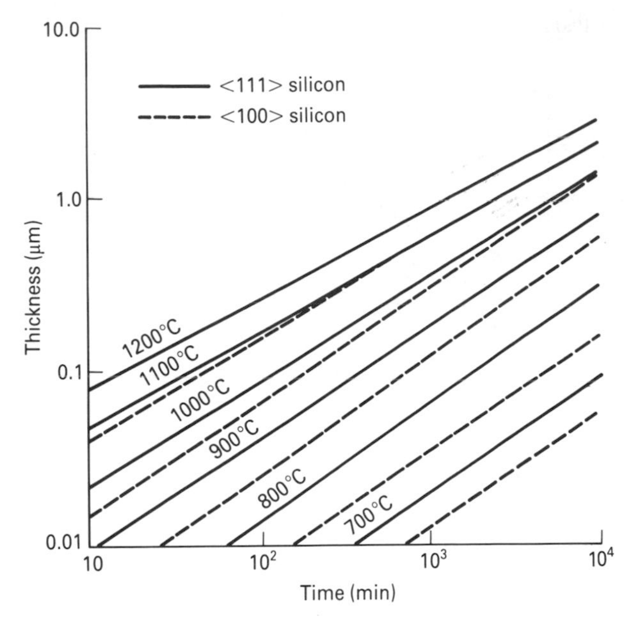

Figure 2.1 Oxide thickness versus time for silicon in dry oxygen. [After Hess and Deal[1,2] and reprinted by permission of the publisher, The Electrochemical Society, Inc.]

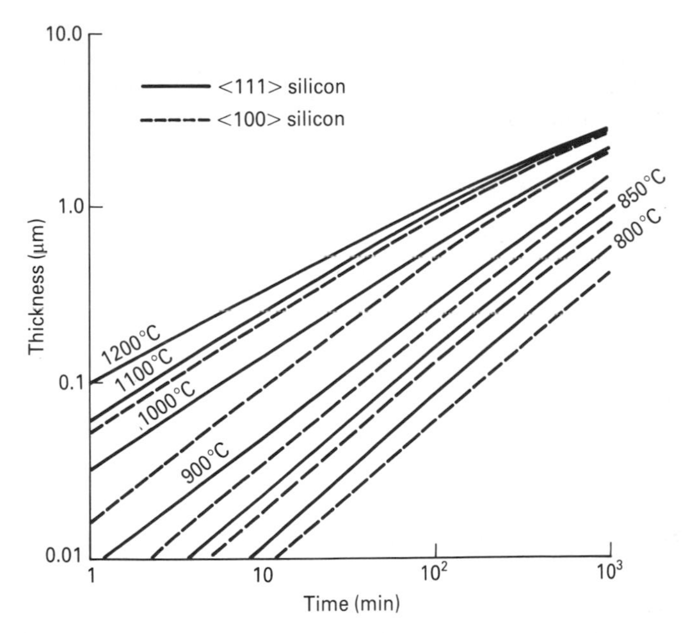

Figure 2.2 Oxide thickness versus time for silicon in steam. [After Deal[3] and reprinted by permission of the publisher, The Electrochemical Society, Inc.]

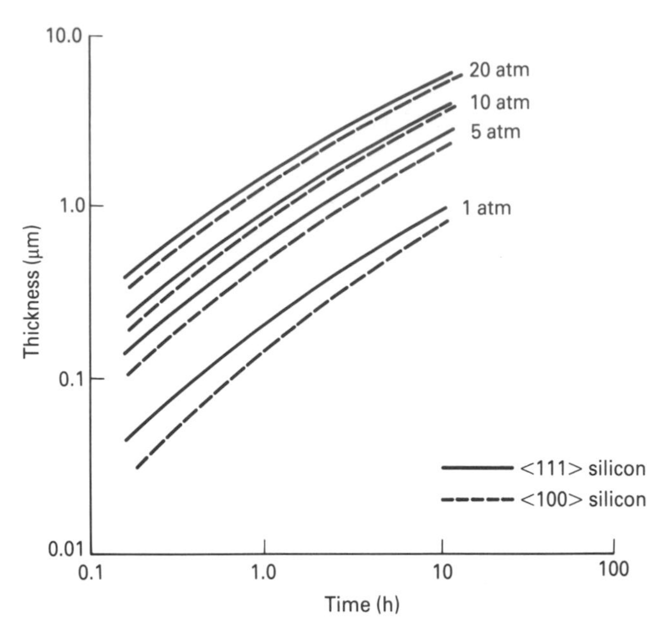

Figure 2.3 Oxidation thickness versus time for silicon at 900 °C and at pressures up to 20 atm. [After Razouk[4] and reprinted by permission of the publisher, The Electrochemical Society, Inc.]

High-pressure oxidation is a very attractive procedure for producing thick oxides which may be required for field oxides in MOS (Metal Oxide Semiconductor) circuits or for oxide isolation in high-speed bipolar circuits. The required oxides can be obtained at lower temperatures and for shorter times, which minimizes the movement of previously diffused impurities.

2.2.3 Effect of halogens

Surface pretreatment is important to obtain good quality oxides with consistent electrical properties. Treating the surface with HCl prior to oxidation can remove many unwanted impurities which may otherwise become incorporated into the oxide. It is also beneficial to include HCl in the oxidizing ambient during oxidation. Typical concentrations of HCl in O_2 are in the range 1–5 %. HCl is also used to clean the oxidation equipment prior to oxidation when impurities which may have diffused through the walls of the silica oxidation tubes are converted to chlorides and flushed out of the system.

2.3 Masking properties

An important requirement of silicon dioxide is that it acts as a mask against dopants such as phosphorus or boron which are used to produce n- and p-type regions in the silicon. The diffusion constants for phosphorus and boron in silicon and silicon dioxide at 1200 °C are given in Table 2.1. It can be seen that the diffusion of impurities in the oxide is several orders of magnitude slower than in silicon. Thus a relatively thin layer of oxide on the surface of the silicon is sufficient to restrict the impurities to clearly defined regions. Typical thicknesses of oxide for masking impurities for bipolar and MOS integrated circuits are 0.5–0.7 m.

2.4 Oxide charges

When an oxide is produced on silicon there is a transition from the very regular arrangement of atoms within the silicon to the non-regular amorphous structure of the oxide. This change of structure has certain implications for the electrical properties of devices formed in the silicon beneath the oxide. At the interface between the silicon and the oxide there is a region where not all of the valence electrons associated with the silicon atoms are bound to neighbouring silicon atoms and thus there is a charge imbalance. This imbalance of electrostatic charge associated with the unbound valence electrons is partly neutralized by combination of silicon atoms with oxygen atoms, but not completely. The result is a net positive charge from the silicon nucleus of those atoms with incomplete bonding. Further charges may also be present in the oxide, either as a result of the oxidation process or as a result of application-specific effects such as external radiation or avalanche breakdown. The number of charges may be described by $N_s = Q_s/q$, where Q_s is the net effective charge per unit area, N_s is the number of charges per unit area and q is the charge of an electron. The type and location of these charges are illustrated in a very simplistic fashion by Figure 2.4

Located at the interface are interface-trapped charges (Q_{it}). These charges, which are caused by the oxidation process, can be minimized by a low-temperature anneal (450 °C) in hydrogen. A fixed positive charge (Q_f) is located in the oxide at the interface and is associated with the unbound valence electrons. It is related to the silicon orientation being lowest for $\langle 100 \rangle$ material and highest for $\langle 111 \rangle$ material. This component of charge is very dependent on the processing conditions and is lowest for a dry oxygen oxidation. For this reason an oxidation process often ends with a dry oxygen step, i.e. the complete oxidation may involve a dry–wet–dry sequence of steps. It has a value which ranges from 10^{10} Coulomb/cm^2 to 10^{12} Coulomb/cm^2.

Table 2.1 Diffusion constants at 1200 °C

Impurity	Silicon	Silicon dioxide
phosphorus	2×10^{-12} cm^2/s	2×10^{-16} cm^2/s
boron	2×10^{-12} cm^2/s	6×10^{-15} cm^2/s

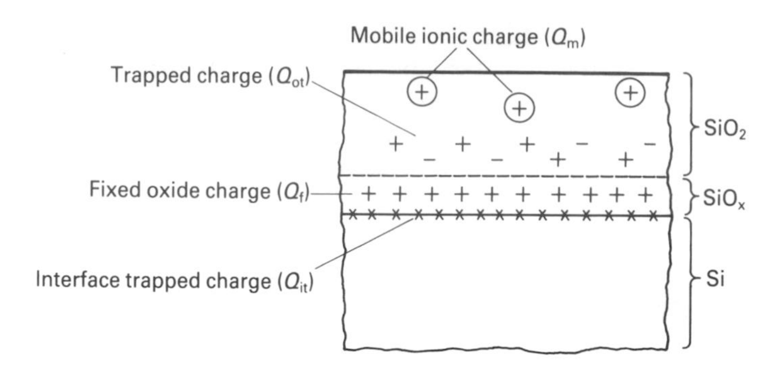

Figure 2.4 Charges in thermally oxidized silicon.

Within the oxide there may be mobile ionic charges which are usually attributed to alkali ions such as potassium and sodium. These elements are usually present in the materials used to construct the furnaces and at the temperatures used for oxidation are able to diffuse through the quartz furnace tubes and be carried to the oxide by the gases flowing through the tube. These ions are even mobile within the oxide at room temperature when an electric field is present and can seriously degrade the electrical properties of devices. Cleaning the furnace tubes with HCl prior to oxidation can greatly reduce the concentration of these ions.

Trapped-oxide charge (Q_{ot}) may be either positive or negative and results from holes or electrons which become trapped in the oxide. These charges may be caused by ionizing radiation, avalanche injection from a pn junction beneath the oxide, or current flow in the oxide between an external electrode and the silicon. Ionizing radiation used during the manufacturing process, e.g. electron beam processing, ion implantation or reactive ion etching, can result in trapped-oxide charge. It can be removed by low-temperature anneal. However, trapped-oxide charges which are application-specific (e.g., avalanche breakdown) are likely to degrade the device.

Thc control of oxide charge is difficult, particularly as device geometries are reduced, as this reduction is usually accompanied by a reduction in the thickness of the oxide. MOS devices are particularly sensitive to the presence of oxide charge but leakage currents of bipolar devices are also affected by the presence of charge near underlying pn junctions. Monitoring oxide charge is an important means of evaluating the quality of an oxide.

2.5 Impurity redistribution

During the oxidation process the silicon is consumed and the interface between the silicon and the oxide advances into the silicon. The dopant atoms within the silicon are redistributed between the silicon and the oxide. Three main factors affect this redistribution:

1. Dopant segregation coefficient:

$$m = \frac{C_{si} \text{ (solubility of dopant in silicon)}}{C_{ox} \text{ (solubility of dopant in silicon oxide)}}$$

2. Diffusivity coefficient:

$$\frac{D_{\text{dopant in silicon}}}{D_{\text{dopant in silicon oxide}}}$$

3. Oxidation rate:

$$\frac{\text{oxidation rate}}{\sqrt{D_{\text{dopant in silicon}}}}$$

The solubility of a dopant is different for different materials and thus a balance must be established for the solubility of the dopant in the oxide and the solubility of the dopant in the silicon at the temperature of the oxidation process. This balance is represented by the dopant segregation coefficient. It has already been noted in Table 2.1 that the diffusivity of the dopant in oxide and silicon is different and this also affects the redistribution as represented by the diffusivity coefficient. Finally, the redistribution is affected by the rate of oxidation and the rate of diffusion of the dopant in the silicon.

Each of the above factors contributes to the way in which the dopants are redistributed between the silicon and the oxide. The overall effect is illustrated in Figure 2.5 where the concentration relative to the bulk concentration in the silicon (C_B) is plotted against distance. For boron the surface is depleted and the concentration of boron in the oxide can be greater than in the silicon. However, the difference depends on the rate of oxidation, with the dry oxidation showing a much higher concentration in the oxide. In the case of phosphorus the concentration in the silicon is much greater at the surface since phosphorus is rejected by the oxide. An effective segregation coefficient is usually obtained experimentally by forcing a fit between a mathematical model for oxidation and experimental data of the impurity profile (see Chapter 4 for profile measurement).

2.6 Oxidation systems

The equipment used for thermal oxidation is a resistance heated furnace, as shown in Figure 2.6. The silicon slices are held vertically in a quartz 'boat' and a boat may contain between 100 and 200 slices. The boat is placed inside a quartz tube which provides a sealed environment for the oxidation process. The necessary gases (O_2, N_2, H_2, HCl) are introduced at one end via valves and flow meters and are exhausted at the other end. The temperatures are in the range 800–1200 °C and are maintained to an accuracy of ± 1 °C along the length of the heated zone. In a production furnace the slices would be cleaned, dried, placed in the boat and automatically loaded into the furnace and the temperature ramped up to the operating value. The inert gas nitrogen

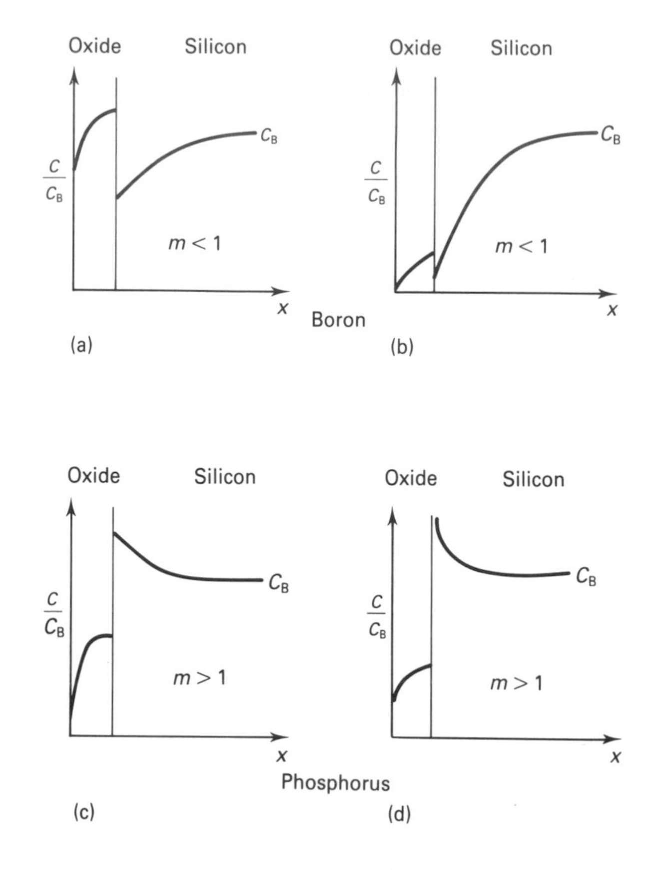

Figure 2.5 Impurity profiles as a result of segregation at the silicon–silicon oxide interface for boron (a) dry and (b) wet, and for phosphorus (c) dry and (d) wet.

(N_2) flows during the initial ramping-up of temperature and again at the end when the temperature is reduced. After the required time at the oxidation temperature the slices would be removed, again automatically and at a controlled rate. Gas flows, temperature ramping, rate of insertion and removal of the boat and oxidation times are all controlled by microprocessor to ensure reproducibility.

Dry oxidation or HCl/dry oxidation is straightforward using metered gas supplies of oxygen and/or HCl. Wet oxidation can be achieved by bubbling a carrier gas (oxygen or nitrogen) through water heated to about 95 °C. This temperature produces a vapour pressure of approximately one atmosphere (640 torr). However, a more likely procedure would involve the direct reaction between oxygen and hydrogen introduced in the correct proportions as gases to produce steam by pyrogenic action. This technique assures very high-purity steam provided that high-purity gases are used.

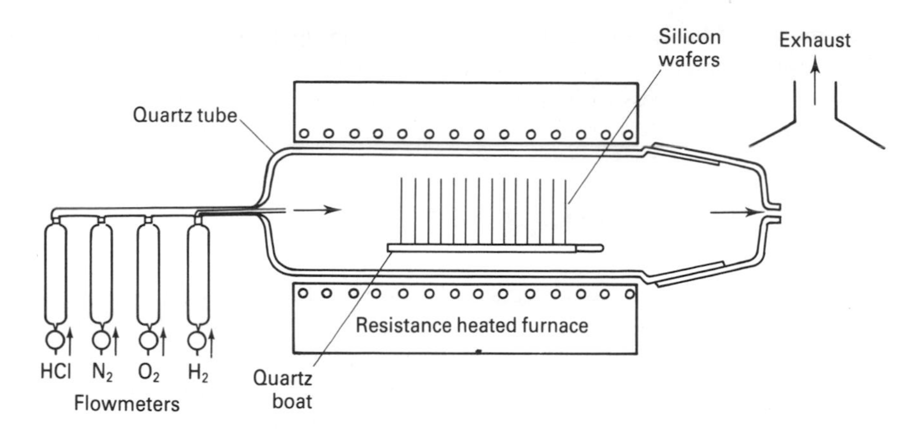

Figure 2.6 Schematic of a thermal oxidation furnace.

2.7 Estimation of oxide thickness

The graphs shown in Figures 2.1 and 2.2 are used to estimate the thickness for a given time and temperature and for a particular crystal orientation for either dry oxygen or steam or to estimate the time required to produce an oxide of a given thickness at a particular oxidation temperature. If no oxide is initially present then the graphs provide a direct reading of time versus thickness (or, alternatively, thickness versus time). However, if a layer of oxide is already present it is necessary to determine an effective time in order to determine the final oxide thickness. Proceed as follows:

1. Determine the time required to produce the layer of oxide that already exists at the temperature of the next oxidation process.
2. Add this time to that of the next oxidation process to produce an effective time.
3. Determine the oxide thickness based on the effective time and the temperature of the present oxidation process.

As an example, assume that a layer of oxide of 100 nm exists and that a second oxide is to be produced in pyrogenic steam at 1000 °C for 30 min. Assume that $\langle 111 \rangle$ silicon is used.

The first step is to determine from Figure 2.2 for wet oxidation the time that would have been required to produce the 100 nm in steam at 1000 °C. From the graph this is approximately 6 min. The 6 min is added to the 30 min to give an effective time for the oxidation in steam of 36 min. From the graph the thickness after 36 min at 1000 °C is approximately 300 nm.

If there was a further oxidation of 15 min at 1100 °C in dry oxygen then Figure 2.1 would be used to estimate the time that would have been required to produce the 300 nm in dry oxygen at 1100 °C. This time would be added to the 15 min to give an effective time for the oxidation process as if the total process had taken place in dry oxygen. From Figure 2.1 300 nm would require approximately 300 min. Thus the effective time would be 315 min and from the graph the thickness for this effective time would be approximately 320 nm. (The change in thickness is very small but this does represent a practical process with the final dry oxidation step being used to remove the OH molecule from the oxide rather than to increase the thickness.)

2.8 Evaluation methods

To achieve consistent results it is necessary to monitor the oxidation process. A very good initial evaluation of the quality of the oxide can be made by visual inspection. The thin layers of transparent oxide act as an interference filter and when viewed in daylight exhibit a series of well-defined colours. The colours depend on the thickness and thus a quick check can be made of the oxide thickness by comparing the observed colour against a table of thickness versus colour. However, the accuracy is not very great, particularly as the colours repeat for different multiples of thickness, and more precise methods based on electronic instruments are necessary for a detailed evaluation. The simplest parameter to measure is the thickness, but for MOS devices surface charge is also important.

Optical methods are usually used for the measurement of thickness, with ellipsometry being the most commonly used method. With this technique, which is shown schematically in Figure 2.7, a beam of single-wavelength light from a laser is passed through a polarizer and directed on to the oxide-covered surface of a silicon wafer. The polarization of the reflected beam is changed by the layer. This change can be detected by passing the beam through an analyzer and to a detector. For particular positions of the polarizer, the analyzer is rotated to produce a minimum in the light intensity observed by the detector. The position of both the polarizer and analyzer are noted for this minimum. The measurement is repeated for two positions of the polarizer. The four sets of readings (two for the polarizer, two for the analyzer) are a function of the thickness of the layer on the silicon. A microcomputer is used to evaluate the thickness from these readings to an accuracy of a few Angstroms (1 Å $= 10^{-10}$ m). Commercial equipment is now available which automates the complete process.

The measurement of surface charge is obtained from the variation of capacitance

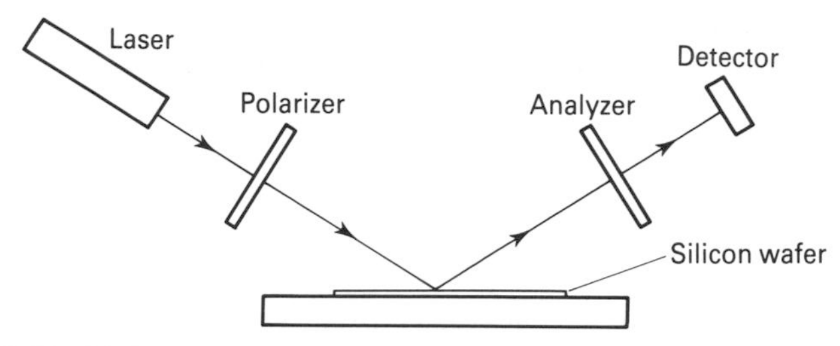

Figure 2.7 Light path geometry for thickness measurement based on ellipsometry.

of a MOS capacitor with applied dc voltage. A typical $C-V$ plot is shown in Figure 2.8. A detailed explanation of the variation of capacitance with voltage is to be found in Section 14.6 on the MOS capacitor (Chapter 14). Special-purpose microcomputer-based equipment is used to make the measurement and to extract the value of surface charge.

2.9 Deposited oxides

As an alternative to thermal oxidation, which involves temperatures in excess of 800 °C, deposited oxides can be used to produce a range of insulating layers which can be formed at temperatures of less than 400 °C.

Deposited layers of dielectric material can be used to provide insulation between multi-layered interconnections, as a capping layer for VLSI circuits to provide additional protection against humidity and as a means of depositing materials which cannot be produced by thermal oxidation. One such material is silicon nitride.

Silicon nitride is an important material for VLSI manufacture. It is impervious to sodium and can be used as a gate dielectric in MOS transistors or as a capping layer. It can also act as a mask during thermal oxidation and is used to selectively mask the silicon for oxide isolation.

The most commonly used method for depositing dielectric layers is by chemical vapour deposition (CVD). A suitable gas is passed over the heated slices so that it decomposes to produce the dielectric film. One such reaction involves silane and oxygen at about 400 °C:

$$SiH_4 + O_2 \rightarrow SiO_2 + 2H_2$$

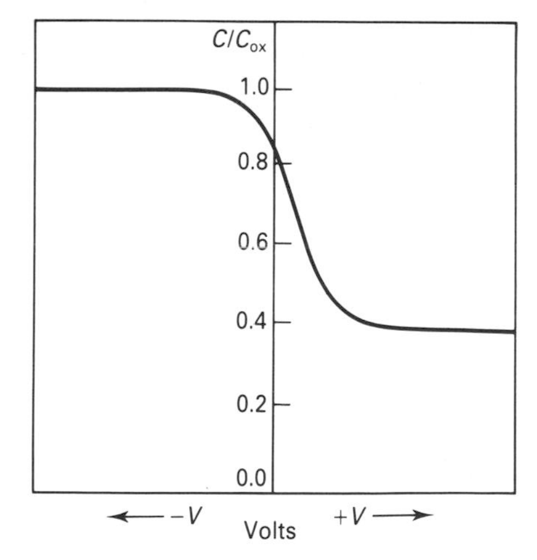

Figure 2.8 Variation of oxide capacitance with applied voltage for a MOS capacitor.

This reaction results in high-quality silicon dioxide with hydrogen as a by-product. There is a linear relationship between thickness and time and, compared with thermal oxidation, the rate of growth is very rapid. Phosphorus is often added in the form of phosphine gas to produce a phosphorus doped oxide. This oxide is useful for multi-level interconnections because during the deposition process the oxide flows over underlying steps which have been formed by previous processing steps, to produce a smooth topography which improves step coverage during the metallization stage. Alternative materials for silicon dioxide are tetraethoxysilane ($Si(OC_2H_5)_4$) and dichlorosilane ($SiCl_2H_2$). However, these materials require temperatures in the range 600–900 °C.

For silicon nitride, silane and ammonia can be used:

$$3SiH_4 + 4NH_3 \rightarrow Si_3N_4 + 12H_2$$

For good-quality films it is necessary to ensure that the correct stoichiometry of the constituent gases is maintained. The quality of the film can be monitored from the refractive index and the etch rate in buffered hydrofluoric acid. A high refractive index indicates a silicon-rich film, while low indices indicate the presence of oxygen. Oxygen impurities increase the etch rate.

2.10 Oxide isolation

The use of silicon nitride for creating recessed regions of oxide in silicon is an important manufacturing step. The process takes advantage of the low diffusivity of oxygen in silicon nitride, even at thermal oxidation temperatures. There are two variations of the process, one in which the nitride layer is used as a mask to etch a recess in the silicon and then as a mask while the exposed silicon is oxidized. In the other the silicon is oxidized without first forming a recess. In the former the resulting surface remains almost flat but in the latter the surface of the oxide is raised above that of the original silicon surface, which is protected by the nitride.

The most commonly used method uses the nitride as an oxidation mask. About 200 nm of nitride is usually sufficient to act as a mask during high-pressure steam oxidation to produce the thick oxide necessary to isolate regions of silicon. Prior to depositing the nitride a thin layer of silicon dioxide (10 nm) is grown by thermal oxidation. This layer is necessary to provide a stress relief barrier between the silicon and the nitride. Without this barrier there is a danger of the nitride film shattering during the heating and cooling for the oxidation step. Photolithography is used to create a pattern in the nitride and it is removed by plasma etching. The exposed silicon is then oxidized. During oxidation, silicon is consumed and the silicon surface is recessed beneath the original surface. A problem arises at the edge of the nitride since the oxidized region moves beneath the nitride. The amount of lateral oxidation is a function of the thickness of the stress relief layer of SiO_2: the thicker the layer the greater the lateral movement. This lateral movement produces a 'bird's beak', which is shown as a computer simulation from SUPREM IV in Figure 2.9. The lateral movement encroaches into the active device area, e.g. with a 20 nm stress relief oxide the bird's beak is in excess of 1000 nm.

While the silicon nitride isolation method is widely used, other methods involving anisotropic etching of silicon to form trenches enable much smaller structures to be produced without the problems of the bird's beak. A cross-section of trench isolation is shown in Figure 2.10.

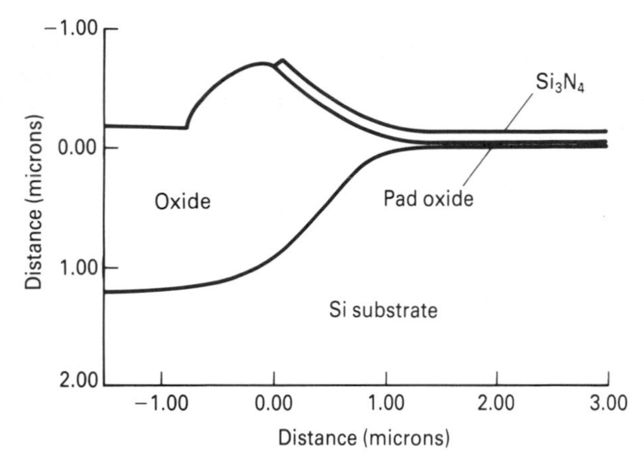

Figure 2.9 Oxide formation known as the 'bird's beak', which results from the lateral oxidation of the silicon beneath the layer of silicon nitride during oxide isolation.

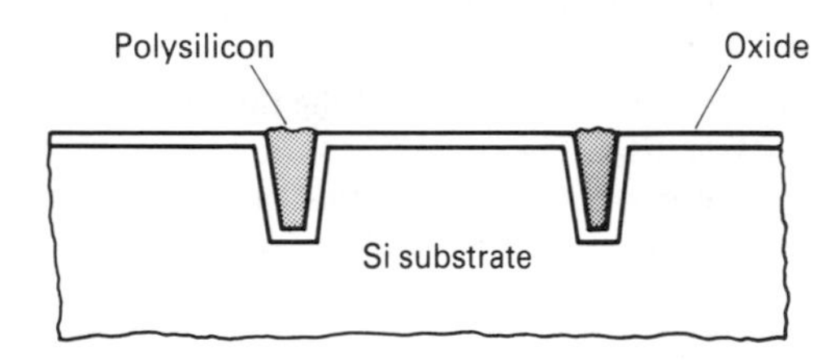

Figure 2.10 Isolation with a silicon trench.

Problems

1. A layer of oxide is grown on ⟨111⟩ silicon in steam at atmospheric pressure at 1000 °C for 60 min:
(a) Determine the oxide thickness.
(b) Determine how long it would take to grow the same thickness at 1200 °C.
(c) Determine how long it would take to grow the same thickness on ⟨100⟩ silicon at 1000 °C in dry oxygen.
[0.4 μm 15 min 1500 min.]

2. A ⟨100⟩ silicon wafer is oxidized using the following sequence:
(a) 30 min at 1200 °C in dry oxygen;
(b) 30 min at 1200 °C in steam;
(c) Remove part of the oxide to expose bare silicon;
(d) a further 30 min at 1000 °C in dry oxygen.
Determine the oxide thickness after each step.
[140 nm, 580 nm, 30 nm, and 580 nm.]

3. If the amount of silicon consumed is $0.45 \times t_{ox}$ determine how far above the original surface of a silicon wafer the top of a 300 nm-thick oxide layer is.
[165 nm.]

4. A ⟨100⟩ silicon wafer undergoes the following sequence of oxidation and etching steps. Determine the oxide thickness following each step in the etched region and in the non-etched region:
(a) 1000 °C dry oxygen for 40 min;
(b) 1000 °C steam for 60 min;
(c) Etch 0.3 m of oxide;
(d) 1100 °C dry oxygen for 20 min;
(e) 1100 °C steam for 50 min.
[0.036 nm, 0.35 nm, 0.36 nm, and 0.085 nm, 0.7 nm and 0.58 nm.]

References

[1] D. W. Hess and B. E. Deal (1975), 'Kinetics of the thermal oxidation of silicon in O_2/N_2 mixtures at 1200 °C', *J. Electrochem. Soc.*, **122**, 579.
[2] D. W. Hess and B. E. Deal (1977), 'Kinetics of the thermal oxidation of silicon in O_2/HCl mixtures', *J. Electrochem. Soc.*, **124**, 735.
[3] B. E. Deal (1978), 'Thermal oxidation kinetics of silicon in pyrogenic H_2O and 5% HCl/H_2O mixtures', *J. Electrochem. Soc.*, **125**, 576.
[4] R. R. Razouk, L. N. Lie, and B. E. Deal (1981), 'Kinetics of high pressure oxidation of silicon in pyrogenic steam', *J. Electrochem. Soc.*, **128**, 2214.

CHAPTER 3

Semiconducting layers

3.1 Introduction

Integrated circuits are fabricated in a silicon wafer by a process of diffusion of n- and p-type dopants in regions defined by creating patterns in silicon oxide. The complexity of the circuit elements can be greatly increased by depositing layers of silicon over previously diffused regions and over layers of oxide. The layers can be produced either by the process of chemical vapour deposition similar to that used for dielectric layers or by the process of epitaxy.

Single-crystal epitaxial layers can also be grown on single-crystal dielectric substrates such as sapphire. Although the lattice structure of sapphire is different from that of silicon the dimensions are sufficiently close to enable the silicon atoms to align and form a single-crystal silicon film. The silicon films can be used to fabricate MOS devices. Silicon On Insulator (SOI) films have important applications for devices which are to be used for high-frequency applications. A special feature of epitaxy is the ability to grow single-crystal semiconductor layers on single-crystal substrate. For example, when a gas containing silicon atoms is passed over a heated silicon substrate the atoms separate from the gas and align themselves with the atoms on the substrate to form a continuation of the crystal lattice. The crystalline properties of the layer are indistinguishable from those of the substrate. Dopants such as phosphorus, arsenic or boron can be added to the gas so that the dopants also become incorporated into the layer and it can have different conductivity type and value to that of the substrate. The layer can be grown over previously diffused regions to allow complex circuit elements to be manufactured.

When silicon is deposited on an amorphous substrate, for example silicon oxide, the atoms form random clusters which are single-crystal on a microscopic scale, but the crystallites are randomly orientated with respect to each other. The layer is said to be polycrystalline. These polycrystalline layers can be used as conductors and have important uses in MOS technology for self-aligned gate electrodes and as multi-level interconnections.

3.2 Single-crystal deposition

The epitaxial growth of single-crystal silicon involves a reaction vessel, usually of quartz, a susceptor to support the substrates, usually graphite-based and gas and temperature control equipment. The susceptor supports the wafers either horizontally or vertically and is usually heated inductively by radio frequency energy. The walls of the reaction vessel are relatively cool.

A typical sequence may be as follows:

1. Remove any oxide from the wafers with hydrofluoric acid (HF).
2. Place wafers in the reaction vessel and purge with hydrogen. Heat to etching temperature (1200 °C).
3. Etch a thin layer of silicon with HCl gas to expose an undamaged layer of silicon.
4. Purge the HCl from the reaction chamber and lower the temperature to the growth value (900–1100 °C).
5. Introduce the silicon-containing gas (and dopant gas if required) for an appropriate time to produce the required layer.
6. Purge the reaction vessel with hydrogen and lower the temperature (600 °C).
7. Purge the reaction vessel with dry nitrogen; cool and unload the wafers.

Dopants are added to the silicon-containing gas to control the conductivity type and conductivity value of the layer. By this means n-type layers (or p-type) can be grown on p-type (or n-type) substrates and lightly (or heavily) doped layers can be grown on heavily (or lightly) doped substrates. The layers have the same crystal orientation and physical properties as the substrate.

3.2.1 Silicon-containing compounds

There are four commonly used silicon-containing compounds. These are silicon tetrachloride ($SiCl_4$), dichlorosilane (SiH_2Cl_2), trichlorosilane ($SiHCl_3$) and silane (SiH_4). Silicon tetrachloride is reduced with hydrogen to form silicon and hydrogen chloride:

$$SiCl_4 + 2H_2 \rightarrow Si + 4HCl$$

This equation represents the final phase but in practice there are many intermediate gaseous phases. The temperature required for silicon tetrachloride is high at between 1150 °C and 1250 °C. The high temperature means that impurity profiles which already exist in the sustrate prior to epitaxial growth are modified by diffusion. A further problem is auto-doping when dopants enter the gas phase by diffusion and evaporation from the substrate and become reincorporated into the growing layer. Auto-doping is a particular problem during the growth of high-resistivity layers on low-resistivity substrates. It can be reduced by lowering the temperature and by increasing the growth rate.

Dichlorosilane and trichlorosilane also decompose to form solid silicon and gaseous HCl, but at lower temperatures (1100–1200 °C for dichlorosilane and 1050–1150 °C for trichlorosilane). Silane decomposes to silicon and hydrogen and can be used at much lower temperatures (950–1050 °C). However, the rate of growth is lowest for silane at 0.2–0.3 μm/min as compared to 0.4–2.0 μm/min for the other sources. An advantage of silane is that it is a gas whereas the other three sources are liquids. Auto-doping is also reduced with silane because of the absence of HCl in the reaction. When HCl is present it reacts with the silicon and releases any dopants into the gas stream.

3.2.2 Impurity doping

The conductivity type and conductivity value is controlled by the addition of suitable dopants to the gas stream entering the reaction vessel. Convenient sources of n-type dopants are the gases phosphine (PH_3) and arsine (AsH_3). For p-type dopant boron is available as diborane (B_2H_6), which is also a gas. The gases are diluted in an inert carrier gas such as nitrogen or argon to levels of 1–2% of the dopant in carrier gas. Dopant control is achieved by metering the gas into the main gas stream, the bulk of which is hydrogen. The dopant hydrides decompose to release hydrogen while the dopant becomes incorporated into the silicon layer.

Many factors affect the quality of the silicon layer and the distribution of dopants in the layer. Ideally, there should be a step change between the substrate and the layer. In practice, because of diffusion there is a more gradual change, as shown in Figure 3.1, which illustrates the profile that might be expected when a lightly doped layer is grown on a heavily doped substrate. Increasing the rate of growth minimizes the effect of diffusion of impurities from the substrate. However, if HCl is present in the reaction

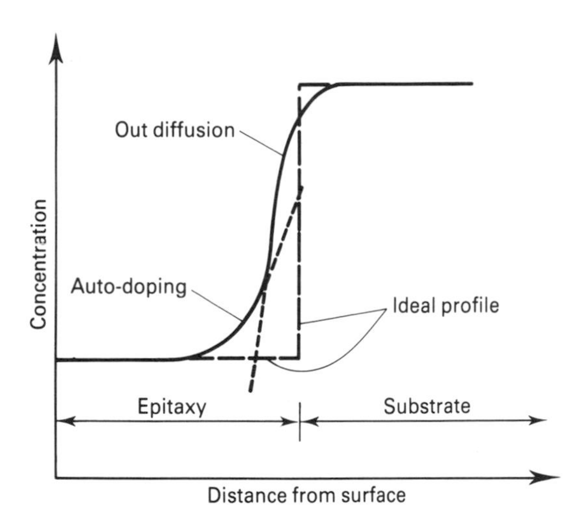

Figure 3.1 An ideal impurity profile (dotted line) and the actual profile for an epitaxial layer on a p-type substrate.

gases then auto-doping from the substrate will have a similar effect to diffusion irrespective of the growth rate.

3.2.3 Process considerations

Two possible configurations for reactor design are shown in Figure 3.2. In one, the slices are placed on the sides of a vertical susceptor, while in the other the susceptor is inclined to the horizontal. In both cases the susceptor is likely to be of graphite which is coated with a layer of silicon carbide by a chemical vapour deposition process. Other coatings include pyrolytic graphite, which is a glassy form of carbon produced by heating the carbon blank in methane at elevated temperatures. The reaction vessel is usually quartz because of its ability to withstand high temperatures without mechanical deformation, and its freedom from impurities.

Energy for the reaction can be provided by a radio frequency source by inductive coupling to the susceptor. The heat is then transported to the slices by conduction and radiation. As an alternative, radiant heating can be used from banks of quartz halogen lamps. In both cases the wall of the reaction vessel remains relatively cool.

Gas flow and temperature are microprocessor controlled to ensure reproducibility. The detailed sequence of operations is determined by a systematic approach of controlled experiments to determine the optimum conditions involving time, temperature and gas flows. The layers are monitored for uniformity of resistivity and thickness across individual wafers, from wafer to wafer in a batch and from batch to batch.

3.2.4 Pattern shift

For many bipolar circuits a low resistivity region is formed beneath the emitter of the transistor to provide a low-resistance path for the current flowing from the emitter to the collector contact. This low-resistance region is formed by diffusion of a suitable

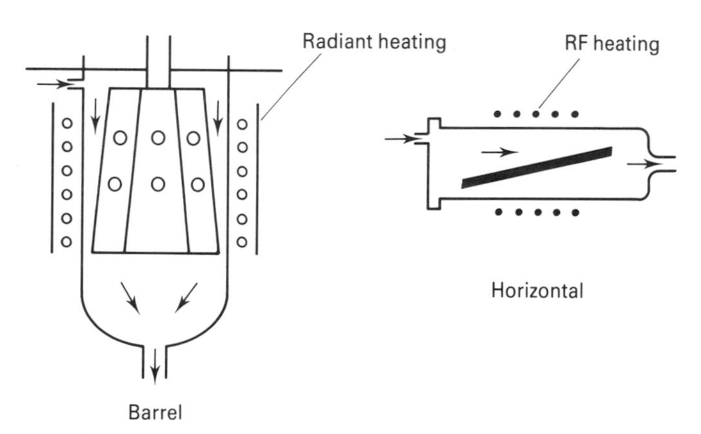

Figure 3.2 Schematics of two examples of epitaxial reactor design.

n-type dopant into the substrate before growing the epitaxial layer. An important requirement is that subsequent masking operations are accurately aligned with the buried low-resistance region, even though it may be covered with several microns of silicon. The presence of the buried region complicates the epitaxial growth with auto-doping and pattern shift.

After the initial diffusion, which involves oxide growth over the buried region followed by removal of all oxide on the surface of the wafer, there is a depression of between 50 and 100 nm in the diffused area. This depression is retained during the growth of the epitaxial layer but, depending on the crystal orientation, it may shift laterally, as illustrated in Figure 3.3. An added complication is pattern washout, which causes the pattern to blur and become distorted.

The crystal orientation has a very marked effect on pattern shift[1] and as a result it can be greatly influenced by cutting the crystal a few degrees (2–5) off the main crystal axis (⟨111⟩ or ⟨100⟩). Pattern shift is reduced at lower growth rates and higher temperatures; it is also reduced at lower reactor pressure. However, pattern distortion exhibits opposite characteristics and increases at higher temperatures.

Diffusion and auto-doping are additional problems associated with the highly doped buried low-resistance region. During the initial phase of layer growth the high concentration of dopants in the diffused region provides a ready source of dopant atoms through evaporation and these become incorporated into the layer above and around the buried region. As the layer thickness increases auto-doping diminishes but the dopants diffuse into the layer. The effect of out-diffusion can be minimized by using a dopant with a low diffusion coefficient such as arsenic, rather than phosphorus.

3.2.5 Epitaxial evaluation

The important properties are thickness and dopant concentration but cosmetic inspection for visible defects is also part of an evaluation routine.

The thickness can be determined using an infrared spectrometer. Infrared radiation is directed on to the surface of the epitaxial layer. The bulk of the radiation is reflected from the surface but some passes into the epitaxial layer since lightly doped silicon is transparent to infrared. Some of this radiation is reflected from the interface between the substrate and the layer. The radiation from the surface, together with that

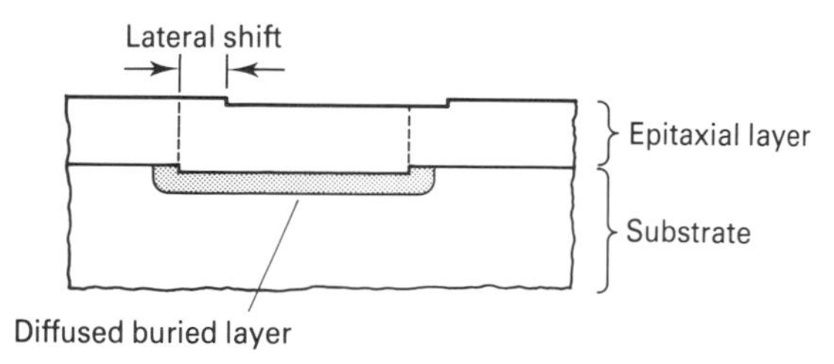

Figure 3.3 Lateral pattern shift after an epitaxial layer has been formed over a diffused region.

from the substrate–layer interface, are combined, which results in destructive interference and the formation of an interference pattern which can be measured using a suitable detector. The location of the maxima and minima can be used to determine the thickness. Equipment is available which automates this process and provides readings of thickness from 1 μm to several hundred micrometers to an accuracy of 50 nm.

Alternative methods involve bevelling and staining to provide visible delineation of the layer and substrate, and spreading resistance (see Chapter 4), which can provide a profile of impurity concentration versus depth.

The dopant concentration can be obtained from electrical measurements such as sheet resistance, capacitance–voltage plots and spreading resistance.

A common technique is to use a control wafer which is lightly doped and of opposite conductivity type to that of the layer. After deposition the sheet resistance is measured by a four-point probe (see Chapter 4). The sheet resistance is converted into carrier concentration from the thickness measurement. The method is not accurate as the control wafer is not the same as the actual wafer and the calculation assumes uniform distribution of dopants which may not be the case for thin layers where auto-doping and out-diffusion can affect the profile.

The capacitance versus voltage of a reverse-biased diode can be used to determine the variation of dopant concentration with distance using the following relationships:

$$N(x) = \frac{C^3}{\mathrm{d}C/\mathrm{d}V q A^2 \varepsilon_s}$$

$$x = \varepsilon_s A/C$$

where N is the doping concentration, x is the distance, C the capacitance, V the voltage, A the diode area and ε_s the dielectric permititivity of silicon. The diode can be formed with a mercury contact which provides a non-destructive way to determine the dopant concentration of a wafer. The measurement is performed using specialized capacitance measurement equipment which measures capacitance while varying the voltage and then plots the impurity profile.

The spreading resistance method can be used to produce a profile of the impurity throughout the layer but this is destructive in that it involves forming a bevel on a test sample of the wafer. However, it can also be used to determine the resistivity from a surface measurement provided the thickness is known, but it also suffers from similar disadvantages to that of the four-point probe measurement.

3.3 Polycrystalline silicon

If instead of a single-crystal substrate an amorphous substrate is used, then silicon is still deposited but the silicon atoms now form clusters of varying orientations and the resultant film is polycrystalline. The amorphous substrate is likely to be a silicon wafer covered with silicon dioxide and the polysilicon is deposited on the oxide.

Polysilicon is used extensively for MOS device fabrication and it is usually formed from the decomposition of silane:

$$SiH_4 \rightarrow Si + 2H_2$$

The layer can be doped with the addition of impurities but it is more usual to rely on subsequent diffusions to dope the polysilicon layer. The diffusion of impurities in polysilicon is much faster than in single-crystal silicon and carrier concentrations equal to the solid solubility limit are easily achieved throughout the thickness of the film. However, the resistivity of polysilicon is higher than single-crystal silicon with similar doping levels. This is because the dopants collect at the boundaries between the crystallites and do not contribute free carriers. In addition, those free carriers that are produced also become trapped at the boundaries and cannot take part in conduction. For boron and arsenic the resistivity saturates at about 2000 ohm/cm, while for phosphorus the limiting resistivity is about 400 ohm/cm. Since the polysilicon is used to form conducting electrodes and interconnections in MOS circuits these resistivity levels can result in serious resistive loss for long tracks.

3.3.1 Practical considerations

Polysilicon is more likely to be produced in a resistance heated furnace than in an inductively heated one. Low pressure is used (0.2–1.0 torr) and the silane is introduced either pure or diluted to about 20% in nitrogen. Batches of 100 or more wafers can be handled per run. The deposition rates are typically 10–20 nm per minute at temperatures of between 600 °C and 650 °C. Too high a temperature results in gas phase reactions occurring, which produce a rough, loosely adhering film. If the temperature is too low the deposition rate is too slow to be practical. Gas flows have to be carefully established to ensure uniform deposition throughout the whole batch. The main problem is the depletion of silane, which results in the wafers at the exhaust end of the furnace having thinner layers than those at the input end.

3.4 Silicon on insulators

When electrical components are fabricated in a silicon wafer they are all electrically connected by the parasitic capacitance associated with pn junction capacitance and the bulk resistance of the substrate. The problem is particularly acute for circuits required to operate at high frequencies. The components only occupy a thin layer near the surface of the substrate and the rest of the substrate simply acts as a mechanical support. From an electrical viewpoint there are obvious attractions in replacing that part of the substrate which acts as a mechanical support with a dielectric substrate.

Device quality silicon can be obtained by depositing silicon on single-crystal insulating substrates such as sapphire (Al_2O_3) or spinel ($MgAl_2O_4$). Because the substrate is different from the layer the process is termed heteroepitaxy.

Silicon-on-insulator is now a well-established process with processing conditions similar to those required for homoepitaxy. Silane is the preferred silicon compound because of its lower deposition temperature which minimizes the auto-doping of aluminium from the substrate. The resultant films are of poorer quality than those produced by homoepitaxy but are suitable for majority carrier MOS devices. The problems arise from the mismatch between the lattice dimensions of the substrate and those of silicon, plus the different temperature coefficients of the two materials.

The film thickness is typically 1 μm and it can be processed in the same way that single-crystal substrates are processed. Good-quality devices can be made and SOI circuits can provide improved frequency performance and also improved resistance to radiation.

References

[1] S. P. Weeks (1981), 'Pattern shift and pattern distortion during CVD epitaxy on $\langle 111 \rangle$ and $\langle 100 \rangle$ silicon', *Solid State Technol.*, **24**, 111.

CHAPTER 4

Diffusion

4.1 Introduction

VLSI components are formed by creating n- and p-type regions in the semiconductor wafer. An n-type region adjacent to a p-type region can form a pn junction diode; a p-type region surrounded by n-type material can be used as a resistor. Solid-state diffusion is an important process for creating these regions.

At temperatures greater than 800 °C dopants such as boron and phosphorus will move from a region of high concentration to a region of low concentration by the process of solid-state diffusion. If the process is allowed to continue for a sufficient time the distribution of dopants throughout the wafer will become uniform. In practice, time is an important process variable and the dopants are only allowed to diffuse a limited distance by controlling the time. Below 600 °C very little movement takes place and the dopants are effectively frozen in place.

Diffusion is used to form pn junctions for bipolar transistors, the source and drain regions of MOS transistors, the plates of junction capacitors and resistors. The dopants or impurities which are used to create these regions can be introduced into a silicon wafer by a variety of methods including transfer from a chemical vapour at a high temperature, from a solid source which is applied to the surface and then heated in a furnace at a high temperature, and by ion implantation. Diffusion distributes the dopants in a controlled manner to produce the required structures for electrical devices.

The distribution of the dopants, and any other impurities, is governed by the laws of diffusion, which are based on Fick's laws. These laws describe the rate of flow of impurities and it is possible to define a diffusion constant which describes the rate at which the impurity moves through the solid. It varies with temperature and, therefore, the rate of movement can be controlled by varying the temperature. In order that impurities can diffuse into a silicon substrate, spaces must exist within the crystal lattice which can be occupied by the impurities' atoms.

4.2 Diffusion in solids

The movement of impurities through a solid is predominantly by vacancy and interstitial diffusion. At high temperatures defects and vacancies are created in the single-crystal solid and an impurity is able to occupy such a vacancy and thus substitute for the host atom. As a result of thermal energy the lattice vibrates and an adjacent host atom may move to another vacancy or to an interstitial site. This movement creates a vacancy which can then be occupied by the impurity. This substitutional (or vacancy) diffusion is illustrated in a simple two-dimensional array in Figure 4.1(a). Here the open circles represent the host atom and the solid circles the impurity. The number of vacancies, even at an elevated temperature, is not high and as a result this form of diffusion is slow. This 'slowness' is reflected in the value of a constant known as the diffusion coefficient.

For impurity atoms which are sufficiently small to fit into the gaps or interstices between the atoms within the lattice, the diffusion can be much more rapid as the interstitial spaces are not dependent on the formation of vacancies. Interstitial diffusion is illustrated in Figure 4.1(b). Many metallic impurities are interstitial diffusers.

4.2.1 Analytical model of diffusion

Diffusion is the result of the random motion of particles. For the silicon wafer these particles are the impurity atoms and the lattice vacancies. If in adition the concentration is non-uniform then there is a tendency for more particles to be moving away from the region of highest concentration rather than towards it. Thus there is a net flow of particles away from the region of highest concentration to the region of lowest concentration. The larger the rate of change, or gradient, between the region of high concentration and low concentration, then the greater the flow. This flow can be described by Fick's first law:

$$f = -D\frac{\mathrm{d}N}{\mathrm{d}x}$$

where f is the number flowing per second through 1 cm^2, D is the diffusion coefficient

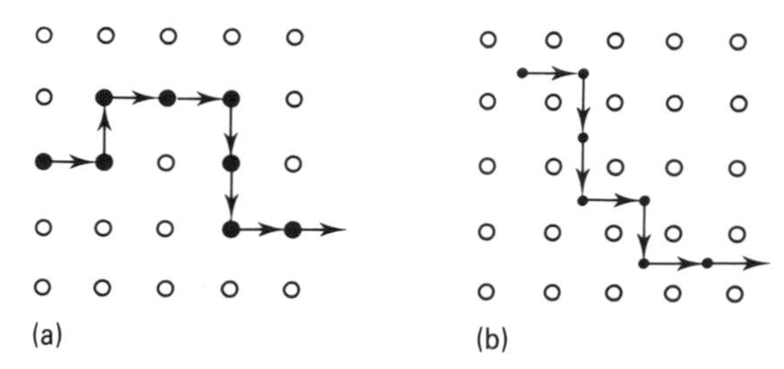

Figure 4.1 Model of substitutional (a) and interstitial (b) diffusion mechanisms.

and has the dimensions of area per unit time and dN/dx is the concentration gradient. The negative sign results from a negative gradient (which decreases from left to right) producing a flow in a positive direction (also from left to right).

The diffusion coefficient varies with temperature and is given by

$$D = D_o \exp(-E_a/kT)$$

where D_o is a constant and E_a is an activation energy. This equation holds for many substitutional and interstitial impurities in silicon. Typical values for D_o and E_a are given in Table 4.1. Values of D may be obtained from the equation or, alternatively, from a graph of D versus $1/T$, as shown in Figure 4.2.

It can be seen from Table 4.1 that there is a considerable difference between the activation energies for the metallic impurities and the n- and p-type substitutional impurities. These differences result in much larger values of diffusion coefficient for the metallic impurities and, as a result, the metallic impurities diffuse very rapidly through silicon. Of the three metals, gold is the only one that has been used to improve the switching speed. However, problems with long-term reliability have generally resulted

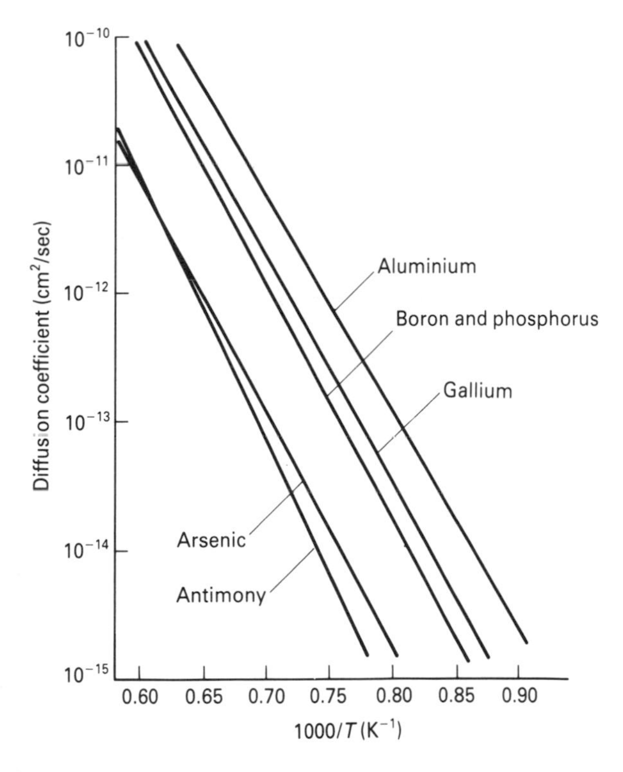

Figure 4.2 Variation of the diffusion coefficient with the reciprocal of temperature for a number of the more commonly used impurities for silicon.

Table 4.1 Diffusion and activation energies

Impurity	D_0 (cm^2/s)	E_a (eV)
Phosphorus	3.85	3.66
Arsenic	0.066	3.44
Boron	0.76	3.46
Gold	1.1×10^{-3}	1.12
Copper	4×10^{-2}	1.0
Iron	6.2×10^{-3}	0.87

in other methods being used to improve switching speed. Copper and iron are generally unwanted impurities and great care is taken to ensure that these and other metallic impurities are not present in any of the gases or chemicals used during the manufacturing process.

Fick's first law can be developed to describe the rate of flow of the impurity and can be expressed as:

$$\frac{dN}{dt} = D\frac{d^2N}{dx^2}$$

where N is the impurity concentration (cm^{-3}). This equation can be solved for a variety of different boundary conditions, the most commonly used of which are described below.

4.2.2 Constant surface concentration

Impurity diffusion for junction formation may be achieved by maintaining a constant source of vapour over the wafer for the duration of the diffusion. The solution of the equation gives:

$$N_{x,t} = N_s\left[\operatorname{erfc}\left(\frac{x}{2\sqrt{Dt}}\right)\right]$$

where $N_{x,t}$ is the impurity concentration (atoms/cm^3) at a distance x (cm) from the surface at a time t (sec), N_s is the constant surface concentration (atoms/cm^3), D the diffusion coefficient, x the distance (cm) into the wafer (i.e. $x = 0$ is the surface), t is the time in seconds and erfc is the complementary error function. The impurity profiles for different times are shown in Figure 4.3. The surface concentration N_s is determined by the solid solubility for the impurity in the semiconductor and is a function of temperature. Graphs of solid solubilities for some of the more commonly used impurities are shown in Figure 4.4[1].

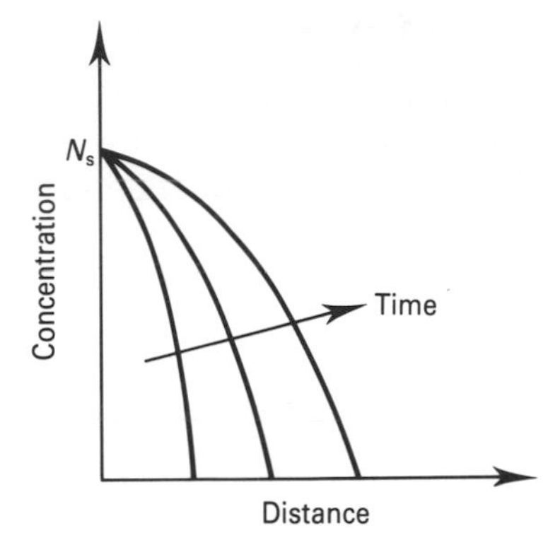

Figure 4.3 Variation of the impurity concentration with distance and time for a constant source.

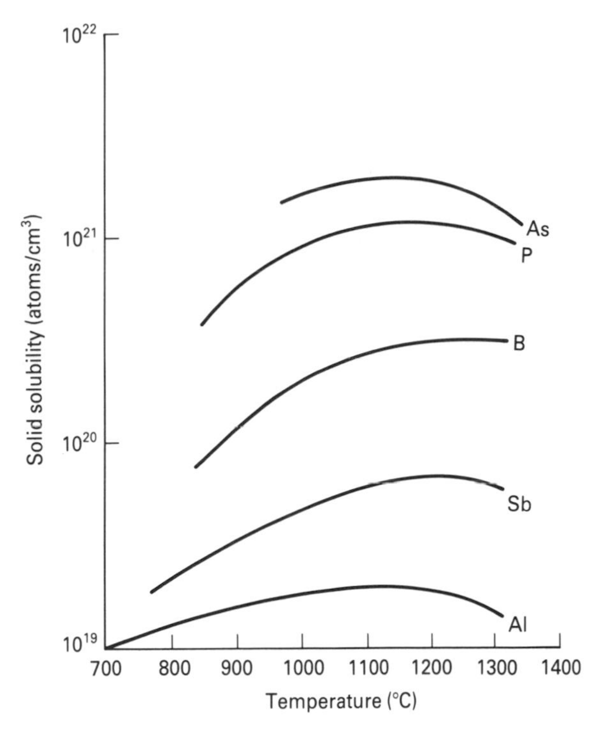

Figure 4.4 Solid solubilities of commonly used impurities in silicon. [After Trumbore[1].]

4.2.3 Diffusion from a finite source

Impurity diffusion may also be achieved by depositing a thin layer of impurity on to the surface of the wafer to produce a finite number of dopant atoms (Q) per unit area. Under these conditions the solution of the equation gives

$$N_{x,t} = \frac{Q}{\sqrt{\pi D t}} \exp - \left[\frac{x}{2\sqrt{Dt}} \right]^2$$

This equation produces a Gaussian distribution and the profiles for different times are shown in Figure 4.5. Notice that the surface concentration decreases with time.

4.3 Junction formation

A junction is formed when the impurity being diffused is of opposite conductivity to that of the substrate. Notice that, unlike the epitaxial layers which are described in Chapter 3, the thickness of the surface is not changed during diffusion. The net impurity concentration $N_d - N_a$ versus distance is shown in Figure 4.6. The junction x_j is formed where $N_{x,t} = N_b$, where N_b is the background concentration of the substrate. The value of x_j can be obtained from the solution of the above equations:

$$N_b = N_s \operatorname{erfc} \left[\frac{x_j}{2\sqrt{Dt}} \right]$$

or

$$N_b = \frac{Q}{\sqrt{\pi D t}} \exp \left[\frac{x_j}{2\sqrt{Dt}} \right]^2$$

The solution of these equations is greatly simplified by using the graphs of $Y = \operatorname{erfc}(X)$ and $Y = \mathrm{cxp} - (X)^2$, as shown in Figure 4.7.

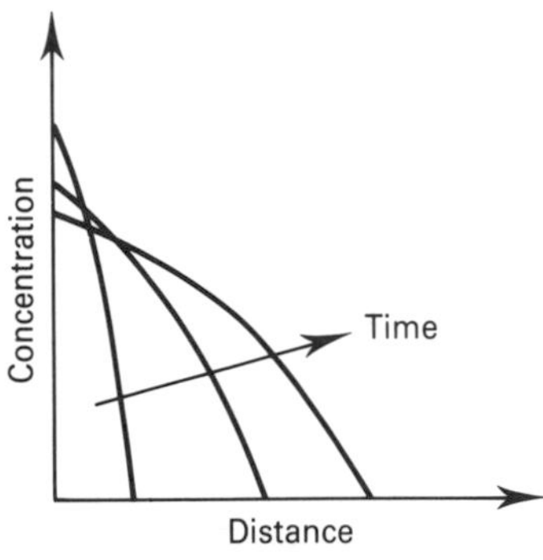

Figure 4.5 Variation of impurity concentration with distance and time for a limited source.

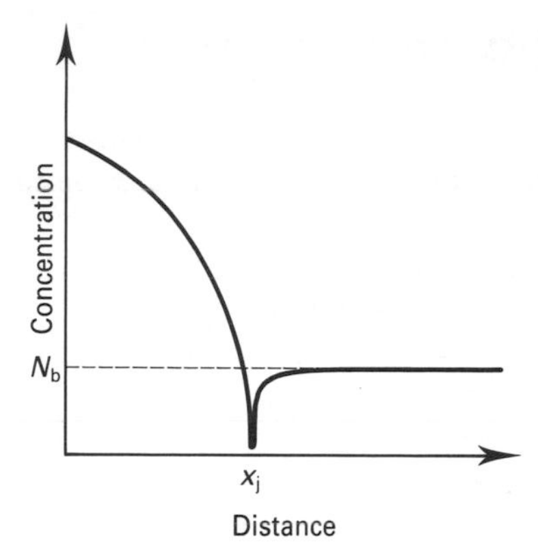

Figure 4.6 Variation of impurity concentration with distance showing the formation of a junction at x_j.

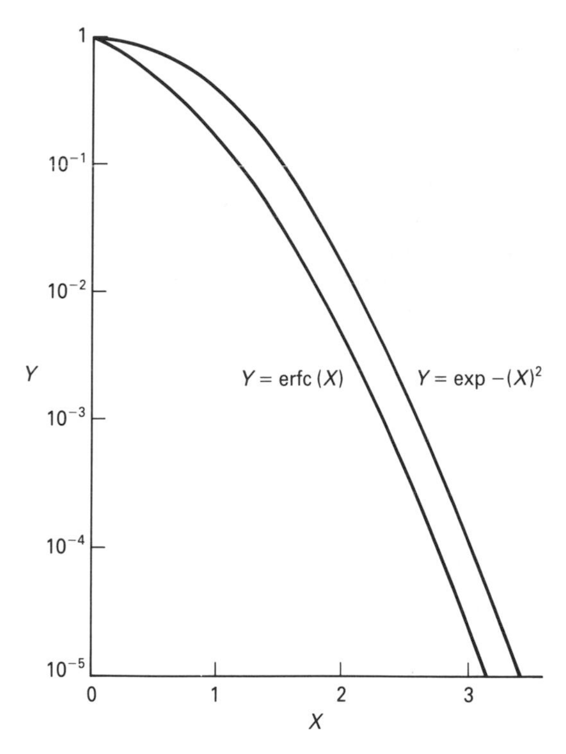

Figure 4.7 Graphs of $Y = \text{erfc}\,(X)$ and $Y = \exp - (X)^2$.

4.4 Redistribution diffusion

In practice, it is found that much better control of the impurity profile is obtained from a two-step diffusion involving a pre-deposition and a drive-in rather than from a single diffusion. The pre-deposition is performed at a relatively low temperature (900–1000 °C) to produce a shallow erfc profile. The surface concentration after this deposition can be obtained from the graphs shown in Figure 4.4. Notice that for the commonly used n- and p-type impurities the surface concentration is in excess of 1×10^{19} atoms/cm^3, even at temperatures as low as 800 °C. For many devices a much lower surface concentration is required, for example 1×10^{17} atoms/cm^3 is required for the base region of an npn transistor. This can only be achieved with a two-step diffusion.

The amount of impurity introduced during the pre-deposition is:

$$Q = N_s[D_1 t_1/\pi]^{1/2}$$

where the subscript 1 refers to the pre-deposition.

After the pre-deposition the wafer is transferred to a second furnace and a drive-in diffusion is performed, usually at a higher temperature (1050–1200 °C). Provided an oxide is not allowed to grow during the diffusion, that is, provided a non-oxidizing atmosphere is used, then the diffusion profile is Gaussian and is described by:

$$N_{x,t2} = \frac{Q}{\sqrt{\pi D_2 t_2}} \exp - \left[\frac{x}{2\sqrt{D_2 t_2}}\right]^2$$

where the subscript 2 refers to the second or drive-in diffusion.

While these equations can be used to estimate the profile and junction depth, in practice they are not very accurate. The assumption that the drive-in takes place in a non-oxidizing atmosphere is unlikely to be valid in practice. It is usual for the drive-in to be performed in an oxidizing atmosphere so that when the diffusion is complete the wafer is protected by an oxide film ready for the next step of the manufacturing process. When oxidization and diffusion are taking place at the same time the mathematical solution is much more complex as there is now a moving boundary at the silicon–silicon oxide interface. An added problem is the segregation of the impurity between the silicon and the oxide at the interface. These problems lend themselves to computer simulation and one such simulation program is SUPREM, which is described in more detail in Chapter 6. An early version of this program (SUPREM II) is available from Stanford University and, while lacking some of the sophistication and accuracy of later versions, is nevertheless a very adequate tool for examining a range of diffusion problems. Some examples of its use will be given in Chapter 6.

4.5 Cumulative heat cycles

During the manufacture of an integrated circuit a diffused region will be subjected to a number of heat cycles during each of which the impurity will diffuse. The cumulative effect of these temperature–time sequences on the diffusion profile can be obtained by calculating an effective *Dt* product as follows:

$$Dt_{\text{eff}} = D_1t_1 + D_2t_2 + D_3t_3 + \cdots$$

The diffusion coefficient for the temperature and time of each heat treatment is determined and an effective *Dt* product obtained. Thus, for example, to fabricate the emitter and base of a bipolar transistor the base is diffused first, followed by the emitter. Therefore, the base region of the transistor is subjected to the heat cycles required for the emitter pre-deposition and the emitter drive-in.

4.6 Lateral diffusion

In practice, the introduction of impurities into the silicon wafer is restricted by an oxide mask. It was noted in Chapter 2 that a relatively thin layer of oxide (0.5–1.0 μm) is sufficient to act as a mask against the diffusion of an impurity from an external source. The impurity that enters the exposed silicon diffuses in all directions, transversely into the silicon and laterally beneath the edge of the oxide window. However, the amount of impurity available at the edge of the window for lateral diffusion at the oxide–silicon interface is limited and the amount of diffusion is reduced. Typically, the lateral diffusion is about 80% of the transverse diffusion. This is shown in Figure 4.8. Lateral difusion must be taken into account when deciding the dimensions for device geometries.

4.7 Evaluation of diffused layers

Diffusion coefficients can be obtained from the measurement of the impurity profile, but this need only be done once for a new impurity or a new semiconductor. The more routine measurements are concerned with the junction depth and the amount of impurity in a diffused layer. This section will deal with the basic evaluation techniques and will also review some of the methods used for profiling.

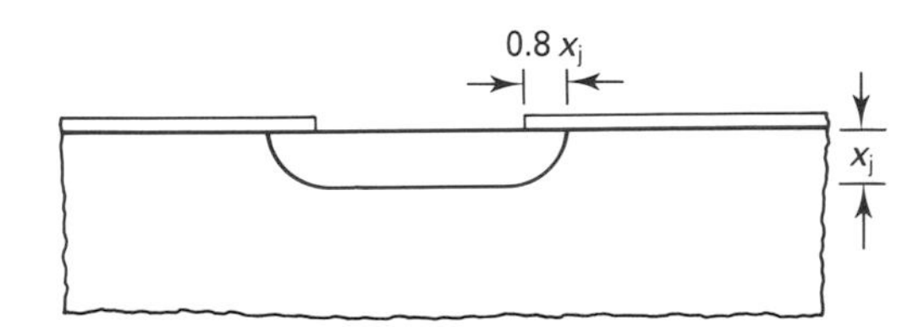

Figure 4.8 Lateral diffusion associated with a pn junction.

4.7.1 Junction depth and sheet resistance

Diffused layers can be evaluated by two simple measurements, the junction depth and the sheet resistance. The junction depth (x_j) is the distance from the surface to the plane where the concentration of the diffused impurity is equal to the background concentration of the substrate. Usually, the diffused region and the substrate are of opposite conductivity types, which enables a simple chemical staining method to be used to locate the position of the junction. A shallow bevel (1–5°) is ground on the edge of a test sample and a chemical stain consisting of HF with a few drops of HNO_3 is applied to the surface of the bevel. Under strong illumination the p-type region is stained darker than the n-type region. It is then a simple matter to use an optical interference method such as Tolansky[2] to measure the junction depth.

The sheet resistance can be measured using a four-point probe. Four spring-loaded steel probes are pressed on to the surface of a test wafer which contains a diffused layer. Current is passed through the outer pair of probes and the voltage is measured across the inner pair, as shown schematically in Figure 4.9. The sheet resistance R_s is given by:

$$R_s = \frac{V}{I} CF$$

where R_s is the sheet resistance in ohm/square, V is the voltage, I the current and CF is a correction factor. The correction factor is a function of the ratio of the diameter (d) of the test wafer containing the diffused layer and the spacing (s) between the probe tips. If d/s is large then CF approaches a value of 4.532[3]. If the current is set to 4.5 mA then the sheet resistance is obtained directly as a voltage reading.

Both junction depth and sheet resistance measurements are made on test wafers which are included with the batch of production wafers during the diffusion. As an alternative or in addition to these test wafers, special electrical test structures are included on the actual circuit wafers to enable measurements to be made directly on the device wafers. The results from these test structures are more valid but they can only be obtained after the wafers have been fully processed.

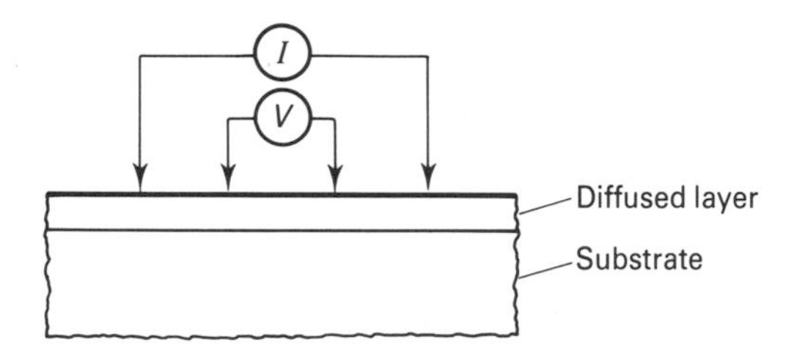

Figure 4.9 Four-point probe used to measure sheet resistance.

4.7.2 Profile measurement

Spreading resistance

The most widely used method for obtaining the impurity profile is the two-point probe spreading resistance method[4]. Since its initial development, the method has been refined and automated and is now used extensively to obtain the impurity profile and the junction depth.

A test wafer is bevelled at a shallow angle and the probes are placed on the surface of the bevel and mechanically stepped along the bevel at very small increments, for example 2.5 μm, as shown in Figure 4.10. At each step the current and voltage are monitored and a value obtained for the spreading resistance. A profile of the form shown in Figure 4.11 is obtained. Since resistance is proportional to resistivity and resistivity is proportional to $1/q\mu N$ where μ is the mobility, q is the electronic charge and N is the carrier concentration, then it is a simple matter to convert the resistance plot to one of concentration versus depth (in making this conversion it is necessary to take account of the variation of mobility with concentration), as shown in Figure 4.12. As can be seen, the junction depth is also obtained.

The measurement is very sensitive to sample preparation and to the condition of the probe tips. It is usual to have a set of silicon samples of known resistivity and to calibrate the instrument against these samples at frequent intervals. The probes have to be 'conditioned' at regular intervals to ensure reproducible results because, unlike the four-point probe, the contact resistance between the probes and the silicon must be considered in calculating the resistance of the underlying silicon.

The spreading resistance method can be used to measure multiple junctions, for example those associated with an npn transistor that includes the epitaxial layer and the buried n^+ layer.

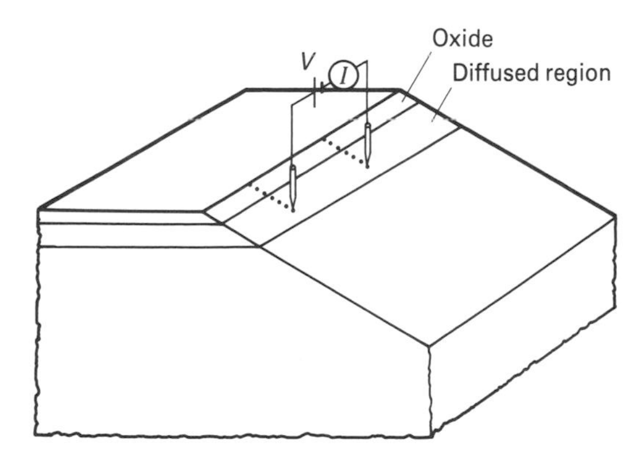

Figure 4.10 Spreading resistance method on a bevelled silicon sample.

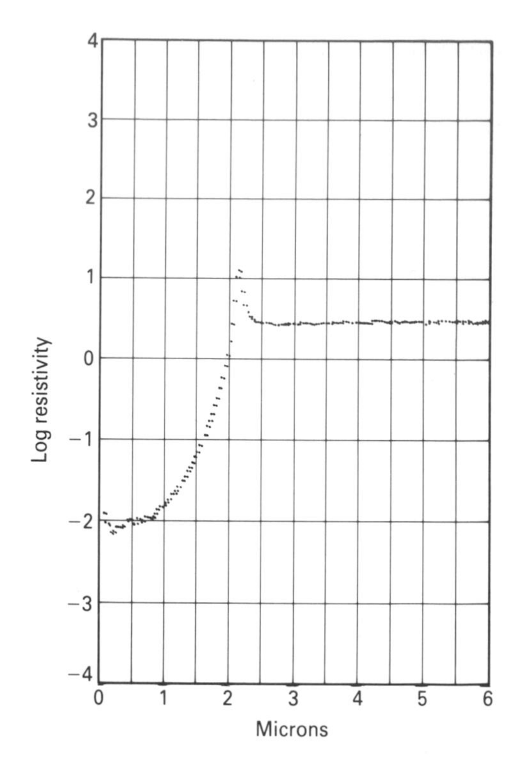

Figure 4.11 Results obtained for spreading resistance versus distance.

To make the measurement, it is first necessary to bevel the sample. A small piece of the test wafer is attached to a steel holder which has the required bevel machined on it. The holder with the sample attached is placed in a jig and rotated against a glass plate which is coated with a grinding medium such as diamond paste. A major difficulty is to ensure that the quality of the bevelled surface is consistent from run to run. To measure the profile on very shallow diffusions it is necessary to produce a bevel at a very small angle, for example 1°. The horizontal displacement of the probe tips is converted to a vertical depth by using the tan of the angle and thus the accuracy with which the angle can be measured is an additional factor in determining the accuracy of the profile.

C–V method

For an abrupt pn junction diode there is a relationship between the width of the depletion layer and the applied reverse voltage. As the voltage increases, the depletion layer uncovers more of the impurity ions and, in effect, uncovers the impurity profile. An expression[5] can be derived for the carrier concentration as follows:

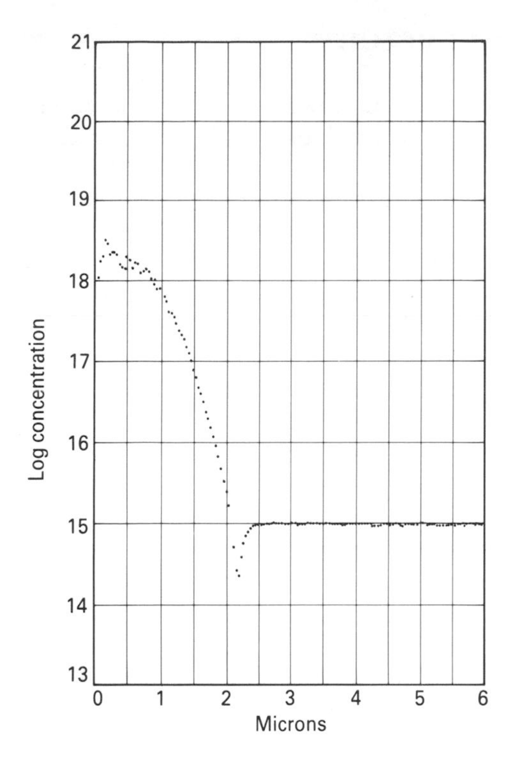

Figure 4.12 Conversion of the spreading resistance to impurity concentration versus distance.

$$N_x = \frac{C^3}{Aq\varepsilon_s} \frac{1}{\mathrm{d}C/\mathrm{d}V}$$

$$x = A\frac{\varepsilon_s}{C}$$

where N_x is the concentration at the edge of the depletion layer, C is the capacitance, V is the voltage, A is the area and ε_s the dielectric constant of silicon. As for the spreading resistance method, commercial equipment is available which incorporates a microprocessor to calculate the gradient of the capacitance versus voltage plot, and by means of the above equations produce the impurity profile of N_x versus x. The diodes can be actual pn junctions or, more usually, Schottky junctions (see Chapter 11) formed between a deposited layer of metal or a mercury probe. The method is well suited to the measurement of shallow junctions. The maximum depth of a few microns is limited by the breakdown voltage of the pn junction or Schottky junction.

4.8 Diffusion systems

The initial pre-deposition of the impurity can be achieved in a variety of ways. Both phosphorus and boron are available as liquid sources (boron tribromide and phosphorus oxychloride) and can be transferred to the slices as a vapour in a carrier gas such as nitrogen. The wafers are placed in a quartz tube furnace similar to that used for oxidation at a temperature of between 900 and 1000 °C. The impurity is transferred from the vapour to the surface of the wafers. The amount of impurity that enters the silicon is controlled by the solid solubility of the impurity in silicon at the temperature of the deposition (see Figure 4.4).

Alternatively, many of the most commonly used impurities are also available as spin-on sources. Oxides of the impurities are finely ground and mixed in a suitable solvent and binder so that they can be applied to the wafers as a liquid at room temperature. The wafers are spun to spread the liquid and the impurity uniformly over the surface. After drying, the wafers are placed in a furnace and heated to between 900 and 1000 °C. The solvents and binding agents are driven off and the impurities enter the wafer.

A further development of the use of the solid oxides of the impurities is to bind the oxide with a ceramic binder and press the material into a disk of the same diameter and thickness as the silicon wafers. The disks are loaded with the silicon wafers into the pre-deposition furnace so that the front surface of each silicon wafer is exposed to the surface of a source wafer (typically the silicon wafers are placed back to back and a source disk is placed between each pair of wafers). At the pre-deposition temperature the impurity vaporizes and is released from the ceramic binder to coat the surfaces of the adjacent silicon wafers.

After pre-deposition the wafers are removed from the furnace and any surplus impurity on the surface is removed by chemical etching. The wafers are then placed in the drive-in furnace and heated to 1000–1200 °C in an oxidizing environment. The impurities redistribute and the surface is reoxidized.

Metallic impurities such as gold can be applied by vacuum evaporation (see Chapter 8) in a layer about 10 nm thick. The impurity is driven into the silicon in the usual manner by heating the silicon in a furnace. However, because of the large value of the diffusion coefficient, at temperatures of 900–1000 °C, the gold is uniformly distributed throughout the wafer after only 10–15 min.

Dopant atoms can also be introduced by ion implantation and this method will be discussed in Chapter 5.

Problems

1. Consider an n-type silicon wafer with a background carrier concentration of 1×10^{15} atoms/cm^3. A constant source of boron of 1×10^{19} atoms/cm^3 is diffused for 45 min at 1100 °C. Determine the junction depth.
[1.8 μm.]

2. A constant source of boron of 4×10^{18} atoms/cm^3 is diffused into an n-type silicon wafer containing 4×10^{16} atoms/cm^3 at 1200 °C:

(a) What is the concentration at a depth of 2 m after 20 min?
(b) What is the junction depth if the difusion is terminated after a further 30 min?

[7.2×10^{16}/cm^3, 3.4 μm.]

3. Boron is diffused into a 0.35 ohm cm n-type epitaxial layer from an initial deposition of 5×10^{13} atoms/cm^2 at 1200 °C for 3 h. This is followed by a second diffusion process at 1180 °C for 40 min and by a third diffusion at 1100 °C for 20 min. Determine the junction depth of the boron in the n-type epitaxial layer.
[5.4 μm.]

4. A p-type diffused resistor is formed by diffusing boron into an n-type epitaxial layer which has an impurity concentration of 5×10^{16} atoms/cm^3. A 15 min constant source pre-deposition at 950 °C results in a surface concentration of 5×10^{19} atoms/cm^3. This is followed by a 30 min drive-in at 1100 °C. Determine the junction depth and the surface concentration after the drive-in.
[1 μm, 3.4×10^{18}/cm^3.]

References

[1] F. A. Trumbore (1960), 'Solid solubilities of impurity elements in geranium and silicon', *Bell System Tech. J.*, **39**, 205.

[2] W. L. Bond and F. M. Smits (1956), 'The use of an interference microscope for the measurement of extremely thin surface layers', *The Bell System Tech.*, **35**, 1209–21.

[3] F. M. Smits (1958), 'Measurement of sheet resistivities with the four-point probe', *The Bell System Tech. J.*, **37**, 711.

[4] R. G. Mazur and D. H. Dickey (1966), 'A spreading resistance technique for resistivity measurement on silicon', *J. Electrochem. Soc.*, **113**, 255.

[5] C. P. Wu, E. C. Douglas and C. W. Mueller (1975), 'Limitations of the CV technique for ion implanted profiles', *IEEE Trans. Electron. Devices*, **ED-22**, 319.

CHAPTER 5

Ion implantation

5.1 Introduction

Ion implantation is a non-thermal method of introducing high-energy impurity ions into silicon. Dopant atoms are ionized, accelerated and directed at the silicon wafer. The high-energy ions enter the crystal lattice, collide with the silicon atoms and finally come to rest. The energies required are in the range 50–500 keV and for ions of the commonly used impurities, such as boron, phosphorus and arsenic, it is possible to implant the ions from 10–1000 nm beneath the silicon surface. The ion implanter is now an essential piece of equipment for the majority of semiconductor manufacturers.

An important advantage of ion implantation is the ability to control the number of implanted impurity atoms accurately, which allows concentration levels from 10^{14} to 10^{20} atoms/cm^3 to be obtained with considerable accuracy. This control is far greater than that which can be achieved using solid-state diffusion, which relies on the solid-solubility concentration of the pre-deposition and a drive-in to redistribute the impurities. A further advantage is that the impurities can be implanted through a layer of oxide. This has important benefits for the manufacture of MOS circuits for gate voltage threshold adjustment. Further advantages result from a much wider range of elements which can be ionized and implanted as compared with the number of elements which can be introduced by high-temperature diffusion, for example oxygen ions can be implanted to form a buried layer of silicon oxide which can be used for component isolation.

Ion implantation takes place in a vacuum and, therefore, is inherently clean. It takes place at room temperature and, therefore, previously implanted or diffused impurities are unaffected, unlike high-temperature diffusions when impurity profiles may change. Ions can be confined to selected regions of the wafer with oxide masks, but in addition polymer masks formed from photo-resist can be used. Impurity concentration is measured directly as an electric current and the profile can be controlled by varying the acceleration voltage. All of these factors contribute to the general acceptance of ion implantation as one of the most versatile tools for the manufacture of

VLSI circuits. As device dimensions diminish and junctions become shallower the n- and p- regions are increasingly being formed by ion implantation rather than by high-temperature diffusion.

5.2 Ion implantation system

A schematic of a typical ion implanter is shown in Figure 5.1. The equipment comprises:

- An ion source where a gas such as arsine, diborane or phosphine is used to generate a plasma at a pressure of about 10^{-3} torr. A dc voltage of about 20 kV is used to attract the positively charged ions and to extract them from the plasma.
- From the many ions produced in the plasma an analyzing magnet selects one species of ion which is passed through to the acceleration tube.
- The acceleration tube supplies the energy required to enable the ions to penetrate beneath the surface of the target.
- After final focussing, the beam is deflected in both X and Y directions by electrostatic deflection plates. The target is positioned slightly off centre so that neutral ions which cannot be deflected or detected by the Faraday cage do not reach the target.
- The raster scanning produced by the X and Y deflections ensures that the beam directs ions uniformly over the surface of the wafer. The vacuum in the acceleration and target chamber is of the order of 10^{-6} torr.
- The target is mounted in a Faraday cage which enables the number of ions to be measured by simply measuring the beam current and integrating with time.

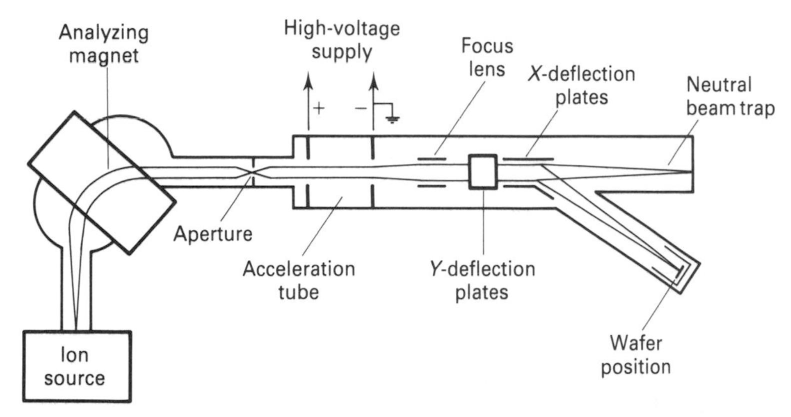

Figure 5.1 Schematic of an ion implantation system.

5.3 Dose control

An important feature of ion implantation is the dose or number of ions per unit area which are implanted in the silicon. The wafer is in good electrical contact with the target holder which is itself part of a Faraday cage (charge integrator) and the dose (Q) can be defined as:

$$Q = \frac{It}{mqA} \text{ atoms/cm}^2$$

where I is the beam current, t is the time, mq represents the charge of an ion and A is the area. For example, consider a beam current of 10^{-9} amps swept over an area of 1 cm^2 for 100 s; then the dose is 6.25×10^{12} atoms/cm^2 for $m = 1$. If the depth of the implanted layer in the silicon is 50 nm then the concentration is 1.25×10^{18} atoms/cm^3. The use of higher current for a similar length of time can realize concentrations of 10^{20} atoms/cm^3 on large-diameter wafers.

5.4 Distribution profiles

The high-energy ions enter the target and collide with the electrons and atoms and lose energy until they eventually come to rest. The directions taken by the ions change at each collision and the resulting paths are completely random. Because of the statistical nature of the collision process different ions come to rest at different positions. The path length is called the projected range R_p (Figure 5.2(a)) and represents an average for a large number of ions. However, some ions come to rest sooner and some later, giving a spread dR_p. The final distribution for a large number of ions can be represented by a Gaussian curve with a peak concentration at a depth R_p and a standard deviation dR$_p$, as shown in Figure 5.2(b). The profile can be represented by the equation:

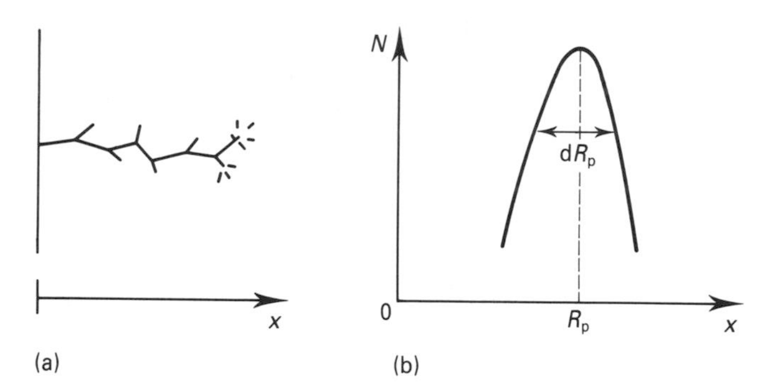

Figure 5.2 (a) Path taken by an ion in coming to rest and (b) the profile of the distribution of a large number of ions with a projected range R_p and a standard deviation dR_p.

$$N_x = \frac{Q \exp[-((x - R_p)^2/2dR_p^2)]}{\sqrt{2\pi}\, dR_p}$$

where $Q/(\sqrt{2\pi}\, dR_p)$ represents the maximum concentration at the position where $x = R_p$.

The Gaussian profile represents a good approximation for ions implanted in an amorphous substrate, but for a single-crystal substrate the profile is more complex with a pronounced skewness and a tail. More complex statistical models have been developed to model these effects. One of these is the Pearson distribution and an example of the distribution as measured and simulated by both Gaussian and Pearson is shown in Figure 5.3.

Graphs of projected range and standard deviation for boron, phosphorus and arsenic are shown in Figure 5.4.

For single-crystal material very large penetrations can occur if the ion enters in a direction which coincides with the open spaces between the rows of atoms. This is known as channelling and allows the ion to travel considerably greater distances than would be the case in amorphous material. Practical application of channelling is limited because of the very critical control of the orientation of the crystal with respect to the ion flight path. Generally, for implantation to be successful channelling is avoided by tilting the wafer a few degrees off the main crystal axis or by implanting through an amorphous material such as SiO_2.

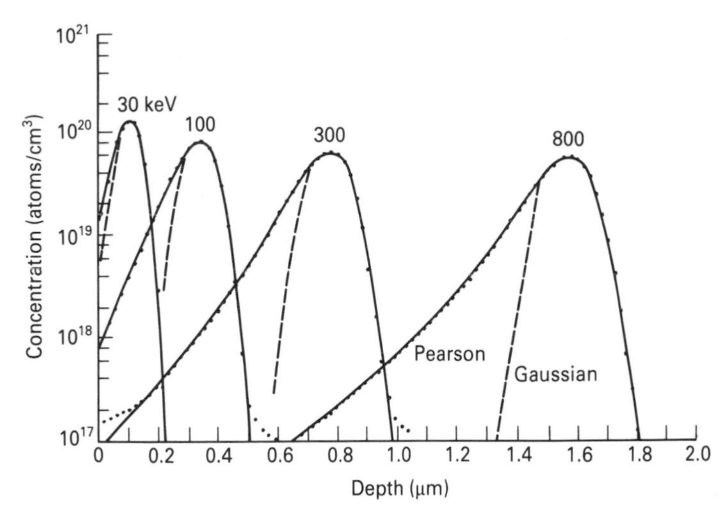

Figure 5.3 Comparison of the measured distribution of boron in silicon with the Gaussian and the Pearson profiles.[1] [Reprinted by permission of Elsevier Science Publishers Ltd.]

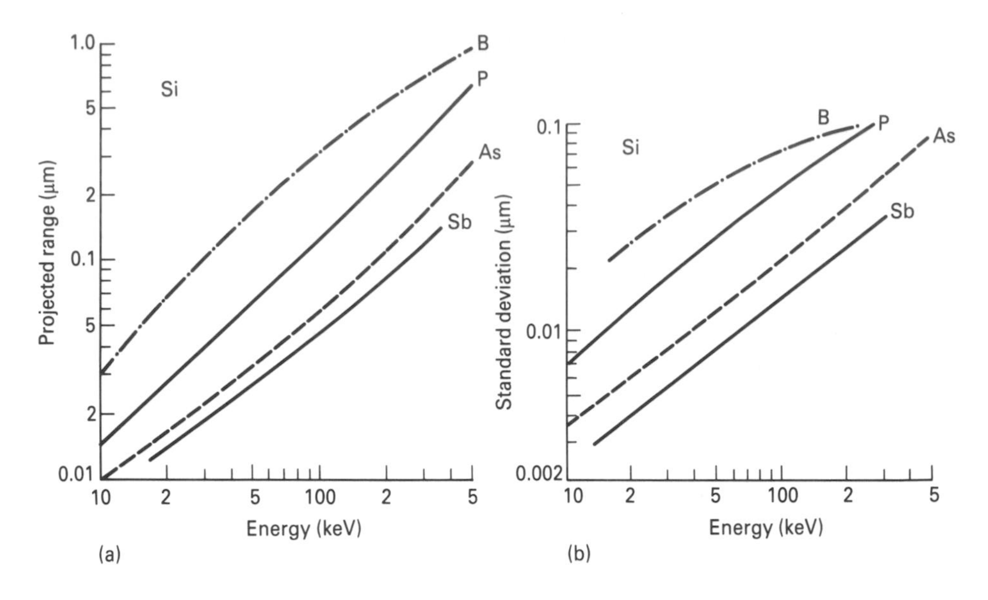

Figure 5.4 (a) Projected range of boron, phosphorus and arsenic in silicon and (b) the standard deviation.

5.5 Annealing of implanted impurities

The ions come to rest by a combination of electronic and nuclear scattering events, with the nuclear events resulting in displaced substrate atoms. Light and heavy ions behave differently. Light ions, such as boron, initially undergo electronic scattering, continually losing energy and then finally coming to rest as a result of nuclear scattering. Thus the light ion leaves a large number of displaced atoms near the peak concentration. In contrast, the heavy ion, for example phosphorus or arsenic, enters the surface and immediately undergoes nuclear scattering. These ions displace large numbers of atoms near the surface, resulting in a much broader spread of displaced atoms in the substrate.

With increasing dose the individual regions of damage merge together and with a sufficiently large dose the material becomes amorphous. The lighter the ion the larger the dose required to produce this amorphous condition. As the material becomes amorphous the profile more closely approximates to a Gaussian profile.

Initially, only a small number of ions will occupy substitutional sites and as a result a large bulk of the ions is electrically inactive. To remove the damage and to make the ions active it is necessary to anneal the substrate by applying heat for an appropriate time.

Annealing is complex and depends on whether an amorphous region has been formed. Generally, temperatures of 400–600 °C for several minutes are required to bring about electrical activity but higher temperatures and/or longer times may be required for full recovery of mobility and lifetime. Ion implantation is often used in place of a high-temperature pre-deposition and the implanted impurities are redistributed with a drive-in diffusion, for example the well for CMOS circuits. The temperature may typically be in excess of 1000 °C and all damage is removed and all of the ions are electrically active. However, for ion implantation which is used for threshold voltage adjustment through the gate oxide of an MOS transistor the annealing temperature must be limited to about 400 °C to avoid damage to the oxide.

As device geometries shrink, the movement of the impurities as a result of diffusion which takes place during the anneal limits the minimum dimensions that may be achieved. This can be avoided by using rapid annealing. Laser beams have been used for this purpose. The energy is concentrated in the surface layer and temperatures of 800 °C can be achieved in seconds which is enough to anneal the damage but with very little diffusion. High-energy beams and infrared radiation are capable of producing rapid annealing and much work is still being done in this area.

5.6 Masking

Because ion implantation takes place at room temperature, light-sensitive polymer films can be used for masking. The photosensitive films can be applied and selected regions removed where implantation is required. Provided the remaining film is of adequate thickness then it stops the ions entering the substrate. This means that ions can be implanted in selected areas. The films are removed after implantation and the slices proceed to the next stage in the process without the need for high-temperature oxide masks which are required for high-temperature diffusions.

5.7 Applications and future trends

Ion implantation is now used extensively for both bipolar and MOS VLSI circuits. For bipolar circuits ion implantation is used for the buried n^+ layer because higher concentrations can be achieved than with thermal pre-depositions. The base region is often formed with two implants; a low concentration implant followed by a shallower and higher concentration implant. This results in a more optimum profile than can be achieved with thermal diffusion alone. Finally, a shallow high-concentration emitter is produced by implanting arsenic rather than phosphorus. This results in a much improved profile with a more abrupt change of concentration between the emitter and base regions and a resultant improvement in the operation of the transistor.

The ability to implant very shallow layers with high concentrations is used to produce low-temperature coefficient high-value resistors. Sheet resistance values of 4000 ohm/square with temperature coefficients of less than 300 ppm/°C greatly extend the applications for operational amplifiers.

For MOS circuits gate threshold voltage adjustment is now a standard procedure which could not be achieved without ion implantation. The n- and p-type wells for CMOS can be produced with low concentrations and with almost uniform distribution because of the ability to implant deep beneath the surface. Avalanche breakdown and latchup can be better controlled with lightly doped drains which are produced with selective ion implantation with a high-concentration region for a low-resistance contact and a low-concentration region adjacent to the channel to minimize hot-carrier injection into the gate oxide.

Ion implantation provides greater control of both concentration and distribution of a very wide range of impurities and will become increasingly important as device dimensions continue to shrink.

References

[1] W. K. Hofker (1975), 'Implantation of boron in silicon', *Philips Res. Repts. Suppl.*, No.8.

CHAPTER 6

Process simulation

6.1 Introduction

The processes of epitaxy, oxidation, diffusion and ion implantation are complex. While the equations and graphs presented in the earlier chapters provide the means for making an initial estimate of a layer thickness or a junction depth for a single step, the task of predicting the effects of a series of process steps, each following one after the other, is extremely complex. For this reason computer simulation was developed during the 1970s and now virtually all aspects of VLSI design and manufacture can be simulated.

SUPREM[1] (Stanford University PRocess Engineering Model) was described in 1978 and is now widely used by many manufacturers. An early version of the program – SUPREM II – is in the public domain (or may be obtained from the Software Distribution Centre at Stanford[2]) and is an excellent tool for investigating the processes which have been described in the earlier chapters. It lacks the sophistication of the later version – SUPREM III – or the two-dimensional version – SUPREM IV, but nevertheless provides the means to study oxidation, diffusion, epitaxy and ion-implantation at a very modest cost. It is particularly useful for studying the effect of process changes on the impurity profiles and junction depths.

Computer simulation is a valuable aid for process engineers, allowing them to run trial processes on the computer before committing the production line to silicon. A complete process can be simulated in a few hours on a computer, whereas on the production line it would take several weeks, and further weeks if it is necessary to make corrections to the process conditions. It is ideally suited to running statistical experiments to obtain the optimum process conditions.

This chapter is largely based on SUPREM II, although the author now uses SUPREM III and SUPREM IV. It should not be too difficult to obtain SUPREM II to run on a variety of computers which have Fortran compilers. If funds are available to purchase SUPREM III then so much the better and the information provided here will still be applicable, although the input statements will be slightly different. SUPREM IV is very nice to use in that it provides a two-dimensional image, or a cross-section through a device, but it is more difficult to use, and it requires much more computer time, which makes it difficult to complete class exercises in a reasonable time.

6.2 SUPREM

The general availability of the early version of SUPREM has resulted in it being widely used by industry and academic establishments. It is a one-dimensional simulator which computes the concentration profile of impurities in the oxide and the silicon normal to the surface of the substrate. The spatial resolution of the profile can be selected by the user so that both very shallow profiles and much deeper profiles can be displayed. Because of the one-dimensional nature of the simulation a number of profiles are necessary to fully characterize a complex device structure such as a bipolar transistor. An example of the cross-section of an npn transistor with a buried n^+ layer beneath the emitter and with junction isolation is illustrated in Figure 6.1. The profile AA′ represents the isolation region, BB′ is used for the collector contact, CC′ shows the base region which also corresponds to a diffused resistor, and finally DD′ provides a cross-section of the emitter, base and collector. Note that the profiles CC′ and DD′ can be compared to show the presence of any difference between the external base profile and the base profile beneath the emitter, for example emitter-push, as shown in the diagram (see Chapter 10).

6.3 Input data format

The input data for the program is in the form of a text file describing the process steps, times, temperatures, concentrations, etc. The text file comprises the following line types:

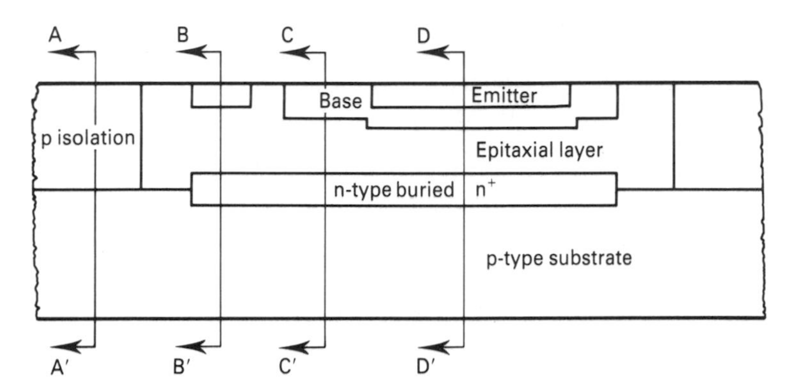

Figure 6.1 Cross-section of an npn transistor with junction isolation and buried n^+ layer.

TITLE
COMMENT
GRID
SUBSTRATE
PRINT
PLOT
STEP
MODEL
END

The name which follows the word TITLE is used as a header for the various output files.

Comments may be included anywhere in the file to make the data easier to interpret. A comment line starts with COMM.

The GRID line provides the means of scaling the output information: fine scaling for shallow junctions and coarser scaling for, say, a deeper isolation region. The parameters required for the GRID line are illustrated in Figure 6.2. The silicon is divided into two regions, one region close to the surface where impurity profiles may change rapidly and a region beneath this with a lower resolution. DYSI specifies the grid spacing in the high-resolution area, which has a depth DPTH. YMAX is the maximum calculation thickness and includes the low-resolution region.

The material properties of the substrate are contained in the SUBSTRATE line, which includes crystal orientation, impurity type and concentration.

The TITLE, GRID and SUBSTRATE lines form the first three lines of data. The PRINT and PLOT lines determine the output format. The process steps are described in the STEP line and the following processes are modelled:

Ion implant
Oxidation
Pre-deposit
Etching
Low-temperature oxide deposition
Epitaxy

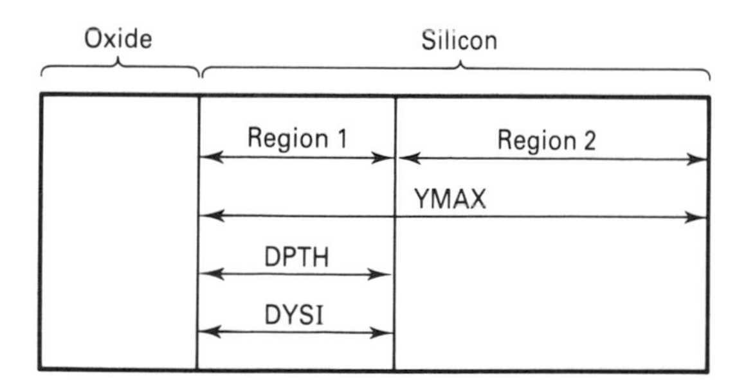

Figure 6.2 Diagram showing the relationship of the three GRID parameters – DPTH, DYSI and YMAX in SUPREM II.

For each process line there are a number of parameters such as time, temperature, element, concentration, ion dose, implant energy, growth rate and a 'model'. The model describes the particular mathematical algorithm which is used to describe the particular process. There are a number of default models which are adequate for many applications, and particularly for educational purposes. More complex models can be included with the MODEL line.

Model information is provided within SUPREM for elements, oxidation and epitaxy as follows:

Element:	Boron MBO#
	Phosphorus MPH#
	Antimony MSB#
	Arsenic MAS#
Oxidation:	Steam STM#
	Wet WET#
	Dry DRY#
	Nitrogen NIT#
Epitaxy:	EPI#

where # has a value from 0 to 5 and identifies variations within each of the models. For the purpose of demonstrating a particular process a value of # = 0 is adequate. If SUPREM is to be used to model an actual process then it would be necessary to select the correct model from those available, or enter data for a new model, to ensure that there is agreement between simulated and measured results.

6.4 Example program

An example of a deposition and drive-in of boron into a phosphorus substrate is shown in Figure 6.3. Notice that the data is terminated with END at line 9. In line 5 the PLOT command specifies a window (WIND) of 1 μm, while in line 7 there is a value of 4 μm. This change of spatial resolution allows the first plot to show the very shallow pre-deposition while the second plot shows the deeper drive-in. The output from SUPREM is shown in Figures 6.4–6.7.

From the printed data in Figure 6.4 notice that the junction depth is 0.34 μm, the sheet resistance of the deposition is 82.5 ohm/square and the surface concentration is 9.9×10^{19} atoms/cm^3. These figures are confirmed in the plot of the profile in Figure 6.5. After the drive-in the junction depth is 1.4 μm and the sheet resistance is 110 ohm/square with a surface concentration of 1.35×10^{19} atoms/cm^3 as described by the output in Figure 6.6 and illustrated graphically in Figure 6.7. Note also from Figure 6.6 that there is an oxide of 0.1 μm. The scaling factor for the oxide region is different from that used for the silicon region.

```
* * * STANFORD UNIVERSITY PROCESS ENGINEERING MODELS PROGRAM * * *

                        * * * VERSION 0-05 * * *

1----TITLE DIFF3
2----GRID YMAX=4, DPTH=2,DYSI=0.01
3----SUBS ELEM=P, CONC=SE15, ORNT=111
4----PRINT HEAD=Y
5----PLOT TOTL=Y, WIND=1
6----STEP TYPE=PDEB, TEMP=950, TIME=45, ELEM=B, CONC=1E20
7----PLOT TOTL=Y,WIND=4
8----STEP TYPE=OXID, TEMP=1100, TIME=45, MODL=DRY1
9----END
```

Figure 6.3 SUPREM II input data for a two-step diffusion.

6.5 Practical considerations

When considering a complex structure such as an npn transistor, as illustrated in Figure 6.1, it is important to include all of the heat cycles for each cross-section. For example, for the npn transistor the manufacturing sequence is as follows:

1. grow an initial oxide;
2. etch the oxide;
3. ion implant arsenic for the buried n^+;
4. drive-in the arsenic;
5. remove any oxide;
6. grow the epitaxial layer;
7. grow an oxide;
8. etch the oxide;
9. deposit phosphorus for the isolation;
10. drive-in the phosphorus;
11. remove any oxide;
12. ion implant boron for the base;
13. drive-in the boron;
14. ion implant arsenic for the emitter;
15. drive-in the emitter.

All of the above steps are required for each cross-section with the exception of some of the ion implants (or pre-deposits if pre-deposition is used in place of ion implantation). For example, for the isolation region represented by section AA′ steps 3, 12 and 14 are not required; for section DD′ step 9 is not required.

For demonstration purposes it is sufficient to simply specify a single temperature and time for a diffusion or oxidation step, but for an actual manufacturing process, particularly one in which the junction depths are sub-micron, the temperature ramp at the start and end of the process must be included, plus the time spent stabilizing the wafers before entering the furnace and after being withdrawn from the high-temperature region. A correction may also be applied for the native oxide which forms

```
GASEOUS PREDEPOSITION
TOTAL STEP TIME      =  45.0 MINUTES
INITIAL TEMPERATURE =   950.000      DEGREES C.
OXIDE THICKNESS      =   0.0000      MICRONS
PREDEPOSITION IMPURITY = BORON
GAS CONC. AT INTERFACE =   1.000000E+20 ATOMS/CC

             I  OXIDE          I  SILICON        I                I  SURFACE        I
             I  DIFFUSION      I  DIFFUSION      I  SEGREGATION   I  TRANSPORT      I
             I  COEFFICIENT    I  COEFFICIENT    I  COEFFICIENT   I  COEFFICIENT    I
--------------------------------------------------------------------------------------
PHOSPHORUS   I   1.72788E-07   I   1.91820E-05   I    10.000      I   5.68325E-03   I
--------------------------------------------------------------------------------------
BORON        I   5.40474E-09   I   2.55658E-05   I   0.20043      I    1.0000       I

 SURFACE CONCENTRATION =   9.983885E+19 ATOMS/CM3

   JUNCTION DEPTH        I       SHEET RESISTANCE
-------------------------I-------------------------
0.338743        MICRONS  I   82.5479      OHMS/SQUARE
                         I   2853.11      OHMS/SQUARE


NET ACTIVE CONCENTRATION

OXIDE    CHARGE =  0.000000          IS    0.000      % OF TOTAL
SILICON  CHARGE =  1.419630E+15      IS     100.      % OF TOTAL
TOTAL    CHARGE =  1.419630E+15      IS    7.098E+04  % OF INITIAL
INITIAL  CHARGE =  2.000000E+12


CHEMICAL CONCENTRATION OF PHOSPHORUS

OXIDE    CHARGE =  0.000000          IS    0.000      % OF TOTAL
SILICON  CHARGE =  1.984968E+12      IS     100.      % OF TOTAL
TOTAL    CHARGE =  1.984968E+12      IS     99.2      % OF INITIAL
INITIAL  CHARGE =  2.000000E+12


CHEMICAL CONCENTRATION OF BORON

OXIDE    CHARGE =  0.000000          IS    0.000      % OF TOTAL
SILICON  CHARGE =  1.417964E+15      IS     100.      % OF TOTAL
TOTAL    CHARGE =  1.417964E+15      IS    0.000      % OF INITIAL
INITIAL  CHARGE -  0.000000
```

Figure 6.4 SUPREM II output for the pre-deposit generated by line 6.

on silicon. For ion implantation into bare silicon the presence of the thin native oxide may affect the impurity profile. In practice there is likely to be a delay between removing the oxide and performing the implantation. During this time a native oxide will form with a thickness of 1–1.5 nm so that for the real process the ions are implanted through the native oxide. This can be included in the simulation by depositing an appropriate thickness of oxide on to the surface at room temperature.

A process simulator uses a large number of mathematical models to represent the various processes, such as oxidation, diffusion and ion implantation. There are often alternative models for the same process. Each of the models relies on constants obtained from experiments on oxidation to determine the growth rate constants, on

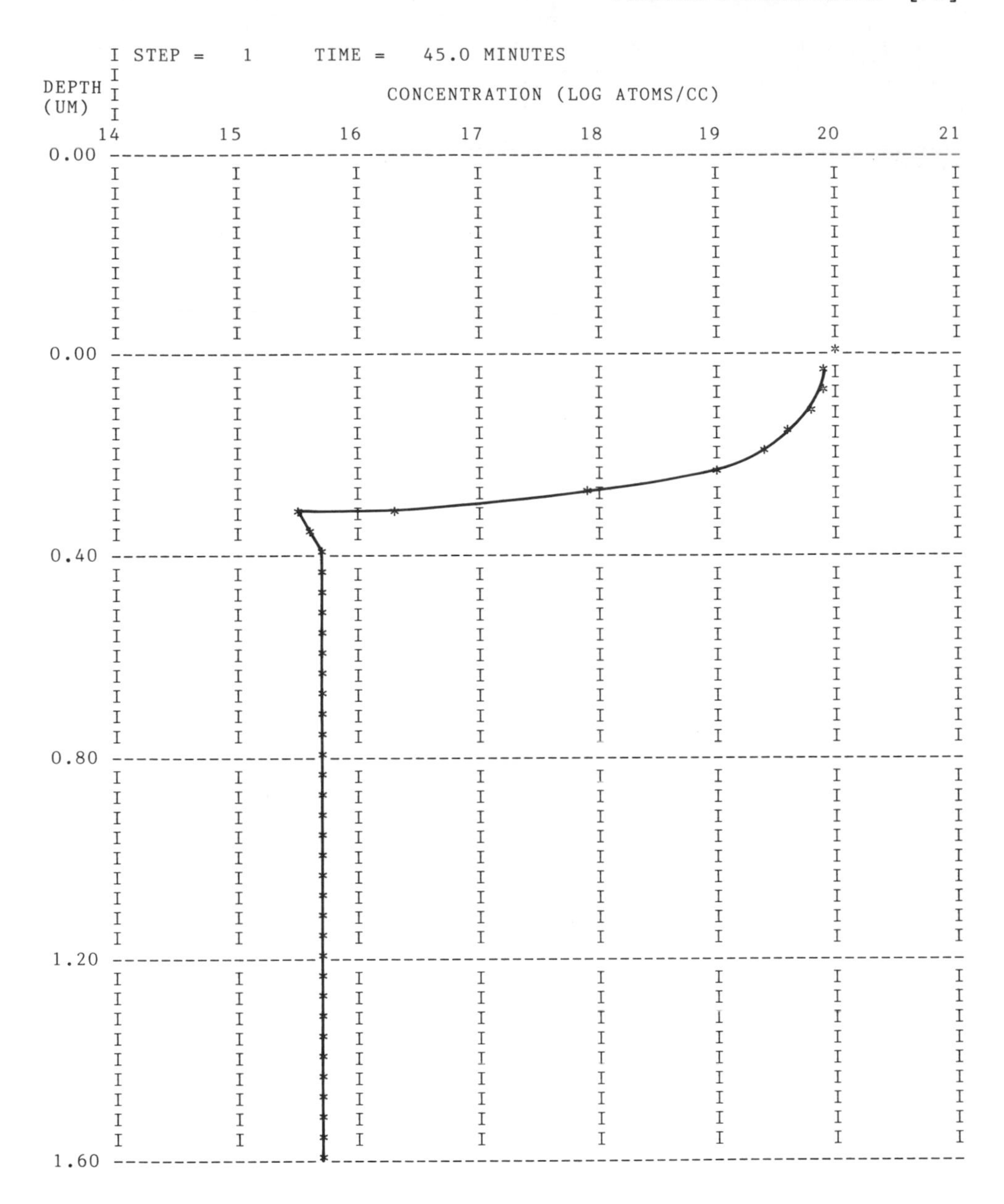

Figure 6.5 SUPREM II graphical output after the pre-deposit.

```
OXIDATION IN DRY OXYGEN
TOTAL STEP TIME      =  45.0 MINUTES
INITIAL TEMPERATURE =   1100.00      DEGREES C.
OXIDE THICKNESS     =    0.1027      MICRONS

LINEAR    OXIDE GROWTH RATE =  4.737751E-03 MICRONS/MINUTE
PARABOLIC OXIDE GROWTH RATE =  3.954384E-04 MICRONS!2/MINUTE
OXIDE GROWTH PRESSURE       =   1.00000     ATMOSPHERES

            I  OXIDE        I  SILICON      I                I  SURFACE       I
            I  DIFFUSION    I  DIFFUSION    I  SEGREGATION   I  TRANSPORT     I
            I  COEFFICIENT  I  COEFFICIENT  I  COEFFICIENT   I  COEFFICIENT   I
-------------------------------------------------------------------------------
PHOSPHORUS  I   6.49936E-06 I   8.51663E-04 I    10.000      I   4.47013E-02 I
-------------------------------------------------------------------------------
BORON       I   2.09718E-07 I   9.81057E-04 I   0.51470      I   1.31132E-02 I

SURFACE CONCENTRATION = 1.358548E+19 ATOMS/CM3

  JUNCTION DEPTH       I      SHEET RESISTANCE
----------------------I-------------------------
 1.42121      MICRONS  I   110.187     OHMS/SQUARE
                       I   4236.52     OHMS/SQUARE


NET ACTIVE CONCENTRATION

OXIDE   CHARGE =  4.095991E+14    IS    32.4       % OF TOTAL
SILICON CHARGE =  8.563965E+14    IS    67.6       % OF TOTAL
TOTAL   CHARGE =  1.265996E+15    IS    89.2       % OF INITIAL
INITIAL CHARGE =  1.419630E+15


CHEMICAL CONCENTRATION OF PHOSPHORUS

OXIDE   CHARGE = 4.349002E+09     IS   0.219       % OF TOTAL
SILICON CHARGE = 1.979730E+12     IS    99.8       % OF TOTAL
TOTAL   CHARGE = 1.984079E+12     IS    100.       % OF INITIAL
INITIAL CHARGE = 1.984968E+12


CHEMICAL CONCENTRATION OF BORON

OXIDE   CHARGE = 4.096034E+14     IS    32.4       % OF TOTAL
SILICON CHARGE = 8.559189E+14     IS    67.6       % OF TOTAL
TOTAL   CHARGE = 1.265522E+15     IS    89.2       % OF INITIAL
INITIAL CHARGE = 1.417964E+15
```

Figure 6.6 SUPREM II output for the drive-in generated by line 8.

diffusion to determine the diffusion coefficients, on activation energies for the different impurities and on ion implantation to determine the range and standard deviation constants. In some cases the growth rates and impurity profiles are calculated from analytical expressions, but in others they are obtained from tables of data with interpolation being used to extract the necessary values where the tabulated data do not coincide with the actual temperature, or energy or dose used in the manufacturing process. While the default values will produce results they may not be sufficiently accurate and the final choice of model and value of a constant is left to the user to vary

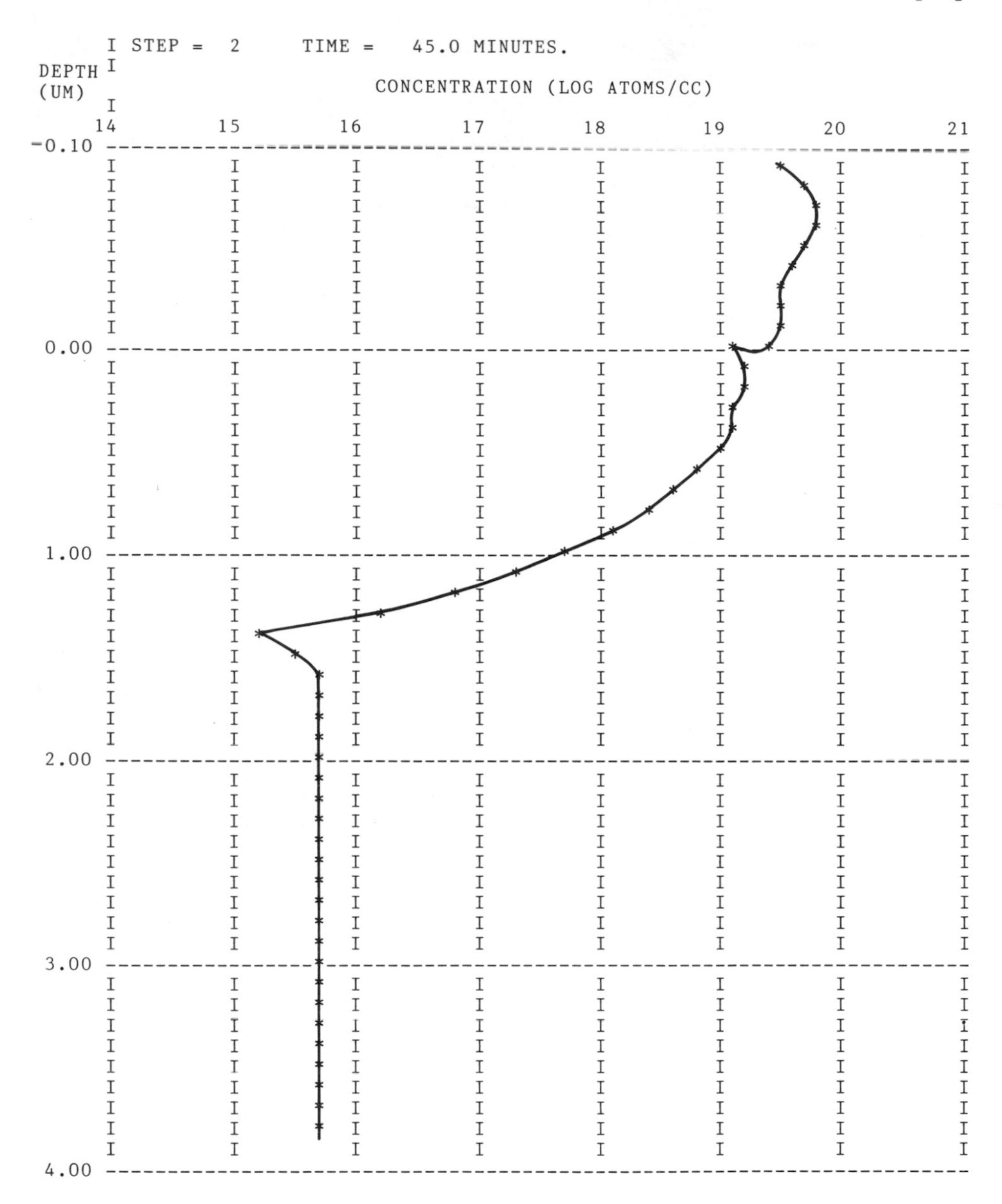

Figure 6.7 SUPREM II graphical output after the drive-in.

to match the manufacturing processes accurately. Accurate calibration of the simulator is very important if it is to be used to quantify changes to a process.

Most simulators, including SUPREM, have default values for many of the constants but because of variations in equipment design, in the silicon being used, in the type and origin of the chemicals, an oxidation or a diffusion carried out under what may appear to be identical conditions by different manufacturers may not result in the same oxide thickness, or junction depth or sheet resistance. An important first step in using a simulator is its calibration. This involves the processing of silicon through oxidation, epitaxy, diffusion, ion implantation, etc., of measuring thicknesses, sheet resistances, junction depths and impurity profiles, and of comparing the measured data with that predicted by the simulator. The simulator is then 'tuned' by making small changes to the appropriate constants to provide the best match between experimental data and simulated data.

It is important to note that the changes should be small and that they should be made to the appropriate constants. Before any changes are made it is necessary to record all the manufacturing processes accurately, for example the presence of a native oxide prior to implantation may be the change required to achieve agreement between experimental and simulated results for ion implantation, rather than trying to change range or standard deviation constants. Small changes to the segregation coefficient for diffusion combined with oxide growth may be more successful than trying to change the diffusion constant in order to obtain agreement between the measured sheet resistance and the simulated value. Each process must be carefully thought through and the constants and models used in the simulator must be understood before any changes are made.

A process simulator such as SUPREM is a valuable tool with which to observe trends, and with careful calibration of the models being used within the simulator it can be extremely useful as a design aid for process and device engineers. For most large manufacturers it is now an essential design and production aid.

Problems

These problems assume the availability of SUPREM or a similar process simulator

1. Create an input data file similar to that shown in Figure 6.3 for SUPREM II. Run the program and compare the output with the data and graphs show in Figures 6.4–6.7. Vary the times and temperatures for the pre-deposit (line 6) and the drive-in (line 8) and observe the results.

2. As a design assignment consider the cross-section of the npn transistor shown in Figure 6.1 and generate input data files for each of the cross-sections shown in the diagram based on the following data:

Substrate is doped with boron with a carrier concentration of $2 \times 10^{15}/\text{cm}^3$ and with a crystal orientation of ⟨111⟩.

Buried n^+ layer: Ion implant with arsenic at dose of $1–5 \times 10^{15}/\text{cm}^2$ and an energy of 30 keV. Drive-in in dry oxygen at 1200 °C for between 60 and 180 min to achieve a sheet resistance of 30–40 ohm/square.

Epitaxial layer: Arsenic doped to give 5 ohm cm (use Irvin curves to determine concentration) with a thickness of 5 μm. Assume a growth temperature of 1000 °C and a growth rate of 0.5 μm/min.

Isolation: Grow an oxide in steam to provide a thickness of between 500 and 800 nm (use graphs of oxide thickness versus time to estimate time and temperature).

Remove the oxide.

Pre-deposit boron at 1100 °C for between 40 and 60 min (use the solid-solubility curves to determine the concentration). Drive-in the boron at 1200 °C in dry oxygen for between 60 and 120 min.

The aim is to drive the boron through the epitaxial layer and also to achieve a sheet resistance of between 20 and 50 ohm cm.

Base: NB the data for the cross-section for the active transistor region must include the time and temperature cycles for the isolation region, but must not include the boron pre-deposition. Pre-deposit boron at 950 °C (solid-solubility curves for the concentration) for between 40 and 50 min. Drive-in the boron at 1100 °C for between 40 and 60 min in dry oxygen to produce a sheet resistance of between 200 and 300 ohm/square.

Emitter: Pre-deposit phosphorus at 950 °C (solid-solubility curves for the concentration) for between 30 and 50 min. Drive-in the phosphorus at between 1000 and 1100 °C for between 30 and 50 min in dry oxygen to produce a sheet resistance of about 10 ohm/square. An important parameter is the base width, which should be less than 1 m.

References

[1] D. A. Antoniadis, S. E. Hanson and R. W. Dutton (1978), 'SUPREM 11 – *A Program for IC Process Modeling and Simulation*', TR No. 5019–2, Stanford Electronic Laboratories, Stanford University, Stanford, California, 1978.

[2] Office of Technology Licensing, Stanford University, 857 Serra Street, Second Floor, Stanford, CA 94305–6225.

CHAPTER 7

Photolithography

7.1 Introduction

Photolithography is the process used to transfer photographic images from a glass mask on to the surface of a silicon wafer. The geometrical shapes which form the images on the glass mask represent the component parts of the circuit elements, such as the emitter, the base and the collector regions, resistors, contacts and interconnections. The images are transferred from the glass mask to a light-sensitive material on the wafer with ultra-violet light. A developing solution is applied to the wafer, and, depending on the type of light-sensitive material, either the exposed or the unexposed material is removed. The material which remains on the surface of the wafer acts as a resist to an etching process used to cut holes in an oxide film, or to remove surplus metal to form interconnections. The photosensitive material (or resist) is removed before further processing of the silicon wafer.

The images are first generated on a computer screen with one of a number of proprietary computer assisted layout programs. These programs are usually part of a suite of programs which are used for the design and simulation of VLSI integrated circuits. Data from the layout program is used to drive a photoplotter or an electron beam machine which generates a photographic master for each of the many layers which form the integrated circuit. The master may be $\times 5$ to $\times 10$ larger than the image which is required on the wafer. This master (or reticle) may be used to generate a multi-image mask at $\times 1$ on a photographic glass plate. Alternatively, it may be used to project images directly on to the wafer.

Masks represent a direct cost for the designer and manufacturer of integrated circuits. The simplest masks for discrete devices, such as diodes and transistors, will be a few tens of dollars per mask, and a mask set may comprise four or five masks. For a small-scale integrated circuit, such as TTL gates, the masks are likely to be a few hundred dollars per mask, and a complete set would require some seven to ten masks. For the advanced, large-scale integrated circuits the individual reticles for a projection system are likely to cost a few thousand dollars per reticle, and there may be between fifteen and twenty masks per set. Masks for contact or proximity printers will need to be replaced at regular intervals as they become damaged through use. The replacements are copied from sub-masters at a relatively small fraction of the original cost of the masters. The reticles for the projection printing systems are more robust and would not need to be replaced very often.

The photosensitive resist is a polymer which reacts to light, usually ultra-violet, in a way which makes it insoluble in a developing solution for negative resists, or soluble for positive resists. Thus, after development, a resist image is left on the surface of the oxide or metal. A suitable etching medium is then used to remove the uncovered oxide or the metal.

Special alignment equipment is necessary to align subsequent masks with previously formed images on the wafer. This equipment uses laser interferometry to position the wafer with respect to the mask and is capable of accuracies to less than $0.5\,\mu m$. Modern alignment equipment is fully automated, with image processing equipment being used to align the fiducial marks which are formed on both the wafer and the mask.

More advanced forms of lithography use electron beams and X-rays in place of ultra-violet light to expose the resist.

7.2 Masks

The first step in producing a photographic mask is to design and simulate the VLSI circuit on a computer using a CAD program. At some stage in the design process the circuit schematic is converted into a layout comprising geometrical shapes which represent the various devices which form the circuit. The data from the layout program is used to drive a photoplotter or an electron beam plotter, either of which create an image of the shapes on a photographic glass plate. The image is usually $\times 10$ larger than the final image and may be formed in a silver-based photographic emulsion or a chrome film covered with a photoresist.

A multi-image $\times 1$ mask is produced by projecting images from the $\times 10$ reticle on to a photographic mask with a set-and-repeat reduction camera. The resulting $\times 1$ mask contains an array of identical images. In a practical step-and-repeat system a multi-barrel camera is used to generate a number of $\times 1$ masks at the same time, with a different $\times 10$ reticle in each camera. This ensures much greater accuracy in the position of related geometrical shapes in the different layers. There is a separate mask for each layer. These multi-image masks are produced on high-quality glass plates using very fine photographic emulsions or chrome films covered in photoresist. They are typically 15 cm, 20 cm or 25 cm square depending on the size of the silicon wafer. Each plate will contain many hundreds of identical images.

To ensure long working life of the $\times 1$ masters, copies or sub-masters are made either on further emulsion plates or on chrome-on-glass plates. The chrome-on-glass plates are more expensive than the emulsion plates but have a longer working life.

The sub-masters are then used in alignment equipment to transfer the images to the surface of the silicon wafers.There are a number of mask-to-wafer alignment strategies which influence both the cost of the alignment equipment and the dimensions of the images produced on the silicon wafer.

7.3 Mask alignment strategies

The simplest and least expensive means of transferring the mask image to the wafer is to place the mask in contact with the wafer and use a flood or collimated beam of ultra-violet light to expose the photosensitive layer which covers the surface of the wafer, as shown in Figure 7.1(a). This method is capable of high throughput and is widely used for discrete devices and small integrated circuits. The disadvantage is that the mask becomes damaged as a result of constantly being brought into contact with the wafer where dust or surface irregularities on the wafer damage the emulsion on the mask. The use of chrome-on-glass masks extends the life of the masks but in both cases the masks have to be replaced regularly. The number of defects introduced as a result of damage to the masks is too great for VLSI circuit fabrication.

An alternative to contact printing is proximity printing, as shown in Figure 7.1(b). With proximity printing there is a gap of some 10–25 μm between the mask and the wafer. While this gap does not eliminate the damage to the mask it is greatly reduced. However, the image is now affected by diffraction, and the optical system must be modified to ensure that the beam is collimated in order to reduce the effects of diffraction. This increases the cost of the equipment.

Finally, with projection printing, as shown in Figure 7.1(c), the mask is placed in the focal plane of the optical system and the image is projected on to the wafer. There is

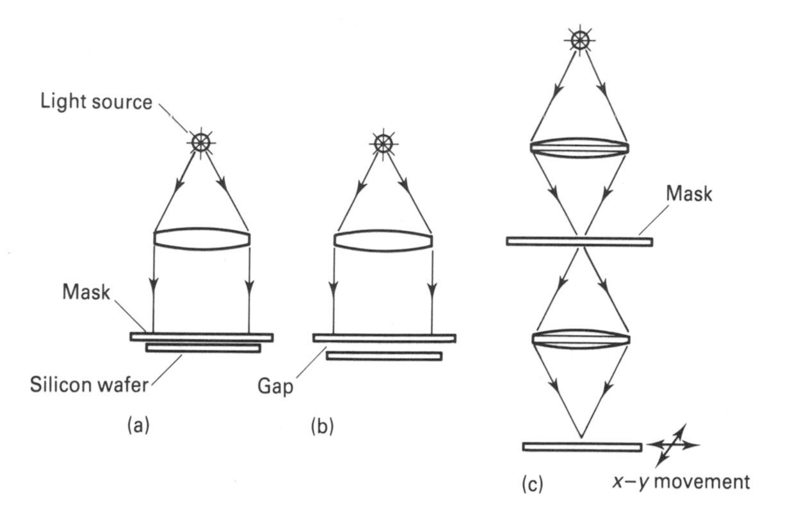

Figure 7.1 Schematic of three mask alignment methods based on (a) contact, (b) proximity and (c) projection.

no danger of the mask being damaged by contact with the wafer. However, it is not practical to project a complete multi-image mask on to the wafer because of the difficulty of producing optical elements that are free of aberrations over such a large area. A more practical system is to project the image from the ×10 reticle and to step-and-repeat the image directly on to the wafer. The reticle may contain a block of four or six ×10 images, rather than a single image, in order to reduce the time required to cover the wafer.

The direct step-on-wafer projection printers are now the standard means for transferring images to silicon wafers for VLSI circuits. There are three important requirements for projection equipment: very good optics, precision measurement of the x and y coordinates of the workstage holding the wafer and automatic control of the focus, since the high-resolution optics have a very small depth of focus. The high-quality optics have been made possible by the use of computers which can solve the complex mathematical equations that are necessary for the design and optimization of lenses. Lenses are now available that can reproduce images with feature sizes of less than 0.5 μm. Laser interferometry is used to position the workstage accurately and automatic focusing systems have been designed that take account of variations in the flatness of the wafer.

Particles of dust can be a major problem for projection printing, particularly if they fall on a clear region of the ×10 reticle because then the image will be transferred to every circuit on the wafer. This problem can be solved by very careful inspection of the mask for any defects and by the use of a 'pellicle'. The pellicle is a transparent protective coating which is placed on either side of the ×10 reticle. It is typically 5 mm thick on either side of the mask. Any dust particles falling on the surface of the pellicle are sufficiently far from the plane of the pattern to be out of focus and will not produce an image on the wafer.

A photograph of a direct step-on-wafer projection printer is shown in Figure 7.2. The wafers are held in two carousels, one for wafers prior to printing and one for wafers which have been processed.

7.4 Photoresists

There are two classifications of resist: negative and positive. Both resists are a mixture of a photoactive material, a resin and a solvent. For the negative resist, exposure to light results in cross-linking of long polymer chains which make them insoluble in the photoresist developing solution. For positive resist the light-sensitive material decomposes into a carbolic acid which is readily soluble in an alkali solution.

Both types of resist are used for VLSI circuit manufacture, with a preference for positive resist for improved resolution. The improved resolution arises because the long polymers of negative resist leave a ragged edge when the unexposed resist is removed. Positive resist does not involve the formation of polymer chains; instead it is a simple change from non-soluble material to soluble material and a much sharper edge is obtained.

Figure 7.2 Optimetrix 8605 direct step-on wafer projection printer. [By courtesy of Edinburgh Microfabrication Facility.]

A typical photomasking sequence is as follows:

1. Apply a resist adhesion promoter such as hexamethyldisilazane to the surface of the wafers. Prior to application the slices may be cleaned and given a 150–180 °C bake to drive off any adsorbed moisture.

2. Apply the resist by high-speed spin coating. A measured amount of resist of known viscosity is applied to the wafer, which is held on a vacuum chuck. The wafer is then spun at between 3000 and 6000 rpm, starting slowly to spread the liquid and then rapidly accelerating to the final speed. This results in uniform distribution of the resist over the surface of the wafer to a controlled thickness.

3. Pre-bake at 80–100 °C to drive off the solvents and to dry the resist.

4. Align and expose the resist to ultraviolet light. Exposure time is usually determined by trial runs and inspection. The optimum time corresponds to that which produces images that have the same dimensions as the images on the mask.

5. Develop the exposed wafers to leave a latent resist image. The developer is usually applied as a spray to the spinning wafer, which is held on a vacuum chuck. The developing solution is removed with a suitable rinse, again applied as a spray to the spinning wafer.

6. Post-bake the resist at about 120 °C to dry the surface and to improve the adhesion of the resist to the wafer. A slight reflow may take place which smoothes the edge profile of the resist.

7. Etch the exposed substrate with either a wet or dry (plasma) etch. A suitable wet etch for SiO_2 would contain a mixture of hydrofluoric acid, nitric acid and deionized water; for aluminium it would contain phosphoric acid, acetic acid, nitric acid (to remove any silicon) and deionized water.

   As an alternative to the wet etch, a dry or plasma etch would involve a low-pressure (10^{-3}–2 torr) ionized gas, which reacts with the exposed substrate to produce volatile compounds which are removed from the reaction chamber by the vacuum pump.

8. Remove the resist with hot acid or an oxygen plasma.

9. Inspect the etched images and check for accuracy and for signs of under- or over-etching.

There are many variables in the complete photolithographic sequence, each of which can affect the image which is produced on the wafer. Because the photolithographic process is repeated many times during the fabrication of an integrated circuit it can have a significant affect on the final yield. Resist thickness, light intensity, exposure time, pre-bake and post-bake temperatures must all be carefully controlled. One of the greatest variables is the profile of the oxide after etching. Typical profiles are shown in Figure 7.3. For wet etching the etch is usually isotropic, with material being removed both horizontally and vertically, that is, there is an undercut beneath the layer of resist,

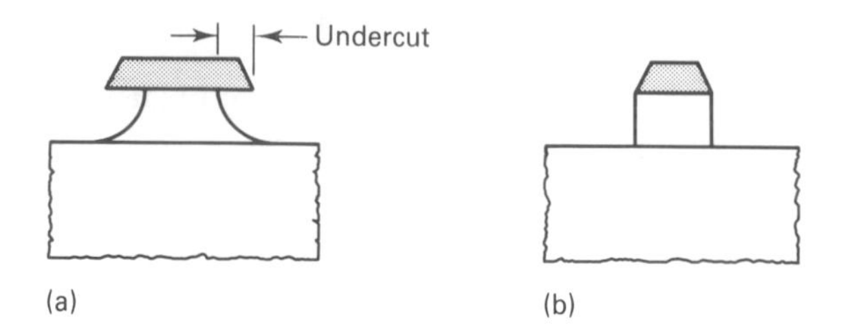

Figure 7.3 Etch profiles resulting from different etch techniques, (a) isotropic and (b) anisotropic.

as shown in Figure 7.3(a). For dry etching it is possible to achieve anisotropic etching which produces the vertical profile shown in Figure 7.3(b). Usually, there is some undercutting with dry etching but it is much less than that obtained with wet etching.

The near vertical edges of the anisotropic etch profile are ideal for masking operations required for diffusion and ion implantation when the tapered edge of the isotropic profile would result in a less well-defined diffusion or implant region. However, the vertical edges are very difficult to cover during the metallization stage and breaks can form in the interconnections at these steps in the oxide.

The degree of anisotropy (A_f) can be described in terms of the lateral (v_l) and vertical (v_v) etch rates:

$$A_f = 1 - \frac{v_l}{v_v}$$

In terms of a feature etched just to completion, as shown in Figure 7.4, the lateral etch rate is given by:

$$v_l = \left(\frac{d_f - d_m}{2}\right) \text{ per second}$$

and the vertical etch rate is given by:

$$v_v = t_f \text{ per second}$$

where t_f is the film thickness.

Thus the degree of anisotropy is:

$$A_f = 1 - \left(\frac{d_f - d_m}{2}\right)\frac{1}{t_f}$$

For isotropic etching the amount of undercut is equal to the film thickness and $A_f = 0$, while $1 \geq A_f > 0$ represents anisotropic etching with $A_f = 1$ representing the extreme case with no undercut and vertical edge profile to the etched feature.

For an amorphous material such as SiO_2 a wet etch results in an $A_f = 0$, whereas using a wet etch for a single-crystal material such as silicon results in $A_f \approx 1$ depending

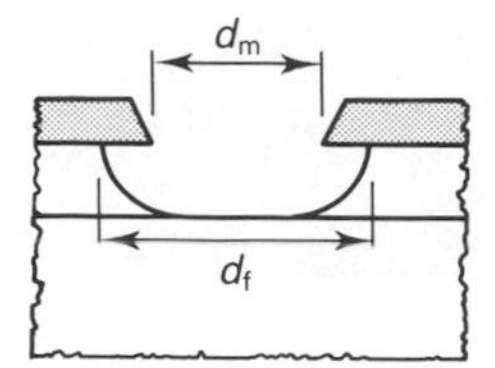

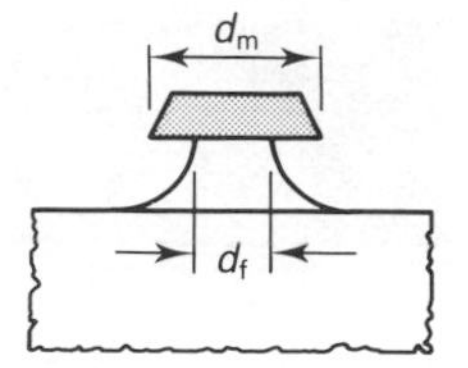

Figure 7.4 Etch conditions as described in terms of the amount of undercut and the mask dimensions.

on the crystal orientation. For plasma etching, A_f can be varied from 0 to 1 depending on the method and the conditions used.

The simplest form of plasma etcher is the barrel etcher where the wafers are held vertically in a barrel, as shown in Figure 7.5(a). A perforated aluminium tube separates the region where the plasma is generated from the region where the etching takes place on the wafers. Barrel etchers are widely used to remove photoresist, as well as for the etching of oxides. In the parallel plate etcher (Figure 7.5(b)) the wafers are placed flat on one of the electrodes of the radio frequency circuit which is used to generate the plasma. A complete range of profiles from isotropic to anisotropic can be obtained with variations of the conditions within a plate etcher.

A photograph of the interior of a plasma barrel asher for the removal of photoresist is shown in Figure 7.6.

7.5 Advanced processing

Optical methods are suitable for the vast majority of pattern transfer operations for VLSI circuit fabrication and are capable of resolving feature sizes down to 0.5 μm. By

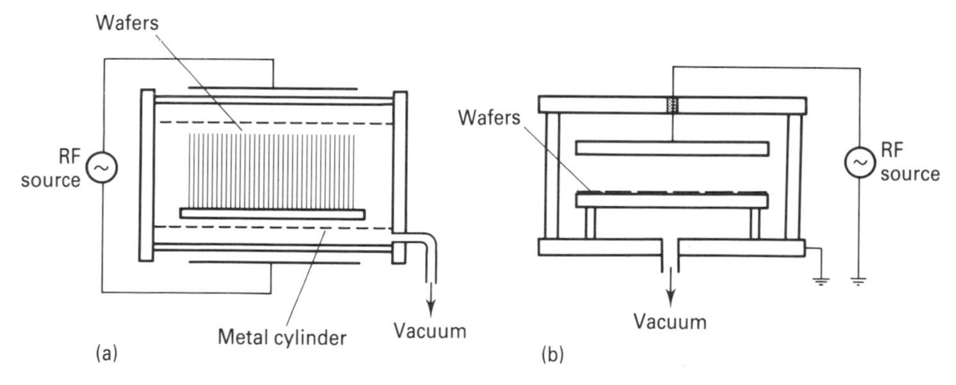

Figure 7.5 Schematic of plasma etching equipment based on (a) the barrel etcher and (b) the parallel plate etcher.

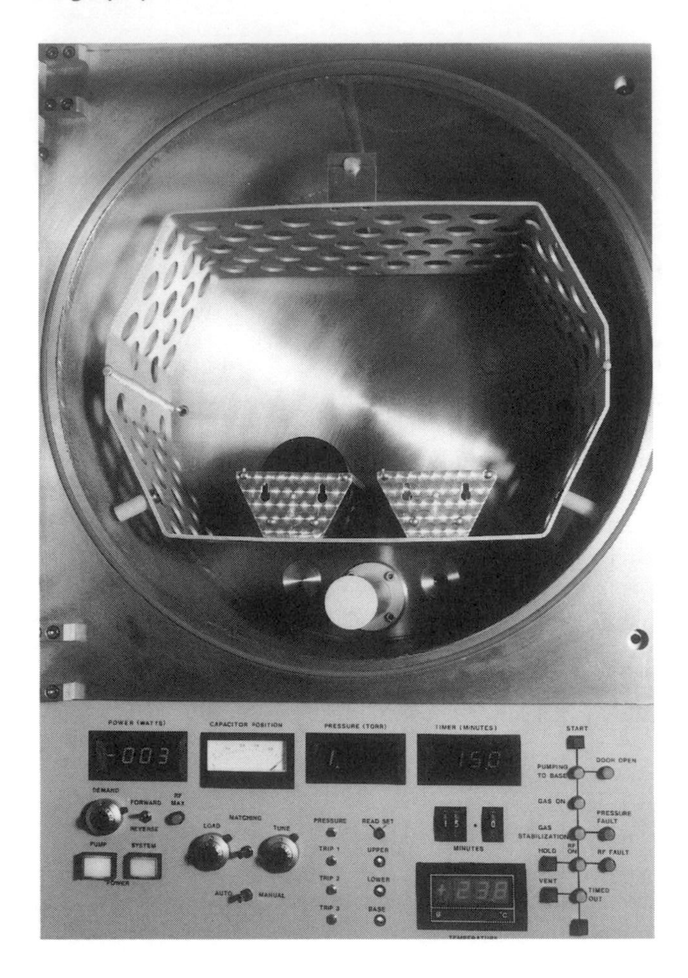

Figure 7.6 Barrel plasma asher from STS Electrotech. [By courtesy of Edinburgh Microfabrication Facility.]

using deep ultra-violet light sources it is possible to reduce the minimum dimension to 0.2 μm. To reduce the dimensions further it is necessary to use an alternative radiation source.

Focussed electron beams can be used to expose specially prepared resists. The beam is raster scanned over the coated wafer and by switching the beam ON and OFF a pattern can be generated in the resist. While electron beams are capable of reproducing feature sizes of less than 0.5 μm, their main attraction is the ability to generate the pattern directly from a computer without the need to produce a photomask. This is particularly attractive for low-volume circuits for which the cost of producing a set of photomasks would be prohibitive. Electron beam lithography is often used for semicustom gate arrays where the electron beam is used to produce the final interconnection pattern over a pre-defined array of logic gates.

For shorter wavelengths X-rays are being used. They have the advantage over electron beams of being able to expose a complete wafer at one step rather than drawing each individual shape as is required with electron beams.

The masks for the X-ray source must be sufficiently rigid to support the opaque patterned layer yet be thin enough to minimize attenuation of the radiation which is being transmitted. The opaque layer must absorb the radiation, and gold is the most commonly used material. The mask must be free of defects and be dimensionally stable.

For the radiation source the long wavelength X-rays of 1–5 nm are strongly absorbed by thin layers of patterning material, and collimated beams can be produced from a synchrotron source. Compact synchrotrons have been produced using superconducting magnets which are capable of producing the very high fields necessary to manufacture a compact synchrotron. These machines are still at the development stage. However, there is the potential to produce feature sizes of less than 0.1 μm.

CHAPTER 8

Metallization

8.1 Introduction

A thin layer of metal deposited on to the surface of the wafer is used to form the interconnections between the many diffused p- and n-type regions in the silicon. The metal layer is separated from the conducting silicon by a layer of oxide, except where contact with the silicon is required. As well as forming interconnections within the boundaries of the chip, the metal layer is also used to provide bonding pads for the wire leads to the integrated circuit package. For very complex circuits two or more layers of interconnections may be required, in which case additional layers of dielectric are deposited over the first layer of metal to provide electrical isolation between the individual layers of metal.

The metal is usually deposited by vacuum techniques and the interconnection patterns are formed by photolithography.

8.2 Interconnect requirements

There are a variety of requirements which a conductive film must satisfy when used for integrated circuit interconnections. Some of these are listed below:

1. The film must conduct current with very little resistive loss.
2. The film must be capable of making good ohmic contact with both n- and p-type silicon.
3. The film must adhere well to both silicon and silicon oxide.
4. It should be possible to pattern the film with conventional photolithography.
5. It must be possible to bond wires to the film to make connection to the integrated circuit package.

6. It must be easy to deposit and the film must be capable of covering steps in the oxide.
7. The film must be stable and must not react with packaging materials or the wires used for bonding.
8. The film should not change with high current densities and should be resistant to electromigration.

No single metal satisfies all these requirements but the one which most nearly meets many of them is aluminium, and many of its deficiencies can be overcome with some additional processing. Gold has high conductivity and is very resistant to corrosion but has poor adhesion to silicon oxide. Refractory metals such as molybdenum, platinum, tantalum, titanium and tungsten have much lower conductivity but can be made to produce very good ohmic contacts to silicon with the formation of silicides.

8.3 Contacts

The contact is the region between the layer of metal and the surface of the semiconductor and it may be classified as being ohmic or rectifying. For certain applications a rectifying characteristic may be required as, for example, in a Schottky diode, but in general an ohmic contact with a low resistance is required. Because two dissimilar materials are used, an electronic barrier exists. For carrier concentrations of less than 10^{17} atoms/cm^3 the conduction of carriers largely results from thermionic emission over the barrier. For concentrations greater than 10^{19}/cm^3 the barrier narrows and carrier conduction is dominated by tunnelling through the barrier. Contact resistance decreases rapidly as the concentration increases beyond 10^{19}/cm^3.

The specific contact resistance[1] may be 1 ohm cm^2 for a carrier concentration of 10^{19} atoms/cm^3 and for a contact of 10 μm × 10 μm the contact resistance is 1 ohm. However, for a contact of 1 μm × 1 μm the resistance would be 100 ohm. Silicides can reduce the specific contact resistance and thus the actual resistance of the contact.

Aluminium is a p-type dopant in silicon with a solid solubility in excess of 10^{19} atoms/cm^3. Provided that the impurity concentration for n-type silicon is greater than 10^{19}/cm^3, then the carrier flow will be dominated by tunnelling and the contact will be ohmic.

8.4 Deposition

The most commonly used method for the deposition of metal layers is vacuum deposition, either by evaporation, or by sputtering in a plasma. Chemical vapour deposition can also be used but vacuum techniques predominate.

The vacuum chamber is a stainless steel cylinder with a closed top and is sealed at the base with a vacuum-tight gasket, as shown in Figure 8.1(a). Air is removed by a

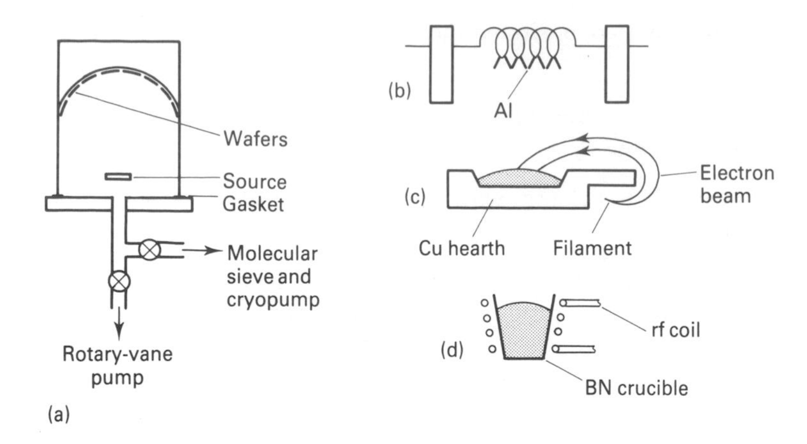

Figure 8.1 (a) Schematic of a vacuum evaporation system with (b) filament heating, (c) electron beam heating and (d) rf inductive heating.

combination of mechanical rotary-vane pump and a liquid nitrogen cooled molecular sieve system and cryopump. The molecular sieve and cryopump have largely replaced the oil diffusion pump because of possible contamination caused by the oil vapour.

Typically, the working pressure in the chamber is 5×10^{-7} torr (6.6×10^{-5} Pa). For plasma sputter deposition systems the pressure is typically $5\text{–}8 \times 10^{-5}$ torr ($6.6 \times 10^{-3}\text{–}1 \times 10^{-2}$ Pa).

8.5 Vacuum evaporation

During vacuum evaporation the source of the material to be deposited is heated until it evaporates. The atoms of the material move away from the source in straight lines in all directions which are not obstructed by the source. The rate of deposition varies with the distance from the source and for a spherically shaped target with the source at its centre the rate of deposition will be uniform over the inner surface of the target. The silicon wafers are placed in the plane of this sphere on a rotating planetary holder.

8.5.1 Resistance heating

The simplest form of heater is a coil of tungsten wire supported between large current carrying supports as shown in Figure 8.1(b). Short lengths of aluminium wire are either placed through the centre of the coil or hung over the turns of the coil. The rate of evaporation is rapid and difficult to control, and film thickness is largely controlled by the weight of the Al placed on the filament.

The main disadvantages are the possible contamination from the filament, short filament life and limited mass of the source material, which limits the thickness of the film deposited.

8.5.2 Electron beam heating

A schematic of the cross-section of an electron beam source is shown in Figure 8.1(c). A water cooled copper hearth contains the source material and a heated wire filament provides a source of electrons. The electrons are accelerated through about 10 kV and are deflected by a magnetic field to strike the source. The energy from the beam can be controlled to vary the rate of evaporation and with a large source the deposition can continue until the required thickness is achieved. Multiple sources can be used to coevaporate a mixture of materials to form alloys.

In addition to Al, many other metals can be deposited such as Pd, Au, Ti, Mo, Pt, W, as well as many dielectric materials.

At 10 kV the electrons have sufficient energy to generate X-rays and the ionizing radiation can penetrate dielectric layers and silicon, where they cause damage. This is a particular problem for MOS devices.

8.5.3 Inductive heating

A crucible made of boron nitride can be used to hold material in an inductively heated source (Figure 8.1(d)). A large amount of source material can be heated but the disadvantage is possible contamination from the crucible.

8.6 Sputter deposition

When a gas such as argon is contained in a chamber at a reduced pressure and is excited with rf energy then a plasma is formed (Figure 8.2). If the ions (Ar^+) are accelerated through a potential gradient to bombard a target (source) or cathode, then, through momentum transfer, atoms near the surface of the target become detached and are

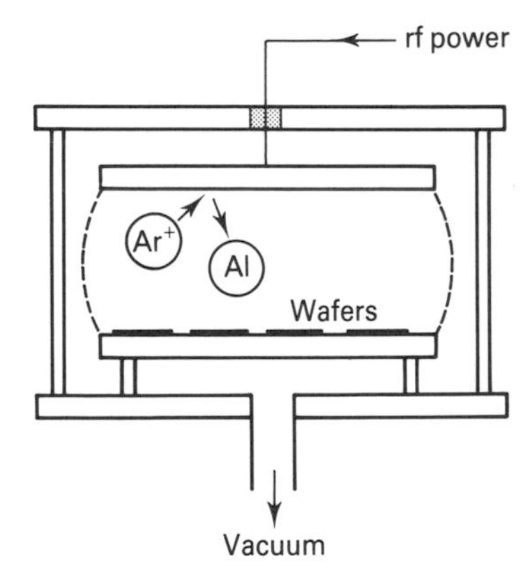

Figure 8.2 Cross-section of a plasma deposition system.

transferred as a vapour to the anode (substrate). At the substrate the film grows by deposition.

Aluminium is a difficult material to sputter because of the presence of residual oxides which inhibit sputtering. A large density of ions is required to sputter aluminium and this can be achieved with magnetron sputtering in which a magnetic field captures electrons in the vicinity of the target and enhances the ion current.

The advantages of sputter deposition are the absence of ionizing radiation, the ability to sputter alloys with a composition similar to the target and improved step coverage.

A photograph of a sputter coater is shown in Figure 8.3. The wafer carrier is seen in the raised position with a number of wafers held in place. In operation the carrier is lowered to provide a vacuum seal and the carrier rotated to provide a more uniform coating.

8.7 Problems encountered with metallization

Aluminium is the most widely used material for VLSI interconnections and while many problems are associated with aluminium it has proved simpler to try to overcome the problems than to use a different metal.

8.7.1 Adhesion

Aluminium adheres well to silicon oxide because it readily oxidizes to form a thin but stable oxide of Al_2O_3 and strong bonds exist between the oxygen in the Al_2O_3 and the oxygen in SiO_2. The strength of this bond is important to ensure reliable wire bonds from the aluminium to the leads of the package.

Gold is an ideal material for interconnections because of its high conductivity and freedom from oxidation and corrosion, but it does not adhere to SiO_2. Good adhesion can be achieved by first depositing a layer of chromium (10 nm) and then coevaporating the chromium and gold so that the two films form a mixture at the interface between the chromium and the gold. Other metals such as W, Pt and Pd also adhere well to SiO_2 but have much lower conductivities than either Au or Al.

8.7.2 Step coverage

The deposition of metal for VLSI interconnections occurs after many process steps have been completed and the surface of the wafer is no longer planar. Therefore, parts of the surface of the substrate are not perpendicular to the trajectory of metal atoms arriving from the source and the growth of the film will be uneven. Step coverage can be improved by heating the substrate (300 °C) to give the arriving atoms greater surface mobility and by rotating the wafers during deposition. The rotation causes the angle between the surface and the trajectory of the atoms to change and, as a consequence, to even out any differences which may have occured in the thickness. The edge profile of

Figure 8.3 Baltzer BAS450 sputter coater. [By courtesy of Edinburgh Microfabrication Facility.]

the oxide can be changed by etching to avoid 90° steps and the surface can be planarized with a low-melting phosphorus doped oxide prior to the deposition of metal. Many of these methods are used but step coverage remains a problem.

8.7.3 Alloying

After deposition and photolithography the wafers are heated to about 450 °C for a few minutes to allow the silicon and the aluminium to form an alloy and hence a low-resistance contact. For large area contacts (10 μm × 10 μm) and deep junctions (2 μm) this has proved to be a very successful means of achieving low resistance and ohmic contacts. However, on closer examination of the contact region the alloying is often patchy and there are often many pits filled with aluminium, as shown in Figure 8.4(a).

During the alloying the silicon dissolves in the aluminium leaving behind a pit which then fills with aluminium. Aluminium spearing[2] is particularly noticeable for very small contacts (2 μm × 2 μm) when a large amount of silicon dissolves because of the excess of aluminium surrounding the contact and the resultant spear or spike can penetrate a shallow junction. The spikes form very rapidly at the alloying temperature and can easily short-circuit a shallow junction. The spikes are inverted three- (111) or four- (100) sided pyramids depending on the crystal orientation. It can be minimized by depositing an alloy of 1–2% by weight of silicon in aluminium. This saturates the aluminium at the alloying temperature and so prevents silicon being taken from the substrate. The method works well for p^+ silicon contact areas but not so well for n^+ areas. This is because the p-type silicon contained in the aluminium (Al is a p-type impurity) precipitates to form a pn junction.

A solution to this problem is to use a barrier layer between the silicon and the

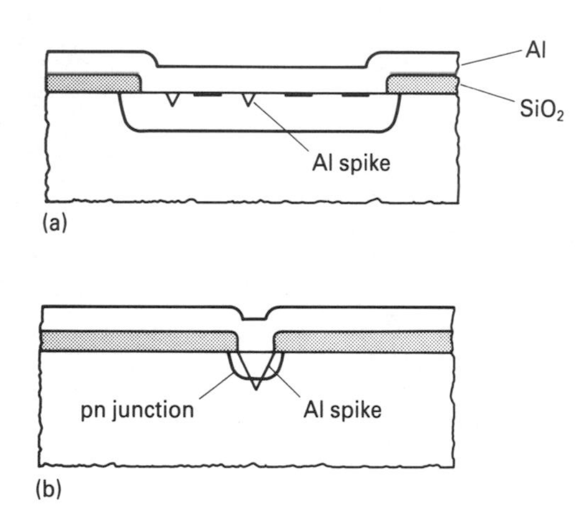

Figure 8.4 Cross-section of aluminium to silicon contacts for (a) a large contact and (b) a small contact.

aluminium. A noble metal such as Pt or Pd is deposited (10 nm) followed by a refractory metal such as W or Cr. The noble metal forms a silicide and a low-resistance contact, while the refractory metal acts as a barrier preventing the aluminium coming into contact with the silicide.

8.8 Electromigration

As the cross-sectional area of the conductor is reduced as a result of scaling, the current density increases. At large current densities there is a transfer of momentum from the current carrying electrons to the metal ions which results in the migration of the metal. The effect is particularly noticeable at oxide steps where the thickness of the metal is likely to be reduced. The movement of the metal eventually results in a break in the track. Microscopic examination usually reveals hillocks adjacent to the break and on the side nearest to the positive supply. The hillocks are formed from the material removed from the cracks.

The resistance to electromigration can be increased by the addition of small amounts of copper to the aluminium.

The mean time to failure of narrow aluminium conductors can be related to current density by:

$$MTF \propto \mathbf{J}^{-2} \exp\left(\frac{E}{kT}\right)$$

where E is an activation energy which has been observed to be 0.5 eV for Al−0.5%Cu[3] for $10^5 \leq \mathbf{J} \leq 2 \times 10^{16}$ A/cm^2.

8.9 Future developments

Metallization was a relatively unimportant process where a single layer of aluminium provided the interconnections between a relatively small number of devices. However, as device dimensions have been reduced and circuit complexity increased, metallization has become a major area for research and development. It is no longer a simple process. Silicides are being used to produce low-contact resistance to shallow junctions. Barrier layers are being used to separate the silicide from the interconnection metal, which is still aluminium. Various combinations of metal and the means of applying them are still being investigated.

The increased circuit complexity means that a single layer of metal is no longer adequate to interconnect all of the devices and two and three layers of metal are now required for this purpose for the more complex circuits. Thus the surface of the silicon chip is becoming a complex three-dimensional electrical interconnection circuit. With such a complex structure many problems are associated with the chemical interaction between the different materials at the process temperatures, the control of surface planarity to ensure good surface coverage over the steps in the dielectric/metal films, and the stress resulting from different temperature coefficients of the different materials. These problems and the resultant increase in the overall manufacturing process means that metallization is now a major process requiring as much effort as the design of the transistors.

References

[1] S. M. Sze (1981), *Physics of Semiconductor Devices* (2nd edn), Wiley, New York, p. 304.
[2] T. E. Price and L. A. Berthoud (1973), 'Aluminium spearing in silicon integrated circuits', *Solid State Electronics*, **16**, 1303–4.
[3] S. Vaidya, D. B. Fraser and A. K. Sinha (1980), 'Electromigration resistance of fine line aluminium', *Proc. 18th Reliability Physics Symposium*, IEEE, New York, p. 165.

CHAPTER 9

Quality and reliability

9.1 Introduction

An understanding of some of the basic factors which affect quality and reliability are important for an appreciation of the problems associated with the manufacture of integrated circuits. The quality of every step of the manufacturing process must be of a high standard. It is measured by the ability to satisfy the required specification for every step of the process without introducing any defects. Any departures from the ideal will affect the end product and with so many process steps the cumulative effect of even small departures can have a very significant effect on the end result. A variety of inspection and monitoring procedures are used to ensure the quality of the process, and statistical methods are used to evaluate the results.

Reliability is a measure of failure-free operation of a component or circuit under a set of stated operating conditions. It is not a precise measurement of how long the component will operate but, rather, a measure of the probability of successful operation over a recognized period of time.

All devices, whether electrical or mechanical, have a finite lifetime. For an integrated circuit the lifetime is closely related to the manufacturing process plus its final operating conditions. For the manufacture of integrated circuits it is the manufacturing conditions which are important: the effect that they have on reliability and the ability to control them to improve the reliability.

Finally, it is important that the manufacturing yield is high, that is, the number of integrated circuits which satisfy the overall design specification at the end of the manufacturing cycle is as large as possible. Yield is an important factor in determining the final cost of a product and for a mature process it is possible to predict yield and hence cost.

9.2 Failure rate

All devices have a finite lifetime and intuitively a device that fails early is said to have a high failure rate, while a device that continues to operate for a long time is said to have a low failure rate. Thus failure rate is inversely proportional to time to failure. Statistically it is not practical to base the measure of failure rate on a single device but, rather, on a number of devices (N), each of which will fail at times $t_1, t_2, \ldots, t_N$. The total number of successful operating hours for the devices is:

$$\sum_{r=1}^{N} t_r$$

and the average failure rate is:

$$f_r = \frac{N}{\sum_{r=1}^{N} t_r}$$

and the mean time to failure is:

$$t_{av} = \sum_{r=1}^{N} \frac{t_r}{N}$$

A practical example of the above relationships is to consider a piece of electronic equipment which contains 100 000 components. What must the failure rate of the individual components be if the equipment is not to fail more than once per year. It may be assumed that the failure of one component is enough to cause the failure of the piece of equipment.

One year is equivalent to 8760 hours and for all devices to survive for one year implies $100\,000 \times 8760 = 8.76 \times 10^8$ device hours per year. If only one device is allowed to fail in one year then the failure rate is:

$$f_r = \frac{1}{8.76 \times 10^8}$$

or

$$f_r = 1.14 \times 10^{-9}$$

or

$$f_r = 1.14 \text{ devices per } 10^9 \text{ hr}$$

This represents a failure rate of one device in approximately 114 000 years! It is impossible to verify this sort of failure rate under normal operating conditions and as a consequence some form of accelerated testing is required.

9.3 Component reliability

Many of the processes which result in the degradation of components are accelerated at increased temperature. It has been noted by practical experiment that the failure rate follows an exponential law which is the same as that which applies to many chemical reactions. The Arrhenius model used to describe failure rate is:

$$f_r = K \exp - \left(\frac{E}{kT}\right)$$

where E is the activation energy, k is Boltzmann's constant, T the absolute temperature and K a constant. Because of the exponential relationship a relatively small increase in temperature produces a very significant change in the failure rate.

The important parameter is the activation energy, which must be determined for each device. This is done by placing a number of devices on test at elevated temperatures, each batch at a different temperature. The devices are connected to electrical supplies so that they conduct current. A particular parameter is then monitored, for example β for a bipolar transistor or V_{TH} for a MOS transistor. Failure is assumed to have occurred when the parameter under test has changed by some prescribed amount. The failure rate for each batch is then plotted against temperature and the activation energy is obtained from the slope of the graph. Typical values of activation energies for integrated circuit components are from 0.5 to 2.0 electron–volts.

An example of an accelerated test for a MOS integrated circuit may produce the following information:

sample size:	300
test time:	500 hr
test temperature:	168 °C
number of failures:	5

With this information it is possible to predict the failure rate at, say, 50 °C.

It is first necessary to obtain from reliability standards data the activation energy for the type of MOS circuit which is being tested. For this example it is 0.65 eV.

The failure rate is:

$$f_r = \frac{5}{(300)(500)} = 3.3 \times 10^{-5}$$

and

$$K = \frac{3.3 \times 10^{-5}}{\exp\left(\dfrac{-1.6 \times 10^{-19} \times 0.65}{1.38 \times 10^{-23} \times 441}\right)}$$

$$K = 881$$

Now the failure rate at 50 °C is:

$$f_r = 881 \exp\left(\frac{-1.6 \times 10^{-19} \times 0.65}{1.38 \times 10^{-23} \times 323}\right)$$

$$f_r = 6.4 \times 10^{-8}$$

or

0.64 devices per 10^9 hr

The temperature of 168 °C is a recognized standard for accelerated life tests. The tests involve placing a number of devices in a heated test chamber with appropriate bias voltages applied so that the device is operating under conditions which are similar to those which would be expected under normal conditions. With current flowing, the junction temperature of diodes and transistors may be at a higher temperature than the ambient. To simulate the stress of switching the device ON and OFF the bias voltages will also be periodically switched.

Component failure follows a 'bathtub' shaped curve, as shown in Figure 9.1. Early in the life of a component there are a large number of failures which are due to defects and weaknesses produced during the manufacturing process. Many of these failures can be removed with stress testing and screening immediately after manufacture.

During the middle period the rate of failure is low and random. This period represents the useful life of the component. Finally, after a 'long' period of occasional failure there is a rapid increase in the failure rate. This is the region where the components wear out as a result of high temperature, slow chemical reactions, a breakdown of the hermeticity or the movement of mobile ions caused by the constant presence of high electric fields.

9.4 Integrated circuit failure mechanisms

A failure mechanism is a physical or chemical process which causes a device to fail. Impurities continue to diffuse, particularly if the junction temperature is high; electromigration of aluminium can result in a thinning and eventually an open circuit of an interconnection; mobile ions in the oxide can move under the influence of electric fields and change the surface conditions and the device characteristics, and moisture may leak into the package containing the circuit. Failure may not be catastrophic but, rather, a gradual change of the operating point until it moves out of specification.

For integrated circuits there are three basic groups of failure mechanisms, namely:

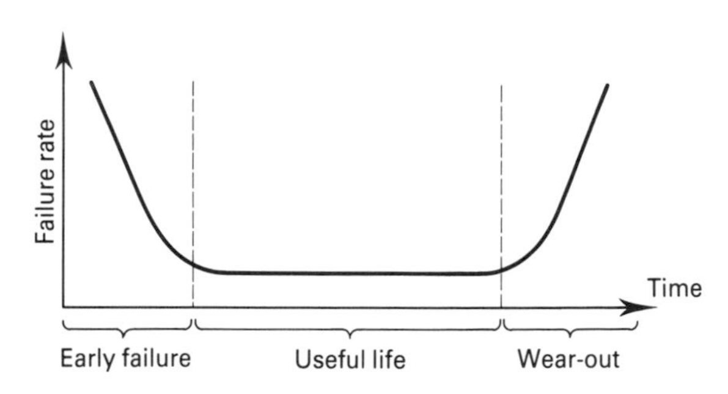

Figure 9.1 Bathtub-shaped curve of component failure rate.

- chip-related, such as oxide defects, metallization defects and problems related diffusion.
- assembly-related problems, such as weak wire bonds, incorrect chip attachment and poor packaging.
- miscellaneous, undetermined or application-related problems.

9.4.1 Diffusion-related

The non-uniform introduction of impurities, which may result from the presence of oxide in the diffusion window after etching, or the presence of crystal defects which may result locally in an increase in the diffusion coefficient, can result in nonuniform junction profiles and low-voltage breakdown, excessive currents and changes in sheet resistance values.

9.4.2 Oxide-related

Oxide contamination either during oxide growth or after growth can result in oxide charge which can seriously affect the surface potential. This can change the threshold voltage of MOS transistors, the leakage current of reverse-biased pn junctions associated with diodes and transistors, and the gain of transistors, particularly at low currents when surface recombination effects are more significant. If the contamination results in mobile ions then the effects described will change with time and temperature.

Defects in the oxide can result in voltage breakdown between the metal tracks and the underlying silicon substrate.

9.4.3 Metallization-related

It is important that a low-resistance contact is formed between the metal and the silicon. This is achieved by the application of heat to create an alloy between the metal and the silicon. Incorrect formation of the alloy can result in high-resistance contacts, while too high a temperature can result in the alloy interface moving towards the junction, which can result in low breakdown and high leakage currents.

The large variations in oxide thickness resulting from field oxides which may be 1 μm thick and oxides over shallow diffused areas which may only be 0.1 μm thick create problems of step coverage over these changes in thickness. Poor control of the deposition process results in thinning of the metal and the formation of cracks at these steps.

Thinning of the metal and the formation of cracks increases the resistance of the interconnection and also the current density. At high current density and at elevated temperature, movement of the metal atoms can occur. The atoms move towards the positive terminal and are deposited in a region where the current density is less. Electromigration occurs when the current density is in excess of 10^5 A/cm^2 and at temperatures in excess of 100 °C. It is important to ensure that the cross-section of power tracks is sufficient to prevent the current density exceeding the value at which electromigration occurs.

9.4.4 Wire bond failures

Interconnections between the chip and the package are made with fine wires of gold or aluminium. A combination of heat and pressure is used to form the bond between gold wire and the aluminium pads on the chip. The two metals are effectively welded together, which results in a strong bond. However, aluminium–gold bonds can fail because of the formation of the intermetallic compound $AuAl_2$. This compound, which is known as purple-plague, forms rapidly at elevated temperatures and leads to a loss of strength and an increase in resistance.

Aluminium wires are bonded to the aluminium pads with a combination of pressure and ultrasonic vibration to break the oxide film which forms naturally on aluminium. If the aluminium pad or the wire have been exposed to the atmosphere for longer than is necessary then the oxide may be thicker than normal and may not completely break down during the bonding process, and a poor bond is likely to be produced.

It is important to test the bonds for mechanical strength regularly in order to maintain the quality of the bonding operation. This can be done by breaking the bonds and measuring the force required.

9.4.5 Package-related

The packages are usually either ceramic with metal lids or injection moulded plastic. In both cases a metal lead frame is sealed into the package. For the ceramic package a low-melting-point glass is used to form the seal between the ceramic and the metal lead frame and the ceramic and the metal lid. The uneven application of the glass frit can result in voids which are open to the atmosphere. For the plastic package the formation of voids or lack of adhesion to the metal lead frame can result in a lack of hermeticity. Variation of temperature can cause the gas within the package voids to expand and leak to the atmosphere through any voids which may exist. When the device cools, the gas contracts and air, which may contain moisture, is drawn into the package. Over a long period the impurities resulting from the moisture will contaminate the surface of the integrated circuit chip and may result in electrical failure.

9.5 Screening

Many of the basic faults which occur during manufacture are detected by visual inspection and electrical test at various stages during the manufacturing process, but some of the faults will remain and, ultimately, the packaged device must undergo some form of screening to eliminate rogue devices.

There are different levels of screening. For consumer applications it may be adequate to subject a sample to a more rigorous form of screening before an electrical end test. However, for devices which are to be used by the military, in space or in telephone repeaters under the sea, it is necessary to screen every device.

A typical screening sequence may involve the following:

stabilization bake
thermal shock
thermal cycle
mechanical shock
centrifuge
hermeticity
electrical test

plus further tests for more critical applications:

burn-in
X-ray inspection
critical electrical test

In designing a screening sequence it is important to ensure that the test conditions are not so harsh as to induce failures which would not otherwise be observed. The purpose of the tests is to simulate the worst conditions that are likely to be met in practice. For the burn-in test it is necessary to select the most appropriate parameter to measure and to decide whether it is the change of the parameter or the rate of change that is important. This requires a physical knowledge of the processes which affect the parameter.

9.6 Yield

Yield provides a measure of the number of good die per wafer at the end of the manufacturing process. As such it is an important factor in determining the economics of manufacturing integrated circuits. Low yield implies that the control of the process is poor and that the costs are likely to be high; high yield implies a lower cost per integrated circuit which can result in improved market share and better profit margins. The basic manufacturing costs are well established and the main problems for the product and marketing engineers when considering a new circuit are the estimation of the yield and hence the market price for a new product.

Yield is closely related to:

1. device area;
2. wafer diameter;
3. inherent defects per unit area;
4. test limits.

Apart from the test limits the other factors are determined by the manufacturing process. The device area is a function of the circuit complexity and the minimum dimensions that can be reproduced by the manufacturing process. Wafer diameters have increased from 50 mm to 100 mm, 150 mm and 200 mm in present-day production lines. For such a complex manufacturing process it is inevitable that defects will be created on the wafer and if they coincide with the active devices of the circuit, then it is more than likely that the circuit will fail during electrical test. For a mature process it is possible to quote an average figure for the number of defects per unit area.

Assuming a circular wafer with a radius R and a chip with an area A, then an estimate of the number of die per wafer is given by:

$$N_d = \frac{\pi(R - A^{1/2})^2}{A}$$

This equation takes account of the fact that circuits around the perimeter of the wafer will not be complete. For example, consider a wafer with a diameter of 100 mm(4 in) then the number of complete die assuming an area of 0.25 cm^2 (0.5 cm × 0.5 cm) is 254, whereas a simple division of wafer area by chip area yields 314. For a die area of 1 cm^2 (1 cm × 1 cm) the number of complete chips is 50.

The yield is defined as the number of good chips (N_o) divided by the total number of chips on the wafer (N_t), that is:

$$Y = \frac{N_o}{N_t} \times 100\%$$

It is useful to be able to predict the potential number of good die based on prior experience and statistical evidence, in order to determine the cost benefit of a design change which may affect the area of the chip, or the move to a larger diameter wafer. Thus:

$$N_o = (N_t)(Y)/100$$

In estimating the yield it is necessary to establish a relationship between the area of the chip (A) and the number of defects per unit area, or the defect density (D). This is illustrated in Figure 9.2 in which a random distribution of defects is shown for a wafer over which is superimposed a grid representing two different chip areas. In Figure 9.2(a) the larger chip area results in twelve good die out of a total of twenty-one producing a yield of 57%. For the smaller chip area shown in Figure 9.2(b) there are seventy-nine good die out of a total of eighty-eight, producing a yield of 89%. In practice, the defect density is not a single number for every wafer manufactured, but, rather, a random variable, and probability theory and statistics must be used to study yield. These studies have produced a number of different models which try to relate yield to chip area and defect density in a simple analytical expression. Three commonly used yield models are as follows:

1. Poisson's model, which assumes that the defect density across the wafer is uniform and is constant from wafer to wafer. The yield for this model is described by:

   $$Y = \exp - (AD)$$

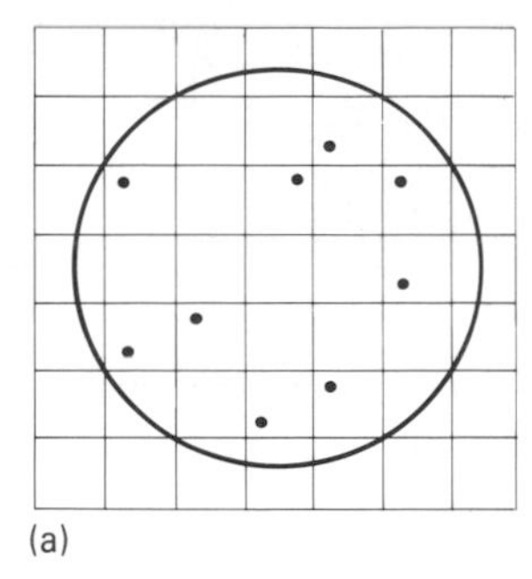

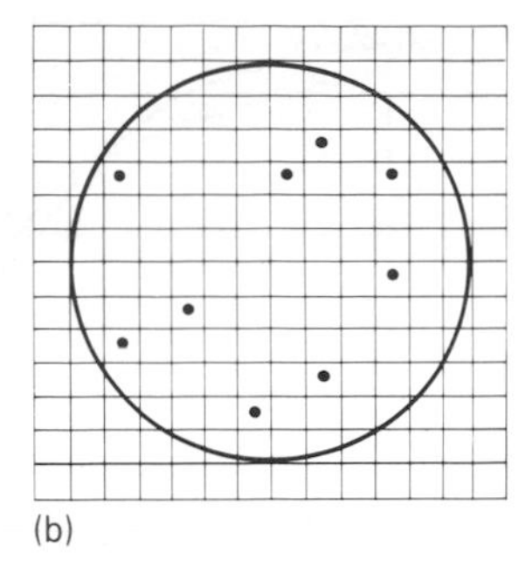

Figure 9.2 Relationship for defect density and (a) the yield for a 1 cm^2 die and (b) a 0.25 cm^2 die.

Based on industrial experience this model only applies for devices of small area, for example discrete devices such as transistors or diodes.

2. Murphy's model, which assumes that the defect density varies across the wafer and from wafer to wafer. The variation across the wafer is assumed to be Gaussian, with the lowest density at the centre and the highest at the perimeter. The model is described by:

$$Y = \left(\frac{1 - \exp - (AD)}{AD}\right)^2$$

For small values of AD this model reduces to the Poisson model.

3. Seed's model also assumes that the defect density varies across the wafer and from wafer to wafer. In addition, it also assumes that the probability of a high defect density is low while the probability of a low defect density is high. The model is described by:

$$Y = \exp - (\sqrt{\mathrm{AD}})$$

While the Poisson model is more suited to discrete devices, the Murphy and Seed models are suitable for integrated circuits.

The variation of the yield for the three models is shown in Figure 9.3. The models can be used to predict possible benefits which may result from changing the design rules to reduce the chip area, or alternatively, to increase the size of the wafer. Table 9.1 illustrates three processes to show how the yield models can be used to predict the effect of changes on the cost per chip.

Process 1 represents a current process with a defect density of $4/cm^2$. The first change that is considered is to shrink the linear dimensions by 10%, as shown for Process 2. This results in a reduction in the area of the chip and thus an increase in the number of die per wafer; the defect density is assumed to be unchanged. For Process 3 the wafer diameter is also increased from 100 mm to 150 mm. The manufacturing cost

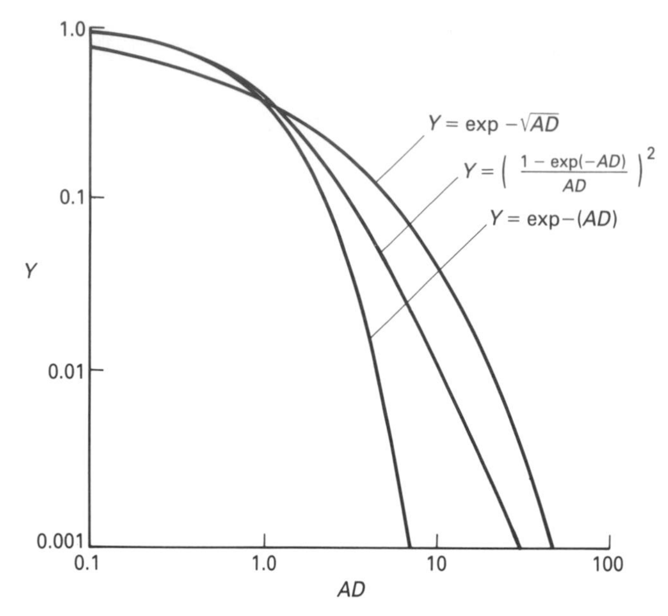

Figure 9.3 Variation of yield (Y) versus AD for the three different yield models.

for the larger wafer is increased but this is offset by the increased number of good die per wafer. The benefits of the changes can be judged in terms of a cost per chip, as shown in the last row of the table.

The calculations of the type which are illustrated in Table 9.1 are important for the integrated circuit manufacturer when considering the cost implications of introducing a new product or when planning major improvements to the manufacturing process.

9.7 Quality control

An important requirement for a complex process is the control of the quality of each stage of the process. All major manufacturers use some form of statistical process control. In its simplest form this may involve the manual generation of process control charts; in its more sophisticated form it involves the use of computers for gathering and monitoring data and, when required, displaying it in graphical format.

The statistical nature of the process results from the fact that for any measurement there are fluctuations from wafer to wafer and from day to day. The different values represent a natural statistical variation and for a very large number of results there is a distribution as shown in Figure 9.4. There is an average value X_{av} with variations on either side of the average of one standard deviation ($\pm\sigma$), two standard deviations ($\pm 2\sigma$) and three standard deviations ($\pm 3\sigma$). An important feature of such a

Table 9.1 Predicted cost per chip based on yield

Parameter	Process 1	Process 2	Process 3
Wafer diam.	100 mm	100 mm	150 mm
D (cm^{-2})	4	4	4
Chip area (mm^2)	7×7 mm = 49	6.3×6.3 mm = 39.7	6.3×6.3 mm = 39.7
Die/wafer	118	151	373
Yield (Seed's)	24%	28%	28%
Good die/wafer	28	42	104
Cost/wafer	$100	$100	$150
Cost/die	$3.57	$2.38	$1.44

distribution is that 68% of the values are within $\pm\sigma$ of the average value, 95% are within $\pm 2\sigma$ and 99% within $\pm 3\sigma$.

The frequency distribution curve of Figure 9.4 is not very convenient for observing day-to-day variations of data and a control chart is more suitable, as illustrated in Figure 9.5. Control charts provide an immediate visual presentation of the state of a particular stage of the process for a particular parameter.

The control charts illustrated in Figure 9.4 are two of the most commonly used charts. They represent the variation of the mean of a sample set of measurements ($X = \Sigma/n$, where n is the sample size), and the range (R), that is, the difference between the maximum and minimum values measured in the sample.

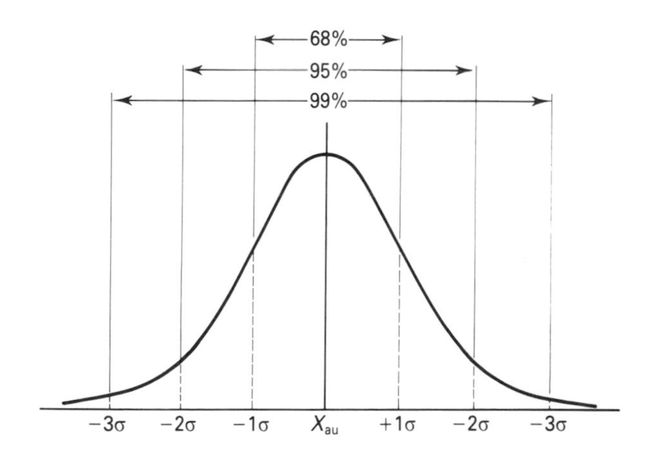

Figure 9.4 Normal distribution with standard deviations of $\pm\sigma$, $\pm 2\sigma$ and $\pm 3\sigma$.

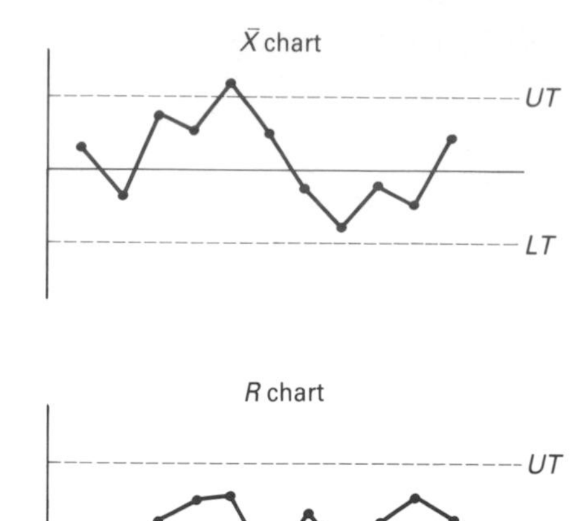

Figure 9.5 *X* and *R* chart for quality control.

Each chart is provided with thresholds, an upper and lower for the mean and an upper level for the range. Readings which are above or below these thresholds represent abnormal readings and demand action if they persist. The position of the thresholds depends on the size of the sample and can be obtained from statistical tables. An example of the threshold calculations is shown in Table 9.2.

In the table $\bar{\bar{X}}$ and $\bar{X}$ represent the mean and range based on a large sample (30–50) of historical data which should display a normal distribution.

For semiconductors, suitable parameters which can be measured are:

C–V flat band voltage
oxide thickness
sheet resistance
oxide voltage breakdown
critical dimensions

Table 9.2 Determination of threshold limits for *X* and *R* charts

X charts $UT = \bar{\bar{X}} + A\bar{\bar{R}}$, $LT = \bar{\bar{X}} - A\bar{R}$ *R* charts $UT = B\bar{R}$

Sample size	A	B
3	1.02	2.57
5	0.58	2.11
10	0.31	1.78

Upon completion, any number of electrical parameters can be measured such as beta, threshold voltage, resistance, etc.

The process is suited to a computer-based statistical process control system with data being entered directly into a computer database. The quality control engineer can then monitor the results at regular intervals. Statistical process control is an important tool for the integrated circuit manufacturer.

9.8 Future developments

The ever increasing complexity of present-day advanced technology processing places ever greater demands on quality control. Many of the observed defects are often caused by the handling of the wafers during processing and many companies are investing in automated wafer handling equipment. Wafers are loaded into cassettes at the beginning of the manufacturing sequence and are transferred to and from these cassettes at the various stages of the process.

Many of the defects introduced during the manufacturing process can be detected by visual inspection, and sophisticated optical equipment is now available which automatically scans wafers and produces a map of the defects. The data can be stored in digital form on a computer and compared with a similar map produced after final electrical test. This process provides the means to correlate visual defects with final test data and provides a means for pinpointing those stages in the process which produce the greatest number of faults.

Many of the problems that were previously associated with diffusion have now been eliminated by the wide-scale introduction of ion-beam implantation, which is cleaner and which provides much improved control of the introduction of impurities.

Wire bonding is now largely automated, which ensures much greater reproducibility, and thus reliability of the connections to the leads of the package.

Computers are now used extensively to process data obtained from inspections and measurements. When this data is combined with data obtained from the production line equipment in the form of times, temperatures, pressures etc., it will be possible to provide much greater control of the complete process.

Twenty years ago it would have been unthinkable that circuits would contain 64 million or more transistors, and yet that is now the prospect, and the growth in circuit complexity and device count is likely to continue.

Problems

1. In an accelerated test for a semiconductor component the sample size is 200, the test time is 400 hr, the test temperature is 200 °C and the number of failures is three. From reliability data for this component the activation energy is 0.68 eV. Determine the failure rate at 70 °C.
[0.68 devices/10^9 hr.]

2. For a mature metal-gate NMOS manufacturing process the cost for 100 mm-diameter wafers is $90 per wafer. The number of defects is 4/cm^2. A new silicon-gate line for 150 mm-diameter wafers has a defect density of 2.5/cm^2 and a cost per wafer of $150. The die size for the old metal-gate process is 5 × 6 mm and it is proposed to transfer the product to the new line with a 15% shrinkage of the dimensions.

Determine the number of good die for each manufacturing process based on Seed's yield model.
[69, 270.]

Determine the manufacturing cost per die for each line.
[$1.3, $0.55.]

3. The final electrical test for a bipolar transistor is obtained from five test sites on each wafer and the collected data for the mean and range for the beta of one of the test transistors located on each test site is shown below. Draw the control charts for *X* and *R* and determine how many times a value falls outside the prescribed limits.

Batch no.	Gain	Range	Batch no.	Gain	Range
1	125	10	11	122	27
2	130	15	12	170	17
3	128	20	13	165	33
4	150	25	14	153	41
5	170	30	15	142	21
6	160	15	16	137	17
7	155	18	17	129	13
8	130	22	18	139	31
9	141	40	19	146	22
10	140	36	20	142	29

CHAPTER 10

VLSI technology overview

10.1 Introduction

The processes of epitaxy, oxidation, diffusion, ion implantation, photolithography and metallization which have been described in the preceding chapters are used in sequence to fabricate VLSI circuits. The main circuit technologies are npn bipolar and complementary metal–oxide–semiconductor (CMOS), but n-channel MOS (NMOS) and bipolar and CMOS (BiCMOS) are also used. Some of the important characteristics of bipolar and CMOS devices are listed in Table 10.1. In the table the '−' sign against a property signifies a poor feature, while the '+' sign signifies good features. The bipolar transistor is a low-impedance, current operated device and has large transconductance, good drive capability and large and linear output. In contrast, CMOS circuits use voltage operated MOS transistors which have very high input impedance, are tolerant of large variations in supply voltage, consume almost no power in a static condition and result in very good noise margins when used in digital circuits. Complementary MOS transistors also form a very good analog switch. Combining the positive features of the two technologies has resulted in BiCMOS circuits. However, the combined technology is more complex than either of the two individual technologies and it is difficult to achieve optimum conditions for both devices with a common set of process parameters.

A sclection of completed wafers ranging from 75 mm (3 in) to 150 mm (6 in) is shown in Figure 10.1.

10.2 The bipolar process

The original integrated circuits (circa 1960) were based on bipolar transistors and many of the present-day small-, medium- and large-scale integrated circuits (SSI, MSI, LSI) are derived from these early circuits. The motive for developing these circuits was to produce smaller and more reliable computers. Present-day bipolar circuits for computers are based on transistor transistor logic (TTL), emitter-coupled logic (ECL)

Table 10.1 Bipolar and CMOS transistor properties

General	NPN	CMOS
Supply voltage range	−	++
Power	−	++
Transductance	++	−
Circuit density	−	+
Drive capability	++	+
Digital		
Speed	++	+
Power consumption	−	++
Noise margins	−	++
Logic swing	−	++
Analog		
Gain	++	−
Bandwidth	++	−
Input impedance	−	++
Power consumption	−	++
Output swing	++	+
Linearity	++	+
Analog switching	−	++

and other variants which use npn transistors, diodes and diffused resistors to produce a wide range of logic functions. Analog circuits often use both npn and pnp transistors. In addition, diffused resistors and junction capacitors are also required to perform a variety of analog functions such as amplification, comparators, multipliers, function generators, etc.

An important requirement for all of these circuits is isolation between the various components, particularly for the npn transistors, which without isolation would all have a common-collector region.

10.2.1 Junction isolation

The most widely used form of isolation for the first generation of integrated circuits used the high impedance of a reverse-biased pn junction. The starting material is a p-type substrate upon which an n-type epitaxial layer is grown. Pockets of the n-type layer are then isolated from each other by diffusing a p-type impurity through the epitaxial layer until it reaches the substrate. A cross-section of the substrate with an isolated region is shown in Figure 10.2. The n-type region is completely surrounded by the p-type diffusion and provided the pn junction is reverse biased there is good dc isolation. However, the isolation provided for ac signals decreases as the frequency of

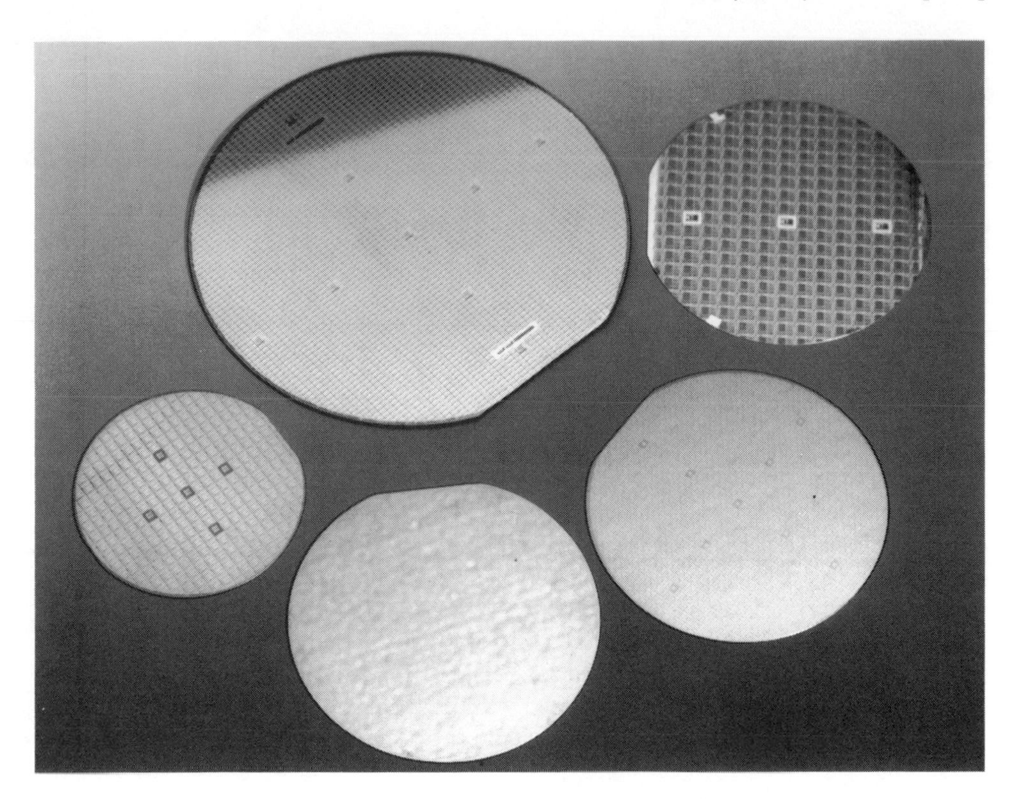

Figure 10.1 Selection of silicon wafers.

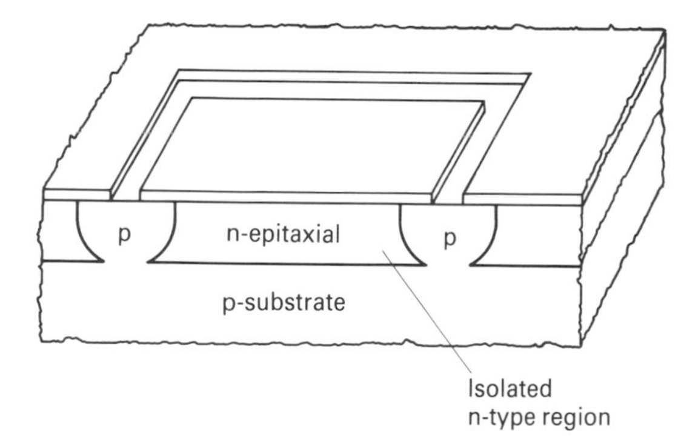

Figure 10.2 Cross-section showing an n-type epitaxial region enclosed by a p-type diffusion.

the ac signal increases, because the junction exhibits capacitance and the reactance of the capacitance decreases with increasing frequency.

Apart from the reduced isolation for ac signals a further disadvantage of junction isolation is the large amount of surface area required for the p-type silicon which is necessary for isolation. The area is not just confined to the p-type diffused region but also to the depletion layer which extends from the reverse-biased junction into the more lightly doped n-type epitaxial layer. The lateral diffusion beneath the oxide from the edge of the oxide window is approximately $0.8x_j$. To this distance must be added the width of the depletion layer. For a step-junction profile this is:

$$w = \sqrt{\frac{2\varepsilon_o \varepsilon_s V}{qN}}$$

where V is the applied voltage and N the impurity concentration of the more lightly doped epitaxial layer. This is shown schematically in Figure 10.3, where it is assumed that the epitaxial layer has a thickness of 5 μm and a resistivity of 5 ohm cm. With 5 V applied to the pn junction the depletion layer is approximately 2.6 μm wide. The combination of the lateral diffusion beneath the oxide (4 μm) and the width of the depletion layer means that the edge of the depletion layer extends some 6.6 μm from the edge of the oxide window. This 6.6 μm zone extends into the isolated n-type pocket on all sides and the active regions associated with any transistors or resistors which are contained within this pocket must not extend into this zone.

10.2.2 Buried n$^+$ layer

A simplified cross-section of an npn transistor is shown in Figure 10.4(a). Note that the three contacts (collector, base and emitter) are all on the top surface of the silicon wafer. Current flows transversely from the emitter, across the base to the collector, but must then flow laterally through the epitaxial layer to the collector contact. It is necessary for

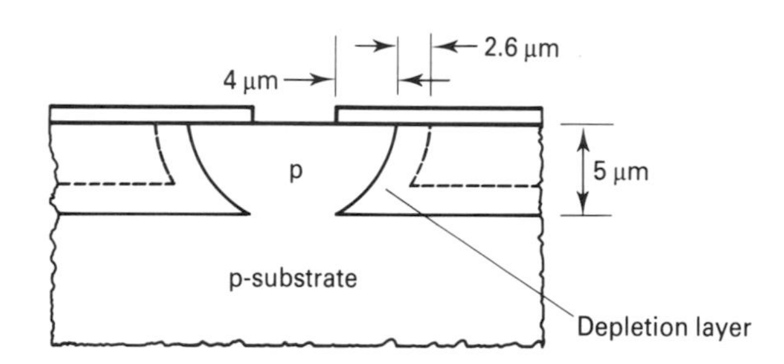

Figure 10.3 The area of the n-type isolated pocket is reduced by lateral diffusion of the p-type impurity and the spread of the depletion layer into the lightly doped n-type region.

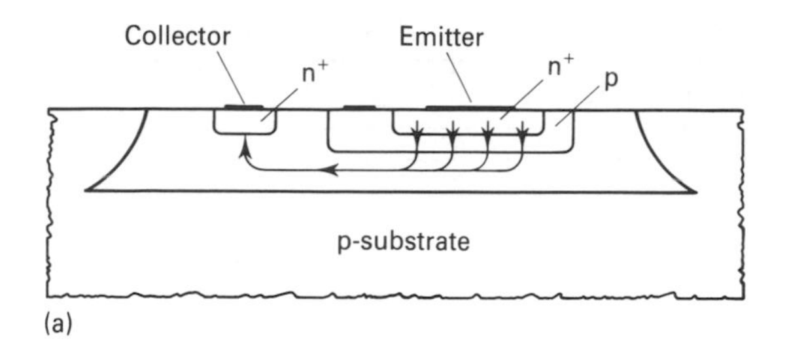

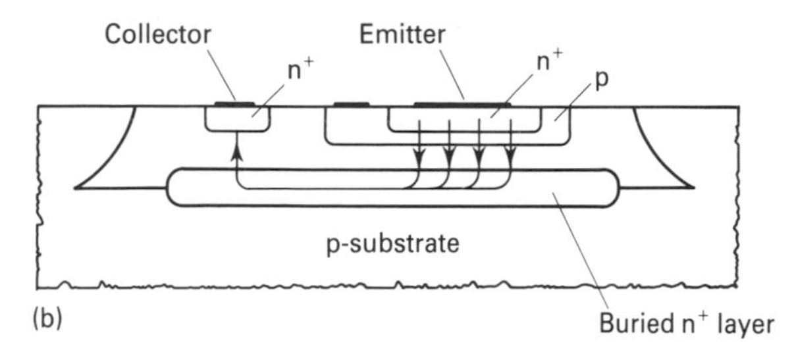

Figure 10.4 Flow of collector current from beneath the emitter to the collector contact (a) without a buried n^+ layer and (b) with a buried n^+ layer.

the epitaxial layer to have a relatively high resistivity and, consequently, there is a large series resistance between the active region beneath the emitter and the collector contact. This resistance affects the performance of the transistor. To maintain the resistivity of the epitaxial layer but at the same time reduce the resistance of the collector current path, a low-resistivity region is created beneath the transistor. This buried n^+ region is formed before the epitaxial layer is grown and is shown in Figure 10.4(b). Notice that, as a result of the many heat cycles, the n-type impurity of the buried n^+ region diffuses up into the epitaxial layer and care must be taken to ensure that it does not reach the collector–base junction. An impurity with a low-diffusion coefficient, such as arsenic, is usually used for this region.

Junction isolation is still widely used for analog bipolar circuits, but for logic circuits it is usually replaced by oxide isolation.

10.2.3 Oxide isolation

Replacing the p-type isolation region with silicon oxide immediately saves silicon because it eliminates the pn junction and hence the depletion layer associated with it. The oxide is a good insulator and the capacitance associated with the vertical walls of the isolated n-type region is reduced.

Capacitance associated with the pn junction at the floor of the isolated n-type pocket still exists. A simplified section of oxide isolation is shown in Figure 10.5(a).

Oxide isolation takes advantage of the fact that silicon is consumed as the oxide is produced such that $x_{si} \approx 0.4x_{ox}$. Thus as the oxidation proceeds, the surface of the

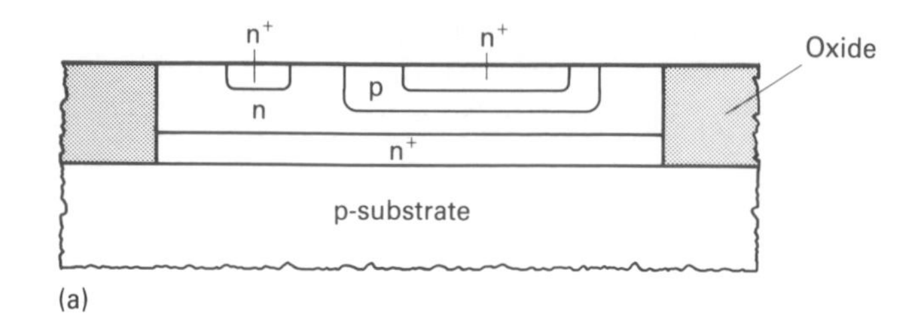

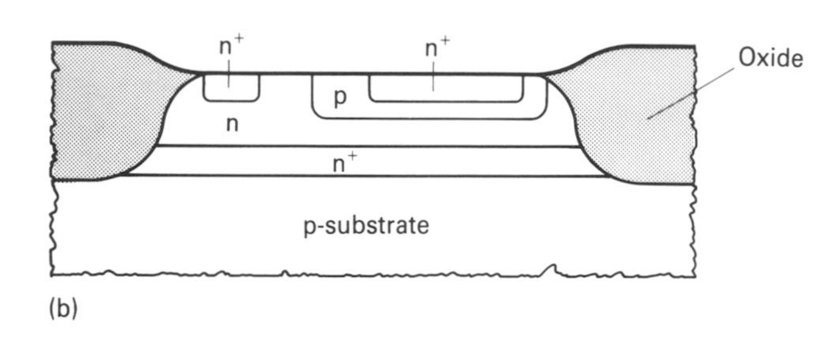

Figure 10.5 (a) An idealized cross-section of oxide isolation and (b) the actual cross-section.

silicon beneath the oxide moves below the original surface position. The oxidation is confined to the isolation channels by a layer of silicon nitride which is deposited on to the surface of the silicon and patterned by means of photolithography. The oxide isolation is performed after growing the epitaxial layer. In practice, the idealized rectangular regions of oxide shown in Figure 10.5(a) are not achieved. Some lateral oxidation takes place beneath the Si_3N_4 forming the so-called 'birds beak' (Chapter 2) shown in Figure 10.5(b). This lateral oxidation encroaches into the active n-type epitaxial layer, but not to the same extent as the depletion layer associated with pn junction isolation. It is usual for the n^+ region to extend over the whole of the base of the n-type pocket. Further reduction of the collector resistance can be achieved with a deep diffusion of an n^+ plug at the collector contact, which diffuses through the epitaxial layer to the n^+ buried layer.

10.2.4 The manufacturing sequence

Bipolar integrated circuits for both analog and digital applications follow a sequence of fabrication steps similar to those shown in Figure 10.6.

For oxide isolation the oxidation step after epitaxy would be replaced with oxidation plus CVD Si_3N_4 and the pre-deposition and following drive-in would be replaced by high-pressure steam oxidation. In practice, there are many more steps, for example there are many etching steps to remove the oxide formed after a pre-deposition, to remove the Si_3N_4, to remove the photoresist; the drive-ins would

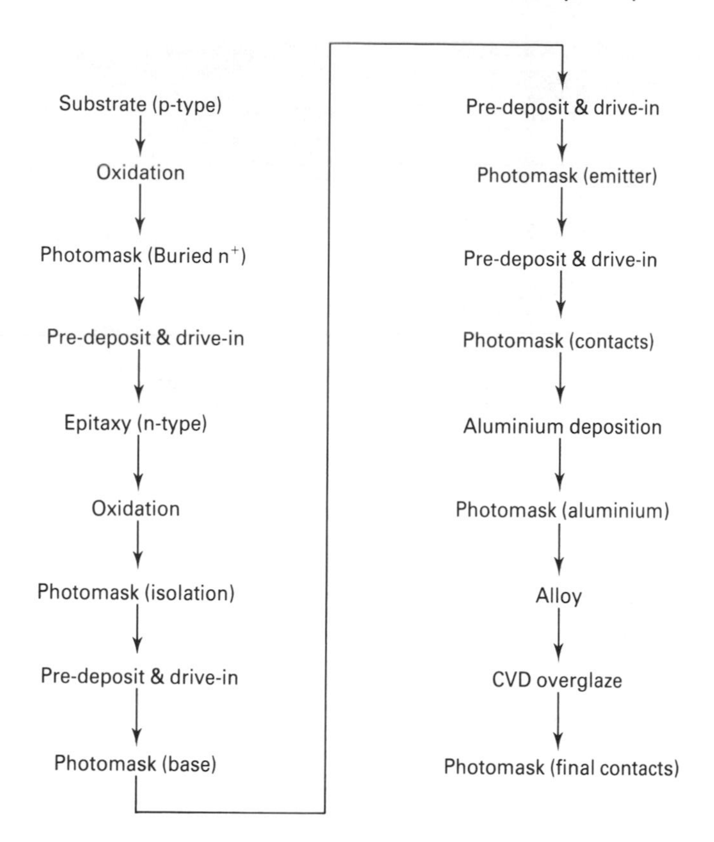

Figure 10.6 A much simplified flow chart for a bipolar integrated circuit.

typically involve temperature–time cycles to ramp the slices up to the working temperature, to grow dry and wet oxides in sequence and finally to ramp the slices back down before withdrawing them from the furnace; there would also be a large number of inspections and quality control tests. Every step of the manufacturing process is recorded on a 'run card' (or stored on computer and accessed by means of a bar-code reader) which is used to monitor the position of the batch of wafers in a manufacturing sequence that may extend over many weeks.

The process steps for analog applications are optimized to achieve higher values of current gain and higher breakdown voltages. This involves deeper junctions, narrower base width, higher resistivity and thicker epitaxial layers. Junction isolation is used because with thicker epitaxial layers oxide isolation is not practical.

Higher packing density is more important for digital circuits and with lower breakdown voltages, thinner epitaxial layers and shallower junctions it is possible to use oxide isolation.

A small section of an analog integrated circuit is shown in Figure 10.7. The bright lines are aluminium interconnections. Transistors can be seen, four on the top left-hand

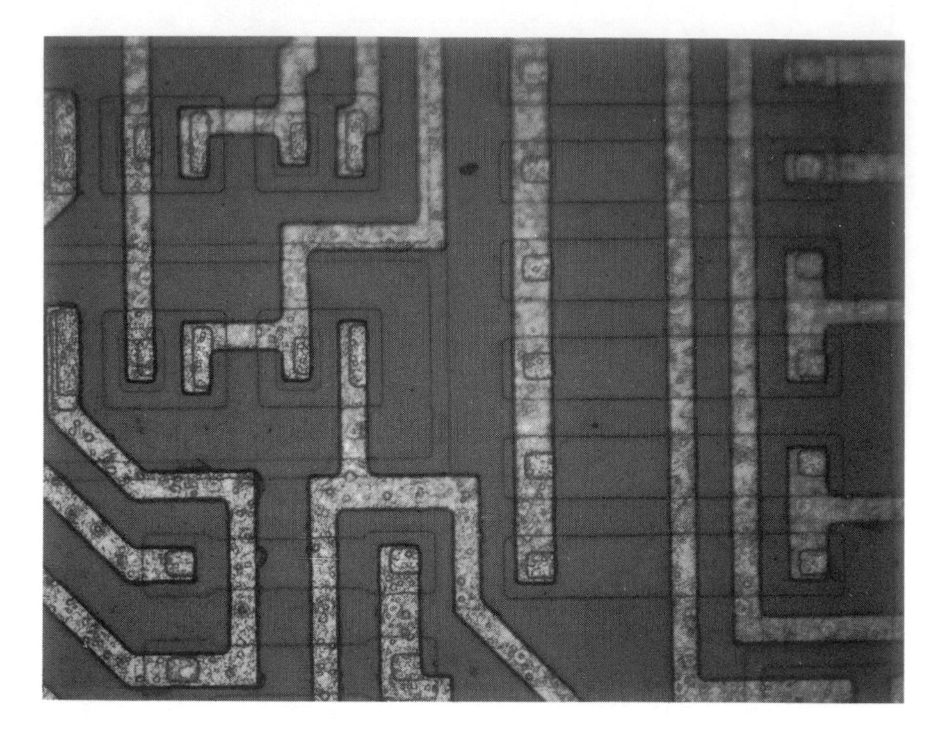

Figure 10.7 Small section of an analog integrated circuit. [By courtesy of National Semiconductor.]

side, together with the narrow isolation diffusion that surrounds them and isolates them from the resistors, which can be seen at the bottom and on the right-hand side.

10.3 The MOS process

The MOS process is used extensively for digital circuits and the fabrication steps are optimized to achieve high packing densities. This is greatly enhanced by the fact that the MOS devices do not require special diffusions or oxidations to provide isolation. The active regions are isolated from each other by reverse-biased pn junctions associated with the actual transistor, and no additional isolation is required. In practice, advanced MOS circuits often employ thick field oxides and implanted p- or n-type guard regions to improve circuit performance. High packing density also results from the fact that MOS digital circuits are composed entirely of MOS transistors without the need for diffused resistors, which usually require a large amount of silicon relative to the amount required for a transistor. All current flow takes place in a very thin layer at the surface of the silicon and there is no need for deep junctions or buried n^+ layers. Also, in principle, there is no need for epitaxial growth but it is often used for other purposes. A vitally important process for MOS manufacture is ion implantation.

While it is now also widely used for bipolar circuits, it is essential for MOS devices to adjust threshold voltages.

The first generation of MOS circuits were based on n-channel enhancement transistors and, later, n-channel enhancement drivers with n-channel depletion loads (Chapter 15). However, it is likely that future generations of MOS circuits will be CMOS with n-channel and p-channel enhancement devices. To fabricate both types of device on a single wafer involves creating a 'well' of opposite conductivity type to that of the surrounding material. This is a form of isolation and does lower the packing density for CMOS in comparison with NMOS.

10.3.1 Well formation

Three well configurations are shown in Figure 10.8. In (a) the starting material is n-type silicon and a p-type well is formed, usually by ion implantation, followed by a drive-in

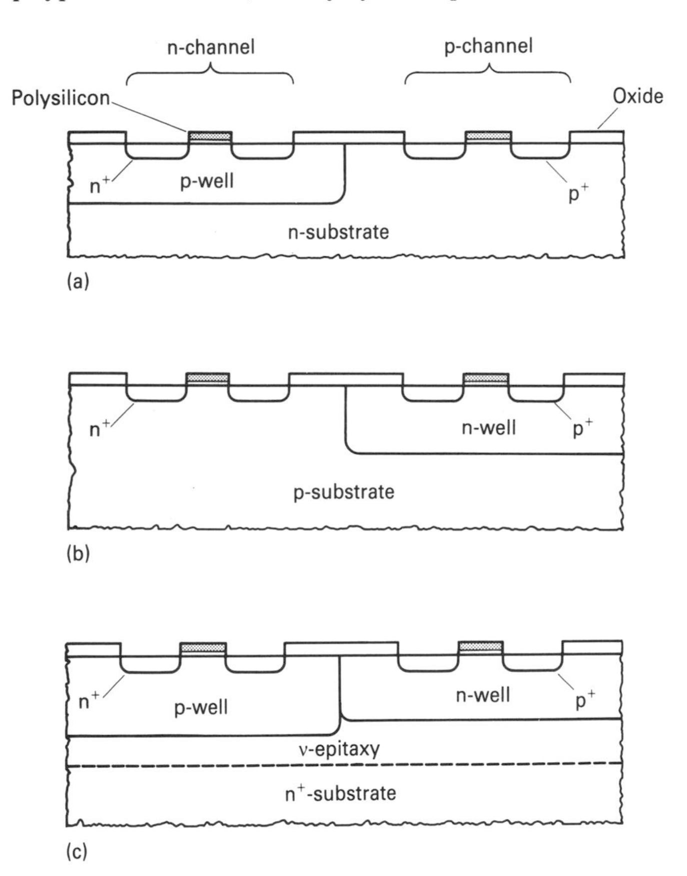

Figure 10.8 Cross-sections of various well configurations for CMOS with (a) a p-well in an n-type substrate, (b) an n-well in a p-type substrate and (c) twin wells in a high-resistivity substrate.

diffusion. The p-channel devices are formed in the original substrate, while the n-channel devices are formed in the p-well. With a p-type starting material an n-type well is formed by implantation and diffusion, as shown in (b). Alternatively, two separate wells are formed in a high-resistivity epitaxial layer, which is grown on an n^+ substrate, as shown in (c). Although the twin well process is more complex it does provide greater flexibility for the design and optimization of the n- and p-type devices.

10.3.2 The gate electrode

The first generation of MOS circuits used a gate electrode of aluminium, which was deposited after the source and drain diffusions had been completed. The disadvantage of this approach is the difficulty of accurately aligning the photomask for the gate electrode with the already formed source and drain regions, so that the gate electrode does not overlap either the drain or source diffusions. Any overlap would result in stray capacitance between the gate and either the source or the drain and a resultant deterioration of circuit performance. Polysilicon has replaced aluminium to produce a self-aligned gate electrode. The polysilicon is deposited before the formation of the source and drain regions. The polysilicon acts as a mask for the channel region and because it is able to withstand the high temperatures required for diffusion it remains in place during and after the diffusion. With this process the source and drain are accurately aligned with the polysilicon gate. Refractory metals such as tungsten and molybdenum have also been used for this purpose. The material used for the gate electrode affects the threshold voltage through the work function, which must then be included when determining the threshold voltages for the n- and p-channel devices of NMOS or CMOS devices.

The successful operation of MOS circuits is very dependent on the ability to control the threshold voltages (V_T). In the case of CMOS devices it is necessary to control two threshold voltages, one of which is positive and the other negative. The accurate control of V_T has been made possible by ion implantation. Implantation usually takes place through the gate oxide before deposition of the polysilicon for the gate electrodes. This ensures that the peak of the impurity ion profile coincides with the surface layer of the silicon where channel conduction takes place.

10.3.3 The manufacturing process

The basic manufacturing steps for a p-well CMOS process are illustrated in Figure 10.9.

There are many more basic process steps for the CMOS process than for bipolar devices and because the active region of the MOS transistor is at the surface of the wafer at the interface between the silicon and the gate oxide, it is very important to eliminate contamination from all parts of the process. MOS devices are much more susceptible to contamination than bipolar devices.

A section of a CMOS circuit showing part of a CPU register block is shown in Figure 10.10. It is impossible to identify individual components because modern digital

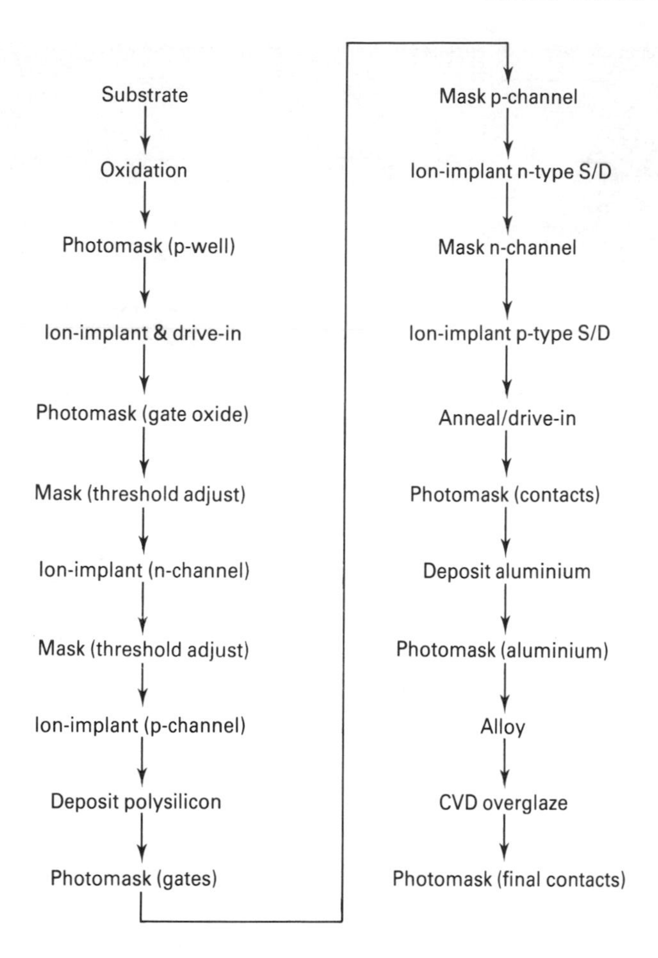

Figure 10.9 A much simplified process flow for a p-well CMOS process.

MOS circuitry is so very small. The regular structure at the lower right is part of a PLA (programmable logic array). This structure would allow the manufacturer to customize the chip for a variety of applications.

10.4 Process monitoring

The fabrication of integrated circuits is very complex and extends over many weeks. It is not enough to rely on an electrical test at the end of the fabrication sequence to determine whether the process has been successful. Each step of the process must be monitored and controlled so that the final electrical tests simply confirm that everything is in order. Many measurements can be made during the manufacturing process, either on test wafers or on the wafers themselves. The application of statistical

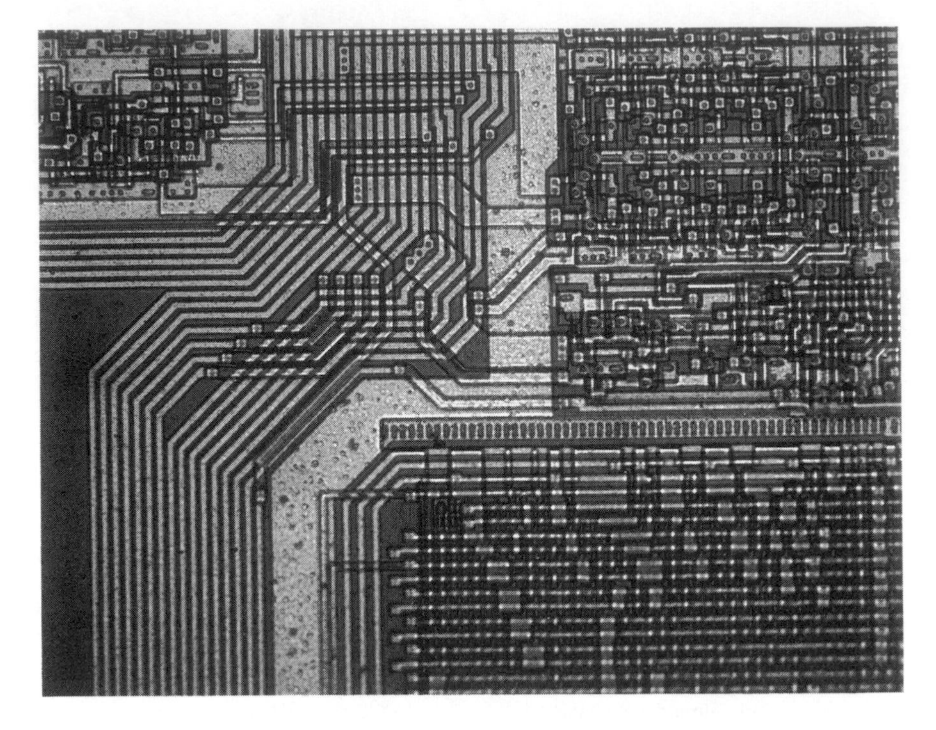

Figure 10.10 Section of a CMOS digital integrated circuit. [Copyright Motorola, reproduced with kind permission.]

process control (SPC) methods to the results of these measurements provides the means for monitoring the quality of the production processes.

Parameters which can readily be measured are:

sheet resistance	oxide thickness
oxide breakdown voltage	critical dimensions
C–V data	epitaxial thickness

One or more of the above parameters is measured at various stages during the production sequence. At each stage, measurements are made on a pre-defined number of samples (3, 5, 10) and the average and minimum and maximum values are calculated. These values are then plotted on 'average' and 'range' charts, as shown in Figure 10.11, to provide a visual indication of the control of the process. Upper and lower thresholds can be established so that a visual indication is provided of any tendency for the process to drift out of control. While such charts may be of value to the individual process engineers, for overall production control the data from every stage of the process would be entered into a computer and SPC methods used to provide a detailed statistical report.

In addition to these process checks, each wafer contains a number of 'test chips' distributed among the actual circuits. Each test chip contains a variety of resistors,

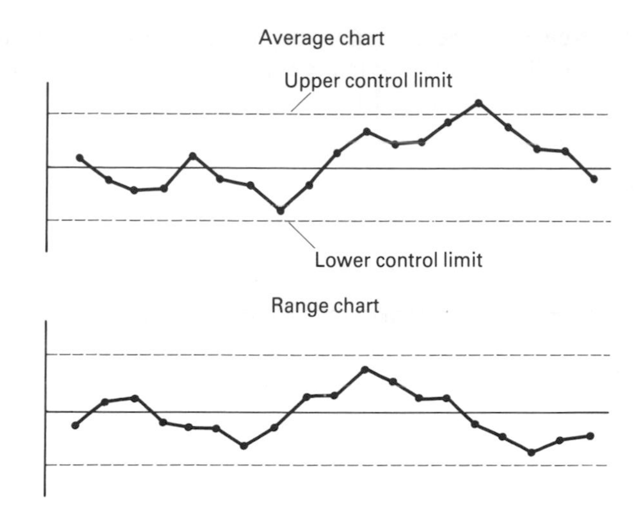

Figure 10.11 Examples of statistical control charts with upper and lower threshold levels for average and range.

transistors, MOS capacitors, devices for measuring contact resistance and interconnection integrity, etc. They enable detailed electrical measurements to be made to assess the quality of the electrical devices, for example resistance values, transistor current gain, threshold voltage, leakage currents, breakdown voltages, etc.

10.5 Future trends

There is continuing pressure to reduce dimensions so that more devices can be packed into a given area. Each reduction results in further demands on process control and for electrical device modelling. Ion-beam implantation is being used much more extensively with a resultant improvement in the control of impurity concentration and hence device characteristics. Deep ultra-violet lithography is being developed for submicron dimensions and, beyond that, X-ray lithography.

Yield of circuits is very dependent on the number of times wafers are exposed to the room ambient and the dust that it contains, even in 'clean rooms'. Consequently, future production lines are likely to make much greater use of robotic handling for the transfer of wafers between different stages of the production process.

Greater use is likely to be made of bipolar and CMOS on the same wafer to provide the circuit designer with the best of both technologies.

Much more use will be made of computer simulation of both the process (SUPREM) and the devices (PISCES[1], MINIMOS[2]) to improve production control and to allow production engineers to simulate process changes before applying the proposed changes to the production line.

Other semiconductors, particularly gallium arsenide, will be more widely used, particularly for extremes of speed or possibly temperature, but silicon is likely to dominate the field for VLSI circuits for many years to come if only because of the large capital investment already committed to it.

References

[1] PISCES (Poisson and Continuity Equation Solver), M. R. Pinto, C. S. Rafferty, H. R. Yeager and R. W. Dutton (1984), Stanford University, Stanford, CA.

[2] MINIMOS V Users Guide, P. Habas, O. Heinreichsberger, P. Lindorfer, P. Pichler, H. Pôtzl, A. Schutz, S. Selberherr, M. Stiftinger and M. Thurner, March 1990, Technische Universitat Wein, Institut Fur Mikroelektronik. Austria.

CHAPTER 11

The bipolar transistor

11.1 Introduction

The design of bipolar integrated circuits involves the interconnection of transistors, diodes, resistors and capacitors. There are no inductors or transformers and, in fact, capacitors are only available in a very limited range of values, up to a few tens of picofarads.

The most important device is the transistor and the design of the integrated circuit is largely based on the ability to control current flow by using transistors with different surface areas. Diodes are formed from one or more of the pn junctions used to make the transistor and small-value junction capacitors are formed from the same junctions when reverse biased. Resistors are isolated regions of n- or p-type silicon, with the isolation being achieved with a pn junction.

Because of surface effects it is easier to fabricate npn transistors than pnp devices. However, pnp devices can be fabricated without adding to the complexity of the npn fabrication sequence, but the current gains are usually less than 50. Field effect devices can also be included but with some additional processing. These devices increase the versatility of the bipolar integrated circuit.

The plan view and cross-section of a low-power npn transistor is shown in Figure 11.1. Oxide isolation is shown which is current practice for the majority of digital integrated circuits, but which may be replaced with junction isolation for analog circuits. The active part of the transistor is the region immediately beneath the emitter. The n^+ emitter injects electrons into the p-type base in response to the bias voltage across the emitter–base junction. The majority of the carriers cross the base to the collector which, under normal operating conditions, is reverse biased.

Some of the important design parameters relate to the dc current, the dc current gain and breakdown voltages. For small-signal operation the frequency response, switching speed, junction capacitance and resistance associated with regions of silicon between the active region and the contacts are important. Much detailed analysis has now been performed on npn transistors and very complex electrical models have been developed for use with computer simulation programs. The explanations provided

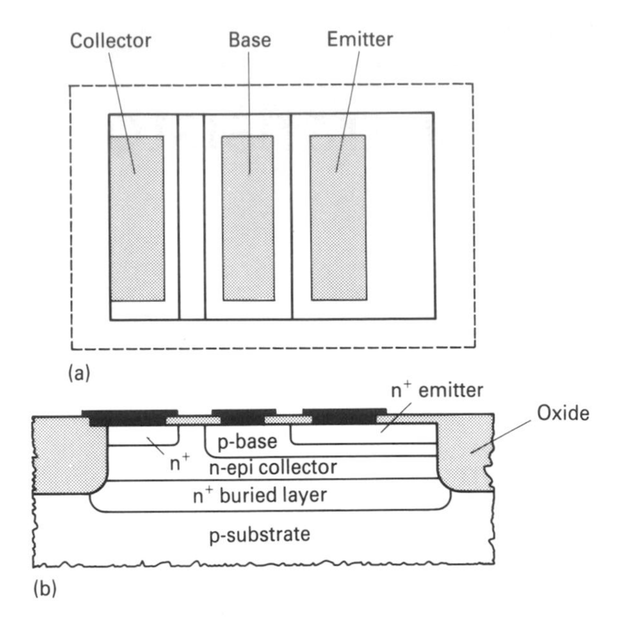

Figure 11.1 Low-power npn transistor with oxide isolation (a) plan view and (b) cross-section.

below are only intended to provide an insight into the various relationships and the problems that may face a device designer.

11.2 Current gain

For normal operation the base–emitter junction is forward biased and the base–collector junction is reverse biased, as shown in the band diagram of Figure 11.2. Forward bias is achieved with a negative voltage applied to the emitter with respect to the base, which lowers the barrier between the emitter and the base. The negative voltage drives the negatively charged electrons in the emitter towards the emitter–base junction, while holes in the base are driven towards the junction by the positive voltage applied to the base. Some of the electrons in the conduction band of the n-type emitter now have sufficient energy to cross the barrier into the conduction band of the p-type base. Similarly, some of the holes in the valence band in the base cross into the valence band in the emitter. A concentration gradient of electrons is created in the base and a gradient of holes is formed in the emitter, as shown in Figure 11.3. The electrons diffuse away from the region of high concentration adjacent to the emitter until they reach the collector. The base–collector junction is reverse biased and the regions adjacent to the metallurgical junction are depleted of free carriers. The electric field associated with the depletion region is in such a direction as to attract any electrons reaching the edge of the depletion layer and to accelerate them across the

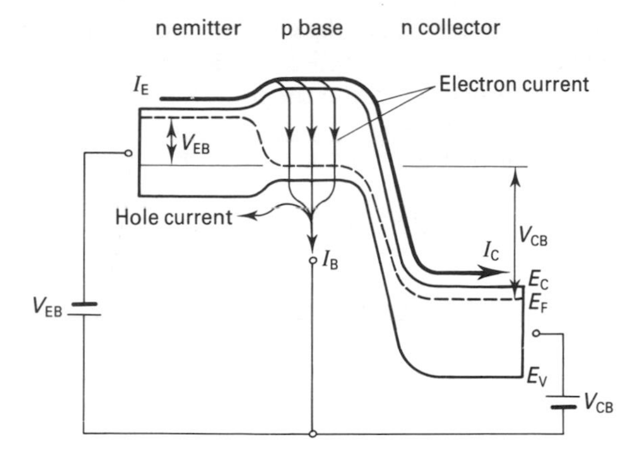

Figure 11.2 Band diagram with forward-biased emitter and reverse-biased collector showing the flow of electrons and holes.

layer to the collector. The emitter acts as a source of electrons and the collector as a sink.

The holes which flow from the base into the emitter constitute base current, which is very undesirable. The base current formed by these holes is a function of the hole concentration relative to the electron concentration injected into the base by the emitter. For large common base current gain (α_F) the efficiency of the emitter is improved if the number of holes is small with respect to the number of electrons.

An additional base current results from the recombination of some of the minority carrier electrons injected into the base with the majority carrier holes which exist in the p-type base region. Thus not all of the electrons reach the collector. The proportion which does is determined by a transportation factor and this also affects α_F. Finally, at the collector junction, the electrons moving from the conduction band in the base to the conduction band of the collector are accelerated by the electric field and can acquire sufficient energy to dislodge electrons from atoms within the depletion layer.

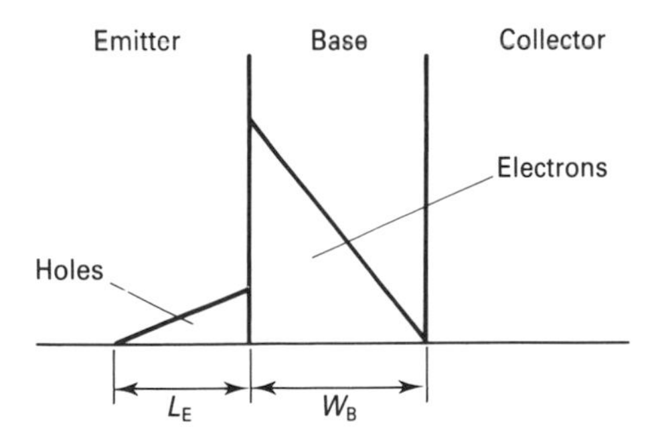

Figure 11.3 Minority carrier distribution in the base and emitter.

Thus the number of electrons arriving at the collector can be greater than the number being supplied by the emitter. This multiplication factor also affects α_F but only near breakdown and since for normal operation the collector voltage is much less than the breakdown voltage it can be ignored.

The common base current gain factor is given by:

$$\alpha_F = \frac{I_C}{I_E} = \frac{I_C}{I_B + I_C} = \frac{\beta_F}{1 + \beta_F}$$

where β_F is the common-emitter current gain.

In terms of the current conduction mechanisms within the transistor α_F is a measure of the injection efficiency of the emitter (γ), the base transportation factor (β^*) and the avalanche multiplication factor of the base–collector depletion layer (M).

$$\alpha_F = \gamma\beta^* M \approx \gamma\beta^*$$

since $M = 1$ for normal operation.

The important mechanism for current flow is diffusion and the electron and hole currents are proportional to concentration gradients in the base and emitter. In the base the gradient extends over a distance W_B, as shown in Figure 11.3 and in the emitter it extends for a distance L_E, which is the diffusion distance for holes in the emitter and represents an average distance before the minority carrier hole combines with a majority carrier electron. The diffusion length is given by:

$$L_E = \sqrt{(D_p \tau_p)}$$

where D_p is the diffusion coefficient of holes (cm^2/s) in the emitter and τ_p is the lifetime of holes in the emitter. The relationship is only true if the width of the emitter (W_E) is much greater than L_E. For many advanced technology transistors this is not the case and W_E must be substituted for L_E.

The diffusion coefficient (D_p) is related to the mobility by Einstein's relationship as $D_p = (kT/q) \times \mu_p$, where $kT/q = V_T$ and is the thermal voltage equal to 0.026 V at 300 K.

The emitter current is equal to the sum of the electron current and the hole current, but the only current which contributes to the collector current is that produced by the electrons. The efficiency of the emitter can be expressed as:

$$\gamma \approx \frac{\text{electron current}}{\text{electron current} + \text{hole current}}$$

or

$$\gamma = \left[1 + \frac{\text{hole current}}{\text{electron current}}\right]^{-1}$$

Each current is determined by the forward bias applied to the base–emitter junction and the concentration gradient. The exponential terms are the same but the gradient terms are different so that in the above ratio the exponential terms cancel. The gradient terms for the holes and electrons are $D_{pE}\,dp/dx$ and $D_{nB}\,dn/dx$ respectively. A full analysis shows that the gradient for the holes is p_n/L_E and for the electrons it is n_p/W_B. The minority carrier concentrations, p_n and n_p can be expressed in terms of the impurity concentrations by using the relationships:

$$p_n N_d = n_i^2$$

$$n_p N_a = n_i^2$$

Substituting N_E for N_d and N_B for N_a, where the subscripts refer to the emitter and base, respectively, then

$$\gamma = \left[1 + \frac{D_{pE} N_B W_B}{D_{nB} N_E L_E}\right]^{-1}$$

In this equation it is assumed that impurity concentration in the base is uniform. In practice, this is not the case. An example of the impurity profile for the emitter, base and collector regions of a transistor is shown in Figure 11.4. It can be seen that the profile for the base is graded. A parameter which is widely used to describe the impurity distribution is the Gummel number[1], which is the number of impurities per cm^2. The Gummel number for the base is defined as:

$$G_b = \int_0^{W_B} N_a(x)\,dx$$

A similar number can be defined for the emitter as:

$$G_e = \int_0^{W_E} N_d(x)\,dx$$

Substituting the Gummel numbers for $N_B W_B$ and $N_E L_E$ in the above equations gives the injection efficiency as:

$$\gamma = \left[1 + \frac{D_{pE} G_b}{D_{nB} G_e}\right]^{-1}$$

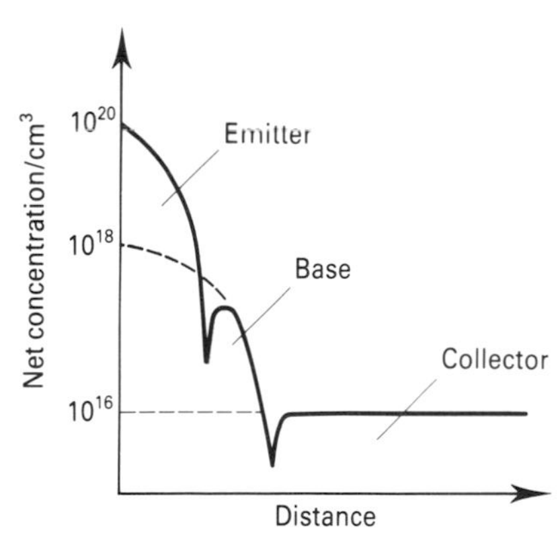

Figure 11.4 Impurity profile for an npn transistor.

As an example assume that the impurity concentration in the emitter is $10^{19}/cm^3$ and the emitter is 0.5 μm wide; assume that the base peak concentration is 10^{17} and that the base is also 0.5 μm wide. Let the collector impurity concentration be $10^{15}/cm^3$. If it is assumed that the profile for the emitter is approximately rectangular, then the Gummel number is $0.5 \times 10^{15}/cm^2$. If the base profile is assumed to vary linearly from 10^{17} at the emitter side to 10^{15} at the collector–base junction, then the Gummel number is approximately $0.25 \times 10^{13}/cm^2$. The diffusion coefficients can be determined from the mobility with the aid of Figure 1.5. For the emitter, D_{pE} is approximately 1.6 cm^2/s and D_{nB} is 18.2 cm^2/s. Then the injection efficiency is:

$$\gamma = \left[1 + \frac{1.6 \times 0.25 \times 10^{13}}{18.2 \times 0.5 \times 10^{15}}\right]^{-1}$$

$$\gamma = 0.9996$$

Minority carrier electrons that are injected into the base by the emitter must cross the base without recombining with the majority holes in the base. If the base width were very large, then all of the minority electrons would recombine with holes and none would reach the collector. The concentration gradient of the electrons would extend from the emitter–base junction a distance L_B, which is the diffusion distance for electrons in the base. At a distance L_B from the emitter–base junction the concentration would be approximately zero. Provided that the base width were very much smaller than L_B, as illustrated in Figure 11.5, then the majority of the electrons would reach the collector before recombining with holes. A detailed analysis yields a value for the base transport factor given by:

$$\beta^* = 1 - \frac{W_B^2}{2L_B^2}$$

The diffusion distance can be expressed in terms of the lifetime (τ_n) and the diffusion coefficient (D_n) of the electrons in the base:

$$L_B = \sqrt{(D_n \tau_n)}$$

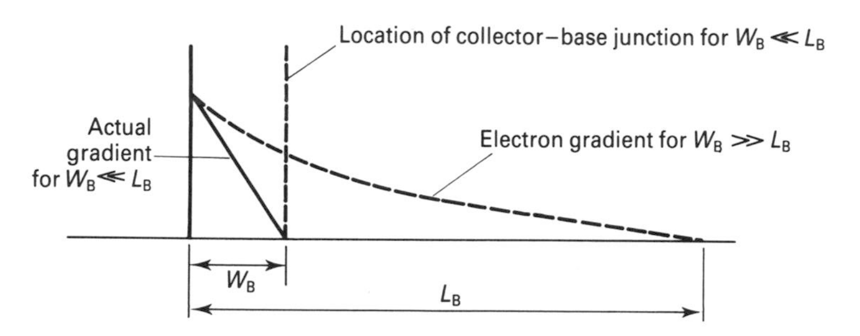

Figure 11.5 An imaginary concentration gradient in the base assuming that $W_B \gg L_B$ and the actual gradient for $W_B \ll L_B$.

It is apparent that for $L_B \gg W_B$ the lifetime of electrons in the base must be as high as possible. For an advanced technology device β^* is of the order of 0.999 93. Thus, the controlling factor for the current gain for many small-area, low-power transistors is the emitter injection efficiency.

The common-emitter current gain (β_F) is given by:

$$\beta_F = \frac{\alpha_F}{1 - \alpha_F}$$

The common-emitter gain can be rewritten in terms of the injection efficiency since $\alpha_F \approx \gamma$. Then

$$\beta_F = \frac{\gamma}{1 - \gamma}$$

or, in terms of the Gummel number, as:

$$\beta_F \approx \frac{D_{nB} G_e}{D_{pE} G_b}$$

For the example given above where the injection efficiency is 0.9996, the value of β_F is 2275. In practice, there are additional effects which reduce this value by at least a factor of ten.

11.3 The collector current

The emitter current can be written as (Sze)

$$I_E = \frac{qAD_B n_i^2}{G_b} \exp \frac{V_{BE}}{V_T}$$

where D_B is the diffusion coefficient of minority carriers in the base, V_T is 26 mV (kT/q) at 27 °C and A is the emitter area. The collector current is:

$$I_C = \alpha_F I_E$$

There is an additional component of collector current which represents the flow of minority carriers across the reverse-biased collector–base junction, and consists of electrons crossing from the base to the collector and holes crossing from the collector to the base.

Thus the collector current is given by

$$I_C = I_S A \exp \frac{V_{BE}}{V_T} + I_{CO}$$

where I_S is a process parameter which is the same for all transistors in a single integrated circuit and I_{CO} is the leakage current associated with the base–collector junction.

With a knowledge of the constant I_S, transistors can be designed for a particular current by selecting the area of the emitter.

11.4 The Early voltage

For the ideal transistor the collector current is independent of the collector–emitter voltage (V_{CE}) but in practice the current increases with increasing V_{CE}, as shown in Figure 11.6. This finite slope of the output characteristics results from the Early effect[2], which describes the modulation of the base width with the collector voltage. As V_{CE} increases, the width of the depletion layer associated with the collector–base junction also increases. Much of the movement takes place in the more lightly doped collector, but there is some movement into the base. This results in a reduction of the effective width of the base. From the equations for emitter injection efficiency and base transport factor, a reduction of the base width increases the current gain and hence the collector current. The slope of the $I_C : V_{CE}$ graphs can be extrapolated back to common intercept, which is known as the Early voltage V_A, as shown in Figure 11.6.

For small-area, low-power transistors, V_A lies between 50 V and 100 V.

11.5 Variation of gain with current

As the collector current changes so does the gain, as shown in Figure 11.7. There are three distinct regions – a low current range, an intermediate range and a high current range.

At low currents the gain decreases with decreasing current as a result of the following:

1. Recombination at the silicon–silicon-oxide interface near the emitter–base depletion layer. Electrons injected into the base by the emitter are more likely to recombine with majority holes near the oxide interface than in the bulk. These

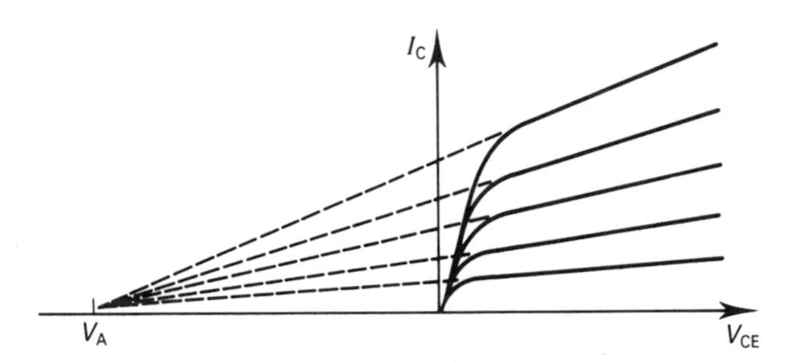

Figure 11.6 Output characteristics for an npn transistor showing the Early voltage as the intercept of the slope of the $V_{CE} : I_C$ curves in the saturation region.

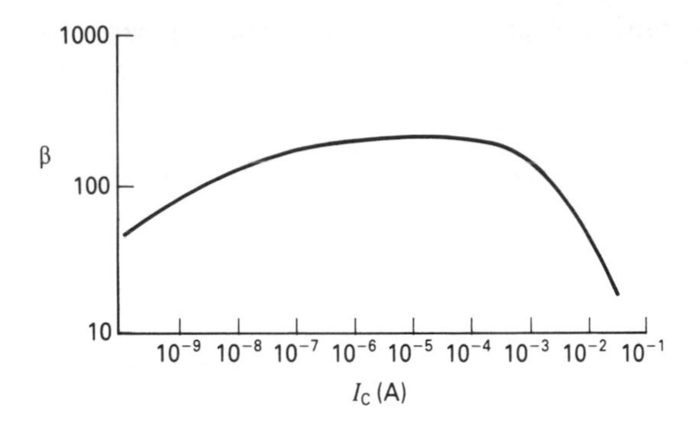

Figure 11.7 Typical variation of common-emitter gain with current for a small-signal transistor.

carriers contribute to the base current and as a result the gain is reduced. The numbers which recombine remain relatively constant throughout the current operating range of the transistor, so that as the current increases, the relative proportion that recombine at the surface with respect to the total current decreases. Thus at higher currents the effect of this surface recombination is insignificant.

2. Recombination throughout the emitter–base depletion layer. The electric field in the emitter–base depletion region opposes the flow of electrons and they spend a relatively long time in this region with, consequently, an increased chance of recombining with majority carriers. The field decreases as the current increases and hence the number recombining reduces, which causes the current gain to increase.

In the intermediate current range the gain remains relatively constant with change of current. This region represents the normal operating condition for the transistor. An important factor for this region is the maximum value of the gain. The gain is largely controlled by the ratio of the carriers in the emitter and the base and it is usual for the emitter to be heavily doped so that $N_E/N_B \gg 1$. However, very high impurity concentrations affect the bandgap and cause it to narrow. The reduction in the bandgap is given by[3]

$$dE_g = 22.5\ (N_E/10^{18})^{1/2}\ \text{meV}$$

where N_E is the emitter doping in atoms/cm^3. The effect becomes significant above carrier concentrations of 10^{19}/cm^3 for which dE_g is approximately 70 meV. The intrinsic carrier concentration in the emitter is now given by:

$$n_{iE}^2 = n_i \exp\left(\frac{dE_g}{kT}\right)$$

where n_i is the intrinsic carrier density without bandgap narrowing. As a result of the increase in the intrinsic carrier density in the emitter there is an increase in the density of the minority holes in the emitter. This has the effect of increasing the number of holes which are injected into the emitter from the base. Thus the ratio of injected electrons to injected holes is reduced which reduces the emitter injection efficiency, and consequently reduces the gain. Typically, bandgap narrowing can reduce the gain by a factor of 2.

At high currents there are three significant factors which produce a reduction in the gain with increasing current:

1. Debiasing of the emitter is caused by the base current flowing through the high-resistance region from beneath the emitter to the base contact, as shown in Figure 11.8. The voltage drop which results reduces the forward bias of that part of the emitter–base junction furthest from the base contact. This debiasing causes the current flow to be concentrated at the edge of the emitter nearest to the base contact, which effectively reduces the area of the emitter and, therefore, reduces the current for a given base–emitter voltage.

2. Conductivity modulation of the base region caused by the increased number of minority carriers. As the current increases the number of minority carriers also increases and there is a corresponding increase in the number of majority carriers to maintain the charge neutrality of the base. This increase in majority carrier concentration represents an increase in base conductivity, which reduces the emitter injection efficiency.

3. For integrated circuit transistors with a lightly doped n-type epitaxial layer there is a relocation of the base–collector depletion region at the metallurgical junction (point A in Figure 11.9) to a region near the buried n^+ layer (point B) under high

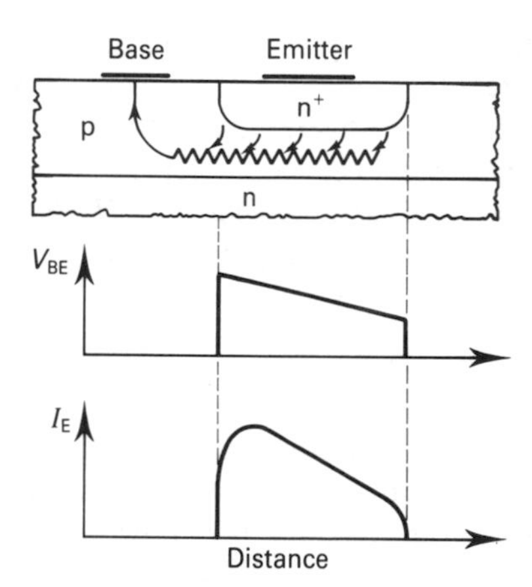

Figure 11.8 Debiasing of the emitter caused by the flow of base current through the lateral base resistance.

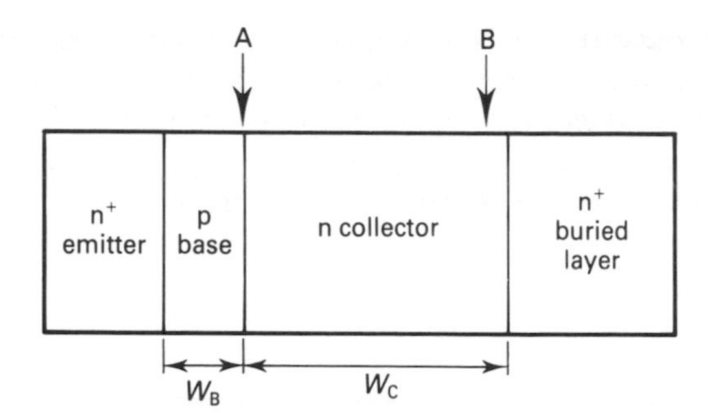

Figure 11.9 Schematic of an npn transistor with a base width W_B and a collector epitaxial layer W_C.

current conditions. This results in an effective increase in the base width from W_B to $W_B + W_C$, where W_C is the thickness of the epitaxial layer. This high-field relocation is known as the Kirk[4] effect and results in an increase in the Gummel number and a reduction in the current gain. It can be shown[5] that the critical current density at which this base widening takes effect is:

$$J_{CR} = qv_s\left[N_C + \frac{2\varepsilon_o\varepsilon_s V_{CB}}{qW_C^2}\right]$$

where v_s is the saturation velocity (10^7 cm/s for silicon at 300 K), N_C is the epitaxial layer carrier concentration, W_C is the layer thickness and V_{CB} is the collector–base voltage.

The critical collector current (I_{cr}) at which the fall in gain is observed is obtained by multiplying the current density by the emitter area.

The debiasing of the emitter has important consequences for the design of power devices for output stages. Since current flow is confined to the emitter periphery at high current levels, power devices are often designed with many long, narrow emitter regions separated by similar long, narrow base regions in order to increase the emitter periphery. The Kirk effect is also important for power devices and the carrier concentration in the epitaxial layer must be sufficient to ensure that the current density does not exceed the critical value given above.

11.6 Simulation

While some simple equations can provide an indication of how different parameters are affected by material properties or physical dimensions, the only effective means of examining the operation of a bipolar transistor in detail is to use simulation. The program SUPREM can be used to model the fabrication steps involving oxidations, diffusions and epitaxial growth. The profiles which SUPREM generates can then be used to model the operation of the transistor. A device simulator, such as PISCES, solves Poisson's equation for electric potential and the continuity equation for current flow, for both holes and electrons, by means of either finite difference or finite element methods. These programs require the generation of a grid within the active region to be investigated. The grid subdivides the material into many small lines (1D) or boxes (2D) across which the solution can be obtained to the appropriate partial differential equation.

An example of the output generated by MEDICI (the TMA[6] version of PISCES) is shown in Figure 11.10.

The device simulated is a small npn transistor typical of that which would be used in low-power Schottky TTL circuits. The impurity profiles are obtained from

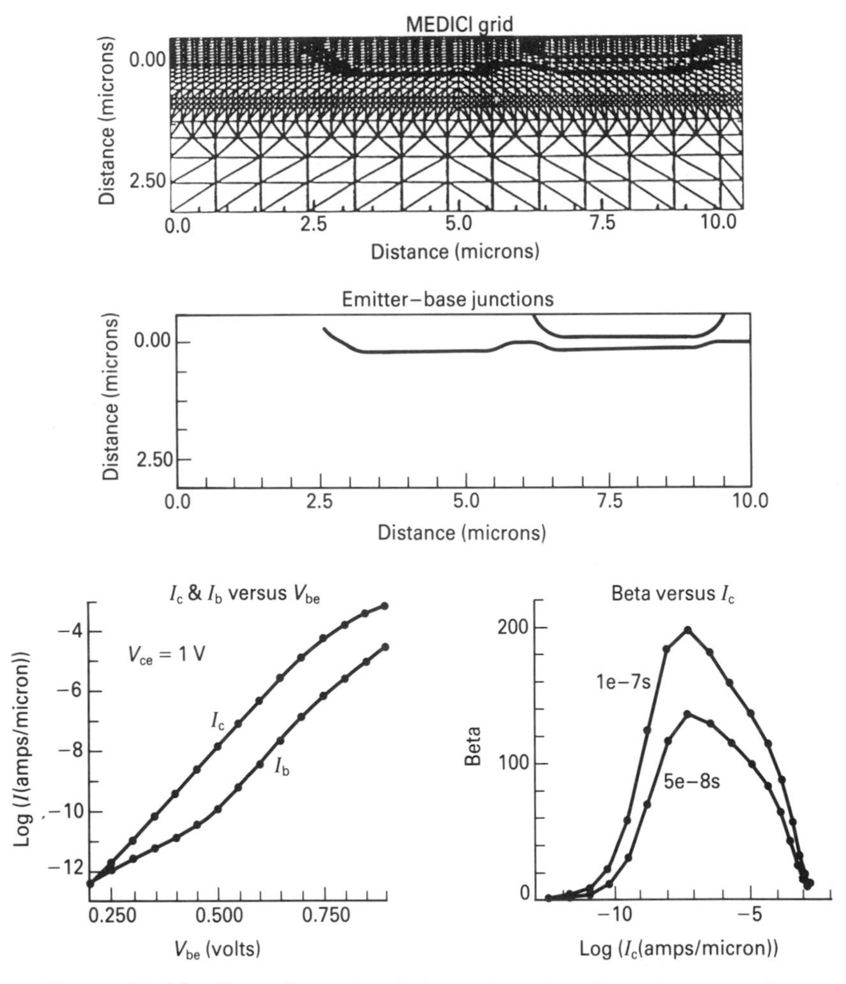

Figure 11.10 Two-dimensional simulation plots for an npn transistor.

TSUPREM4 (TMA version of SUPREM 1V) and are transferred to the MEDICI grid. The one-dimensional profiles are similar to those shown in Figure 11.4, with the addition of a buried n^+ layer in the collector region. During a regridding process a finer grid is created in regions where the impurity and/or the potential change rapidly. This finer grid allows improved accuracy in the solution of the partial differential equations. The resultant grid used in MEDICI is shown in the top diagram of Figure 11.10. The diagram represents a cross-section of the transistor with the emitter and base at the top and the collector at the bottom. The outline of the emitter and base junctions is shown in the middle diagram. The emitter extends from 6.25 μm to 10 μm, with the external base region extending from 2.5 μm to 5.5 μm. The emitter–base junction is approximately 0.4 μm deep with the base–collector at 0.7 μm.

It can be seen from the shape of the lateral base profile beneath the emitter that the base region has diffused further as a result of an 'emitter-push'. This results from the effect of the greater concentration of impurities in the emitter which enhances the diffusion coefficient of the boron immediately beneath the emitter.

The simulator can be used to generate current/voltage characteristics such as those shown in the bottom-left diagram. This shows the variation of the collector and base currents with V_{BE}. These plots are normally referred to as the Gummel plots. The low-current and high-current effects discussed above can be clearly seen. The change of slope of $I_C - V_{BE}$ at high currents indicates the onset of the high-current effects, and similarly the change of slope in the $I_B - V_{BE}$ at low currents shows the start of the low-current effects. It is a simple matter to extract the common-emitter current gain (β_F) from the Gummel plots. The variation of β_F with I_C is shown in the bottom-right diagram. Many other parameters could be investigated, including two-dimensional electron and hole current flow, the electric potential, the variation of the depletion layers, avalanche breakdown and many others.

Once the data has been prepared for the simulators it is a simple matter to vary parameters and to study the effect. One such parameter is the lifetime of the carriers as illustrated in the bottom-right diagram, where two plots of β_F are shown. One is for a lifetime of 50 ns and the other for 100 ns. It can be seen that β_F decreases with decreasing lifetime as the diffusion length decreases, as illustrated in Figure 11.5

11.7 Voltage breakdown

When high voltages are applied to the terminals of a transistor and the voltage limits of the junction are exceeded, then large currents flow and the junction is said to break down. Provided that the current is limited by external resistance then the junction need not be damaged. The current–voltage plots of the breakdown characteristics of a transistor are shown in Figure 11.11. The highest breakdown voltage is exhibited for the collector–base junction with the emitter an open circuit (BV_{CBO}). For the emitter–collector with the base an open circuit (BV_{CEO}) the voltage is much lower and is given approximately by:

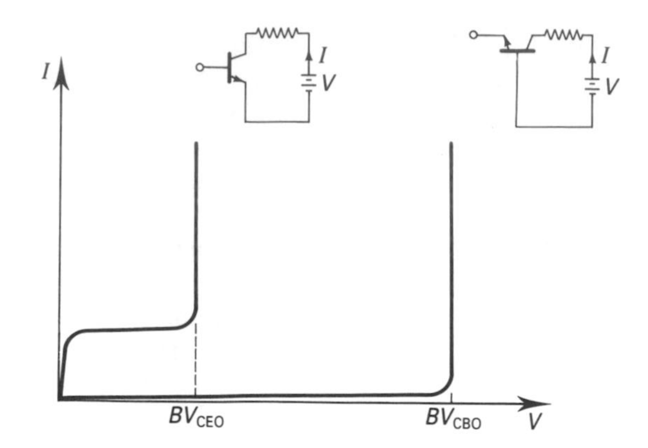

Figure 11.11 Breakdown characteristics of the collector–emitter (BV_{CEO}) and collector–base (BV_{CBO}).

$$BV_{CEO} = \frac{BV_{CBO}}{(\beta_F)^{-1/m}}$$

where $m \approx 4$ for npn and ≈ 6 for pnp.

Avalanche breakdown is responsible for the rapid increase in current beyond the breakdown voltage and it has a positive temperature coefficient of ≈ 300 ppm/°C. The voltage is closely related to the carrier concentration of the lightly doped collector region and the variation of voltage with concentration is shown in Figure 11.12.

11.8 Base and collector resistance

In a practical transistor current flows from the active region beneath the emitter to the base and collector contacts. It passes through silicon which plays no part in the transistor action but which does act as a resistance to current flow and as such must be included in the mathematical model for the transistor. The most important resistive components are those associated with the base and the collector. The base resistance has a significant effect on the high-frequency performance while the collector resistance affects the collector saturation voltage in switching circuits.

The base current for a typical small-signal transistor flows from beneath the emitter through the high-resistance region of the base and to the base contact, as shown in Figure 11.13. The path can be divided into three: one beneath the base contact (r_a) with two equipotential surfaces at right-angles to each other, a rectangular block between the edge of the base contact and the edge of the emitter (r_b) with equipotential surfaces at either end of the block and a third (r_c) beneath the emitter with current entering from the emitter and leaving by an equipotential surface at right-angles to the surface of entry. A detailed analysis of the three regions yields resistance values as follows:

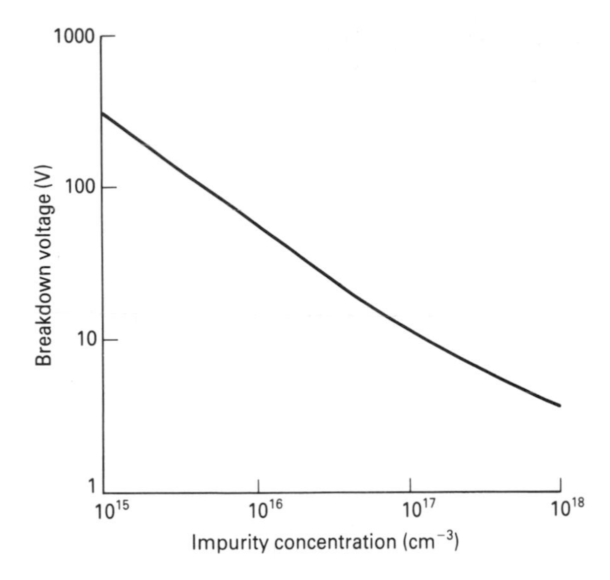

Figure 11.12 Avalanche breakdown voltage versus impurity concentration for a step-junction. [After Sze[7] and by permission of the American Institute of Physics.]

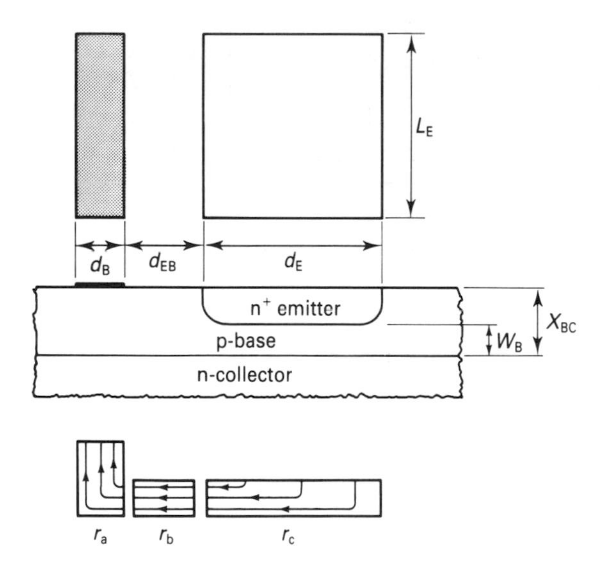

Figure 11.13 Lumped representation of the base resistance between the base contact and the active region beneath the emitter.

$$r_a = \frac{\rho_B x_{BC}}{3 d_B L_E}$$

$$r_b = \frac{\rho_B d_{EB}}{W_B L_E}$$

$$r_c = \frac{\rho_B d_E}{6 W_B L_E}$$

where ρ_B is the average resistivity of the base region and L_E is the length of the base contact (which is also the same as the width of the emitter).

Thus the total resistance for the base is:

$$r_{bb} = r_a + r_b + r_c$$

Typically r_{bb} may have values which range from 50 ohm to 200 ohm.

The collector resistance is obtained in a similar manner. The path for the collector current is divided into a number of separate blocks, as shown in Figure 11.14. A buried n^+ layer is assumed to exist beneath the emitter and to extend under the collector contact.

The three regions yield resistances as follows:

$$r_a = \frac{\rho_C (x_{EPI} - x_E)}{d_C L_C}$$

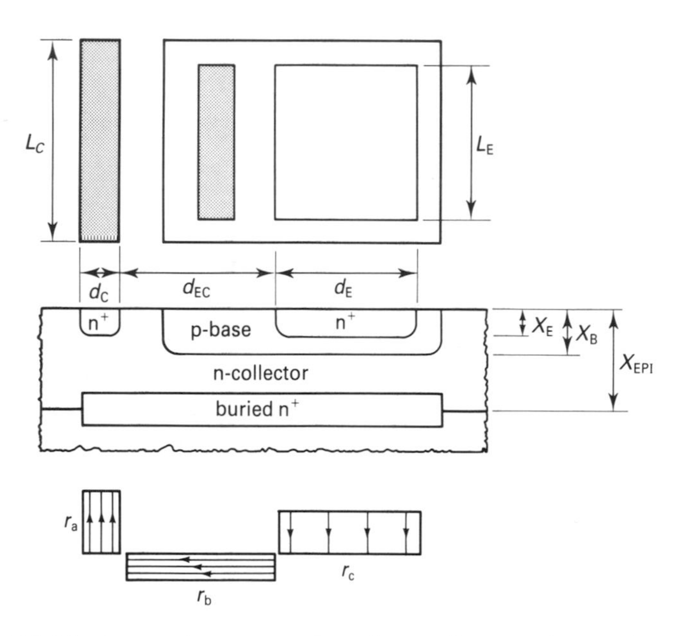

Figure 11.14 Lumped representation of the collector resistance between the collector contact and the region beneath the emitter.

$$r_b = \frac{\rho_S d_{EC}}{(L_E + L_C)/2}$$

$$r_c = \frac{\rho_C (x_{EPI} - x_{BC})}{d_E L_E}$$

where ρ_C is the resistivity of the epitaxial layer and ρ_S is the sheet resistance of the buried n^+ layer. The total collector resistance is:

$$r_{cc} = r_a + r_b + r_c$$

Typically r_{cc} has values which range from 10 ohm to 50 ohm.

11.9 Junction capacitance

A reverse-biased p^+n or n^+p step-junction exhibits a junction capacitance given by

$$C = A\left[\frac{\varepsilon_o \varepsilon_s q N}{2(V + \phi_o)}\right]^{1/2}$$

where N is the impurity concentration in the more lightly doped region and ϕ_o is the built-in voltage given by

$$\phi_o = 2V_T \ln\left(\frac{N}{n_i}\right)$$

Although this equation applies to a step-junction, whereas in practice the impurity profile is graded, it is an adequate approximation for the emitter–base junction, and for normal operating voltages it is also adequate for the base–collector junction.

The capacitance is often expressed as:

$$C = \frac{C_{xo} A}{(1 + V/\phi_o)^{1/2}}$$

where C_{xo} is the capacitance per unit area for zero bias.

A related parameter is the width of the depletion layer which is given by:

$$x = \left[\frac{2\varepsilon_o \varepsilon_s (V + \phi_o)}{qN}\right]^{1/2}$$

Some typical parameters for npn transistors are shown in Table 11.1.

In the table three different processes are represented for low, medium and high voltages. The most important parameter for the different operating voltages is the resistivity of the collector material (refer to Figure 11.12), and this affects the capacitance of the collector–base junction.

Table 11.1 npn device parameters

Parameter	Low voltage	Medium voltage	High voltage
Epi resistivity (ohms cm)	0.8–1.0	2.0–2.5	4.0–5.0
BV_{CEO} (V)	20.0	30.0	45.0
C_{BC} (pF/μm^2 $\times$ 10^{-4})	2.3	1.4	1.0
C_{CS} (pF/μm^2 $\times$ 10^{-4})	1.0	1.0	1.0
C_{BE} (pF/μm^2 $\times$ 10^{-4})	10.0	10.0	10.0

11.10 Frequency response

The variation of gain with frequency is shown in Figure 11.15. With the gain expressed in dB f_β and f_α represent the frequencies at which the gain is reduced by 3 dB from the dc values. At a frequency f_t the common-emitter gain has a value of unity. The relationship between f_β and f_t is:

$$f_t = \beta f_\beta$$

and since $f_t \approx f_\alpha$ then $f_\beta \approx f_\alpha/\beta$.

The cut-off frequency f_α for the common base configuration is given by[8]

$$f_\alpha = \frac{1}{2\pi\tau_{ec}}$$

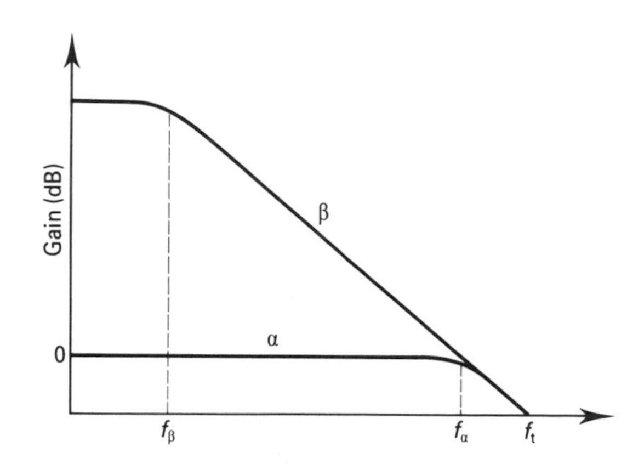

Figure 11.15 Variation of common-emitter gain (β) and common-base gain (α) with frequency.

where τ_{ec} represents the sum of four delays encountered sequentially by the carriers which flow from the emitter to the collector:

$$\tau_{ec} = \tau_E + \tau_B + \tau_C + \tau_D$$

where τ_E represents the charging time of the emitter depletion layer capacitance:

$$\tau_E = r_E C_E = \frac{kT}{qI_E} C_E$$

where I_E is the emitter current. The forward-biased emitter junction exhibits a capacitance of:

$$C_E = A\sqrt{\frac{q\varepsilon_o \varepsilon_s N_B}{2(\phi_o - V_{BE}}\,)}$$

where A is the area of the emitter, ε_s is the dieletric constant of silicon (12.7), N_B is the impurity concentration in the base next to the emitter junction and ϕ_o is the built-in voltage of approximately 0.8 V.

The second delay is associated with the rate of flow of charge across the base. In Figure 11.16 the number of electrons is shown as being distributed linearly across the base from a maximum at the emitter side of $n(o)$ to zero at the collector. The flow is dominated by diffusion and the current is given as:

$$I_n = qAD_{nB}\frac{n(o)}{W_B}$$

where D_{nB} is the diffusion coefficient of the electrons in the base. The total charge in the base is:

$$Q_B = \frac{qAn(o)W_B}{2}$$

$$Q_B = I_n \frac{W_B^2}{2D_{nB}}$$

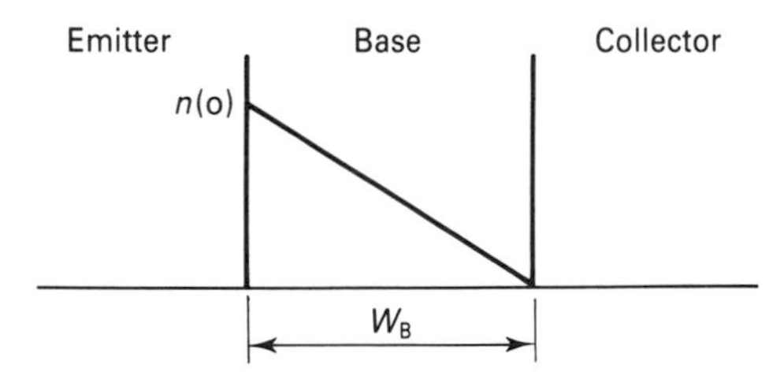

Figure 11.16 Variation of minority carrier charge in the base.

The small-signal component of collector current can be defined as:

$$i_C = i_n = \frac{Q_B}{\tau_B}$$

that is,

$$\tau_B = \frac{W_B^2}{2D_{nB}}$$

In practice τ_B is significantly smaller than the value predicted by the above equation as a result of the graded impurity profile in the base.

The third delay is the collector capacitance charging time:

$$\tau_C = r_{CC}(C_{BC} + C_{CS})$$

where r_{CC} is collector series resistance, C_{BC} is the collector–base capacitance and C_{CS} is the collector–substrate capacitance.

Finally, the fourth delay is associated with the time required to cross the collector depletion layer:

$$\tau_D = \frac{x_C}{v_s}$$

where x_C is the width of the collector depletion layer and v_s is the saturation velocity of the carriers through the depletion layer.

The full expression for the cut-off frequency is:

$$f_t = \left[2\pi\left\{\frac{kTC_E}{qI_E} + \frac{W_B^2}{2D_{nB}} + r_{CC}(C_{BC} + C_{CS}) + \frac{x_C}{v_s}\right\}\right]^{-1}$$

Clearly, it can be seen that, in order to increase the frequency, the transistor should have a small base width, a small collector depletion layer and it should operate at high emitter current. In practice, the choice is not as simple as this and it usually involves various compromises. For example, a small collector depletion layer would imply a low-resistivity collector, but this would reduce the collector–base breakdown voltage.

The emitter time constant is reduced at higher current levels but the emitter injection efficiency decreases as a result of base conductivity modulation and the Kirk effect. For a given set of parameters there will be an optimum emitter current which will maximize the frequency response.

11.11 Switching characteristics

In addition to its use as a linear circuit element the transistor is also an excellent switch. Current flow between emitter and collector can be switched ON and OFF at high speed by application of a signal to the base terminal.

The transistor is normally used in common-emitter configuration and with the application of 0.7 V between base and emitter the maximum current, determined by the external load, flows and the collector voltage drops to between 0.1 V and 0.2 V. Under these conditions the transistor is said to be saturated and the base is more positive than the collector (0.7 V − 0.2 V = +0.5 V), and the collector–base junction becomes forward biased.

When switching from OFF to ON the transistor moves through the normal active region, where the emitter is forward biased and the collector is reverse biased, and into the ON region, as shown in Figure 11.17(a). The minority carrier concentration in the base increases from a very low level corresponding to the OFF condition, when both junctions are reverse biased, to a level where a large number of minority carriers accumulate in the base, as shown in Figure 11.16(b), when both junctions are forward biased.

For a qualitative explanation of the switching action consider a transistor in common-emitter mode with a collector resistance (R_C) and with a pulse generator connected to the base.

When a pulse is applied to the base the base current immediately rises to a value I_{b1}, as shown in Figure 11.18. After a delay time, t_d, required to charge the depletion layer capacitances of the emitter and collector, the collector current begins to rise at t_1. The operating point of the transistor is now at the beginning of the active region of the output characteristics, with the emitter forward biased and injecting carriers into the base. The collector current now rises from the 10% level at t_1 to the 90% level at t_2 of the saturation current given by V_{CC}/R_C. The time taken is the rise time, $t_r = t_2 - t_1$, which is related to the frequency dependence of β.

During the period when the transistor is ON both the emitter–base and the collector–base junctions are forward biased and the emitter and collector both inject carriers into the base. The base fills with minority carriers.

The transistor remains ON for as long as I_{b1} is maintained. If at t_3 the input pulse

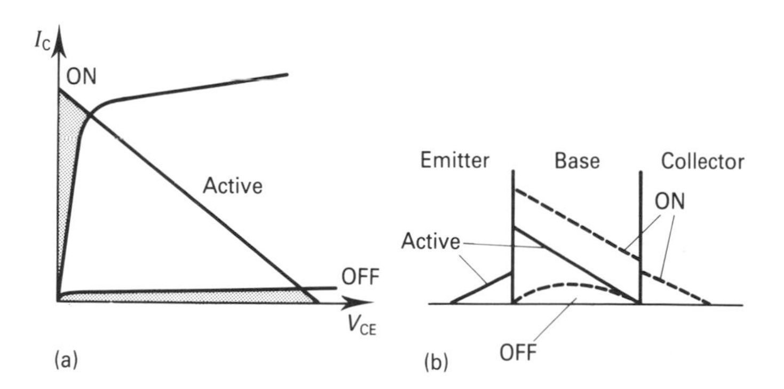

Figure 11.17 Relationships between (a) the transistor operating characteristics and (b) the distribution of minority carrier charge in the emitter, base and collector.

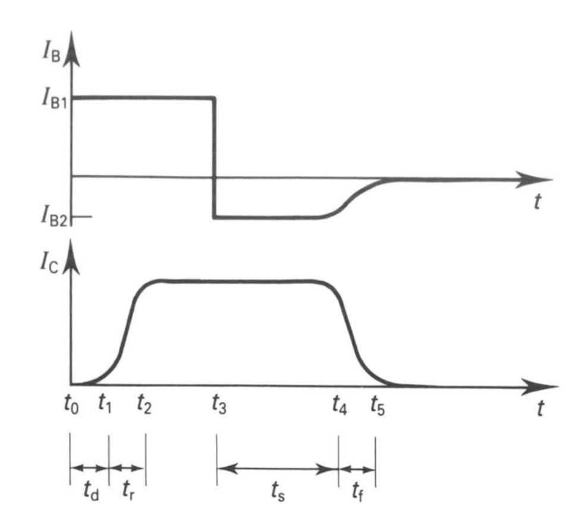

Figure 11.18 Relationship between the base and collector currents for the common-emitter configuration showing the various times associated with the output pulse.

is removed the collector current does not respond immediately. Between t_3 and t_4 the charge stored in the base and collector during the ON period when the transistor is in saturation must now be removed. This storage time, $t_s = t_3 - t_4$, is a measure of how rapidly the minority carriers in the base and collector recombine, that is, it is a measure of the minority carrier lifetime.

During the period t_s a negative base current I_{b2} flows. This current represents the outflow of the stored minority charge, and it maintains the collector current.

Finally, at t_4, the transistor comes out of saturation and the collector operating point moves back through the active region to the OFF condition. The explanation for the time required for the fall time, $t_f = t_5 - t_4$, is largely the same as for the rise time being dependent on the frequency response of β.

Thus the output pulse response can be defined by an ON time t_{ON} and an OFF time t_{OFF} given by:

$$t_{ON} = t_d + t_r$$

$$t_{OFF} = t_s + t_f$$

The critical time is the storage time, t_s. For conventional TTL logic circuits it is usual to include in the processing some means of reducing the minority carrier lifetime. Reducing the lifetime also reduces β but large values of β are not required for switching circuits and the improved switching response is more desirable than the disadvantage of a reduction in gain. Alternatively, circuit modifications are made involving the inclusion of Schottky diodes (Chapter 12) which prevent the collector from becoming forward biased and thus reduce the amount of minority carrier injection into the base during the ON period.

11.12 pnp transistors

A schematic of the lateral pnp is shown in Figure 11.19. The p-type emitter and collector regions are formed together and at the same time as the base region of the npn device. It is called a lateral device because the active region is now the sidewall of the emitter periphery and the current flows parallel to the surface from the emitter to the collector. However, many carriers are injected into the n-type base from the bottom surface of the emitter and do not reach the collector, and the emitter injection efficiency is low. The base width is defined by the photolithographic process rather than the diffusion process as is the case in the conventional npn. Both gain and frequency response are sensitive to base width. To a first order the degradation of the gain is the ratio of the area of the sidewall A_v and the surface area A_s, that is, $\beta = \beta_{ideal} A_v/A_s$.

The advantage of the lateral transistor is that it can be fabricated as part of the conventional npn manufacturing sequence. It provides, at no extra complexity, an additional device for use as an active load, a current source, a level shifter and a device for use in complementary output stages.

An alternative device is the substrate pnp which is shown in Figure 11.20. The device is a conventional vertical pnp transistor where the base of the npn device is now the emitter of the pnp device; the n-type epitaxial layer is the base and the p-type substrate is the collector. Even though the base width is large the gain is greater than for the lateral device. The main limitation of the substrate pnp is that the collector is effectively grounded via the substrate. For a complementary output stage, however, this is not a problem as the collector of the pnp can be connected directly to the negative rail of the supply which is usually connected to the substrate.

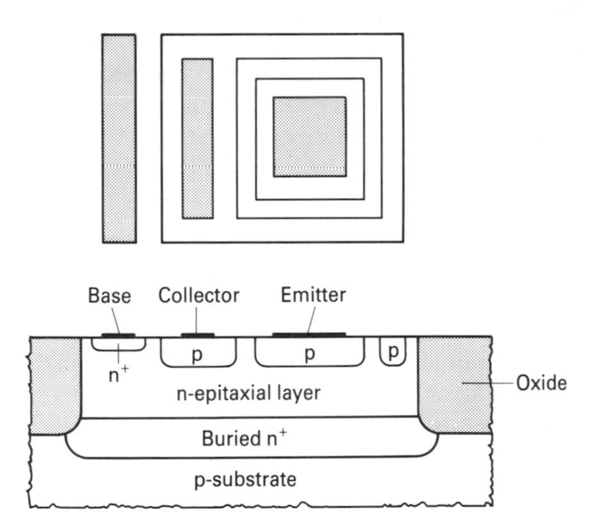

Figure 11.19 Schematic of a lateral pnp transistor.

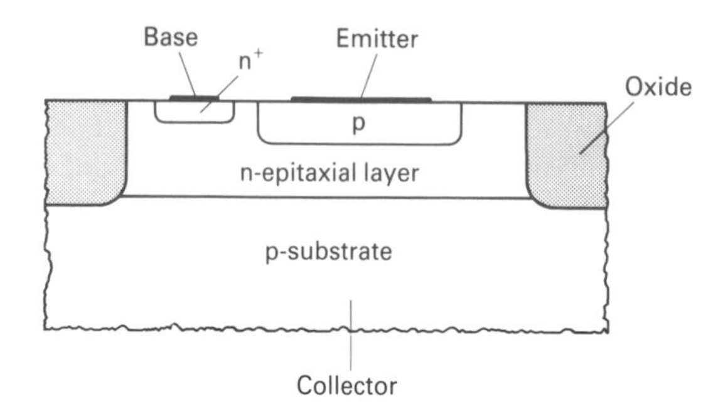

Figure 11.20 Schematic of a substrate pnp transistor.

11.13 Advanced device structures

The advances made in manufacturing technology have resulted in many improvements to the basic npn transistor. This is particularly noticeable in the method used to isolate transistors from one another. A major disadvantage with junction isolation (Section 10.2.1) is the amount of silicon which is required around the active regions to avoid the depletion layer of the isolation junction and that of the collector–base junction merging into each other. The absence of a depletion region when oxide isolation is used eliminates this problem, but oxide isolation allows even greater scaling to be achieved by enabling the diffused regions of both base and emitter to extend to the walls of the oxide. This is illustrated in Figure 11.21. The original junction isolation is shown in Figure 11.21(a). The amount of silicon occupied by the p^+ isolation diffusion can be minimized by reducing the thickness of the epitaxial layer. However, the separation between the base and isolation must still be maintained. When oxide isolation was first introduced the oxide was grown in the area which had previously contained the p^+ isolation diffusion. The dimensions of the n-type isolated pocket were reduced because the depletion layer of the base could now extend up to the walls of the oxide, as shown in Figure 11.21(b). Further scaling can be achieved by allowing both base and emitter regions to extend to the oxide walls, as shown in Figure 11.21(c).

Further developments are aimed at reducing series resistance associated with the base and collector. In Figure 11.22 an additional n^+ diffusion is included for the collector contact. This 'plug' diffusion uses phosphorus ions while arsenic ions are used for the emitter and the buried n^+ layer. The higher diffusion coefficient of phosphorus ensures that in the time required to drive the phosphorus through the epitaxial layer the arsenic in the buried n^+ layer does not move significantly. The base region is also formed with two ion implants to establish a lightly doped active region beneath the emitter and a more heavily doped and deeper external base region where the base contact is formed, as shown in Figure 11.22.

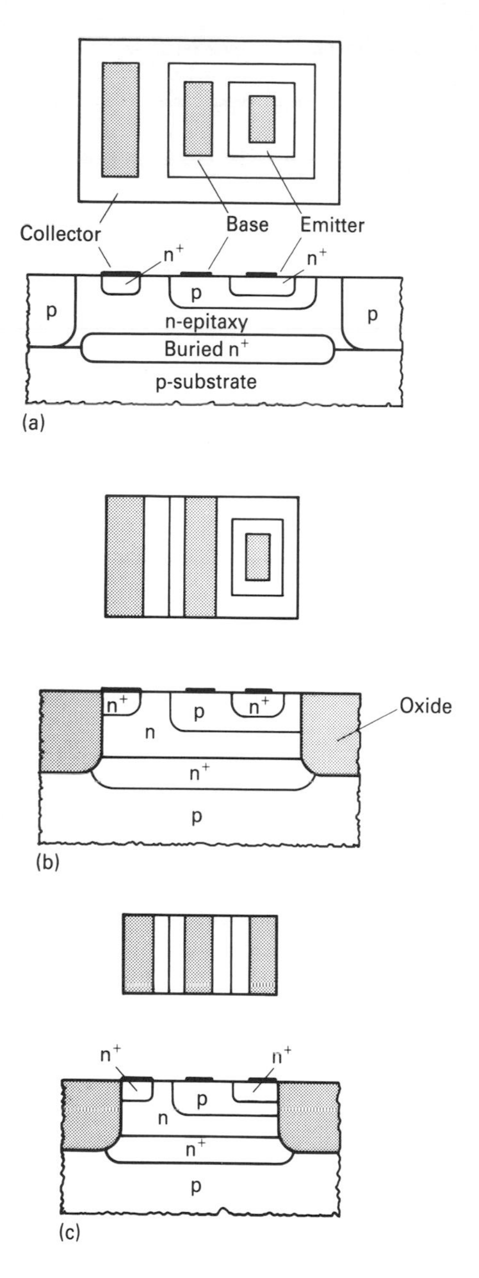

Figure 11.21 Reduction in dimensions produced by changing from (a) junction isolation to (b) oxide isolation and (c) scaled oxide isolation.

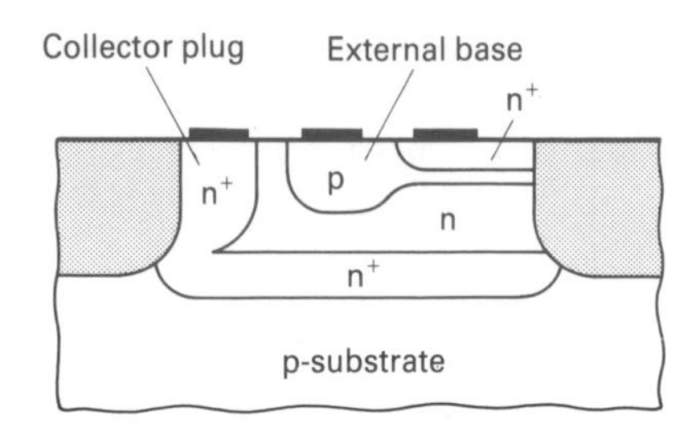

Figure 11.22 Cross-section of a scaled npn transistor showing the use of a collector plug and an external base region.

Other developments include the use of polycrystalline n-type silicon between the n^+ emitter and the aluminium contact. With very shallow emitters produced as a result of scaling there is a marked reduction in the gain. This is caused by the reduction in the thickness of the emitter. The emitter injection efficiency is:

$$\gamma = \left[1 + \frac{D_{pE} N_B W_B}{D_{nB} N_E L_E}\right]^{-1}$$

Normally the diffusion length of holes in the emitter (L_E) is much less than the thickness of the emitter. However, with scaled devices with a very shallow emitter this is no longer the case and the hole diffusion current in the emitter is increased as a result of the increased hole gradient between the edge of the emitter–base depletion layer and the emitter contact. This results in a reduction of the emitter injection efficiency and a corresponding reduction in the current gain (β). The use of polycrystalline silicon produces a barrier[9] to the flow of holes to the emitter contact at the interface of the monocrystalline silicon of the emitter and the polysilicon. This barrier reduces the hole gradient and hence the hole diffusion current. The current gain is, therefore, increased over the value which is obtained with an aluminium contact.

Self-alignment can be used to reduce device dimensions further by using the mask from one process step to define the geometry for the next step. Many of the diffusion steps are now replaced by ion-beam implantations and annealing. Another important area of development is the combination of bipolar transistors with MOS transistors. Such a combination takes advantage of the current drive capabilities of the bipolar device and the high impedance, high noise immunity of the MOS devices.

Problems

1. Sketch the energy-band diagram for a pnp transistor under normal operating conditions of a forward-biased emitter–base junction and reverse-biased collector–base junction.

2. For a pnp transistor the acceptor concentration in the emitter is $1 \times 10^{19}/\text{cm}^3$, the donor concentration in the base is $1 \times 10^{17}/\text{cm}^3$ and the collector acceptor concentration is $1 \times 10^{15}/\text{cm}^3$. With the aid of the relationships derived in Chapter 1, determine the Fermi levels with respect to the intrinsic level for the three regions.
[0.529 V, 0.409 V, 0.289 V.]

3. In an npn transistor the emitter impurity concentration is $2 \times 10^{19}/\text{cm}^3$ and the concentration in a uniformly doped base is $5 \times 10^{17}/\text{cm}^3$; the emitter is 1 μm thick, the base width is 0.6 μm and the diffusion length of carriers is 20 μm. With the aid of graphs of mobility–carrier concentration in Chapter 1 determine the common-base and common-emitter current gains.
[$\alpha = 0.998$, $\beta = 679$.]

4. Estimate the impurity concentration required in a 2.5 μm-thick collector epitaxial layer to prevent base widening at a collector current of 3 mA for a collector voltage of 5 V. Assume an emitter area of 25 μm^2 and a carrier saturation velocity of 1×10^7 cm/s.
[$6.4 \times 10^{15}/\text{cm}^3$.]

5. Determine the base resistance of the transistor shown below if the impurity concentration in the base is assumed to be uniform at $2 \times 10^{17}/\text{cm}^3$.
[895 ohm.]

6. If the impurity concentration for the emitter of the transistor shown in Problem 5 is $5 \times 10^{19}/\text{cm}^3$, determine the built-in junction voltage (ϕ_0) for the emitter–base junction and the junction capacitance for a forward bias of 0.6 V.
[0.998 V, 0.1 pF (bottom), 0.04 pF (sidewall).]

7. If the diffusion coefficient for the electrons for the transistor shown in Problem 5 is 20 cm^2/s, estimate the cut-off frequency based on the emitter and base time constants at an emitter current of 10 μA.
[15 GHz.]

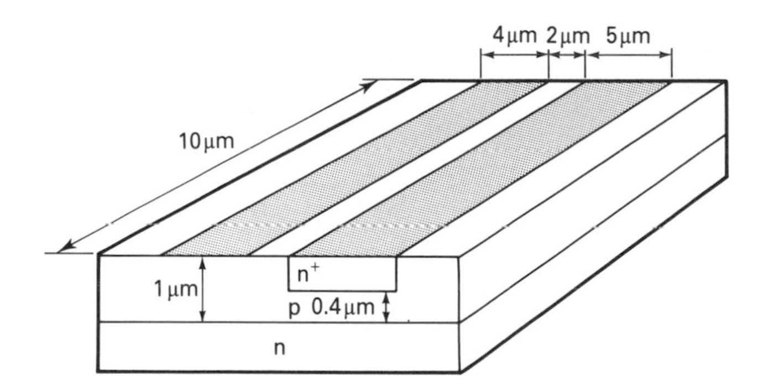

References

[1] S. M. Sze (1981), *Physics of Semiconductor Devices* (2nd edn), Wiley.
[2] J. M. Early (1959), ‘Effects of space charge layer widening in junction transistors’, *Proc. IRE*, **40**, 1401.

[3] H. P. D. Lanyon and R. A. Tuft (1978), 'Bandgap narrowing in heavily doped silicon', *IEEE Tech. Digest, Int. Electron Devices Meeting*, p. 316

[4] C. T. Kirk (1969), 'A theory of transistor cutoff frequency falloff at high current densities', *IRE Tran. Electron. Devices*, **ED 9**, 164–74.

[5] S. K. Ghandi (1977), *Semiconductor Power Devices*, Wiley, New York.

[6] Technology Modelling Associates, Inc., 300 Hamilton Avenue, Third Floor, Palo Alto, California 94301.

[7] S. M. Sze and G. Gibbons (1966), 'Avalanche breakdown voltage of abrupt and linearly graded p-n junctions in Ge, Si, GaAs and GaP', *App. Phys. Lett.*, **8**, 111.

[8] R. L. Pritchard, J. B. Angell, R. B. Adler, J. M. Early and W. M. Webster (1961), 'Transistor internal parameters for small-signal representation', *Proc. IRE*, **49**, 725.

[9] C. C. Ng and E. S. Yang (1986), 'A thermionic-diffusion model of the polysilicon emitter', *Proc. IEDM*, **36**.

CHAPTER 12

Diodes, resistors and capacitors

12.1 Introduction

The range of components available to the integrated circuit designer is limited to resistors, diodes and small-value capacitors; the range of resistor values is limited to a few tens of ohms at the low end of the range and to a few tens of thousands of ohms at the upper end. Higher values can be achieved but the tolerance and the temperature coefficient are often poor. For an integrated circuit the amount of surface area occupied by the components is very important and since the transistor is generally much smaller than resistors or capacitors, designs have evolved which require large numbers of transistors. This approach is in direct contrast to the design methods used for discrete component circuits where the transistors are generally the more expensive items, and efforts are made to minimize their number.

Because of the importance of the transistor the manufacturing process for an integrated circuit is optimized for producing transistors, and any other components must ideally be capable of being manufactured with the same sequence of process steps. For special applications additional steps can be introduced, but at an increase in the overall cost.

Integrated circuit diodes are readily produced from one or more of the pn junctions which form the transistor. The same junctions when reverse-biased can be used as capacitors. Resistors are formed from regions of n- or p-type material, with the value of the resistor being proportional to the length and width of the diffused region. Ion implantation can be used to produce high-resistivity regions for high-value resistors.

An important design feature for integrated circuits is the ability to produce closely matched components, with the added ability for accurate scaling, for example it is a simple matter to produce transistors with accurately scaled currents by simply scaling the area of the emitter. Similarly, very accurate resistor ratios can be obtained by scaling the length-to-width ratios. This scaling is achieved as a result of the precision of the photographic methods used to reproduce the geometrical shapes required for the components.

12.2 Diodes

Diodes are formed from one or more of the pn junctions which are present in the transistor, and as a result a number of different configurations are possible depending on the way in which the junctions within the transistor are used. These are shown in Figure 12.1.

The diodes D_{BE} and D_{BC} represent the base–emitter and base–collector diodes, respectively, while D_{SC} is a parasitic diode between the n-type epitaxial layer of the collector and the p-type substrate. The resistors r_{bb} and r_{cc} are the series resistors associated with the external base and collector contacts.

The basic diode equation for an abrupt junction is:

$$I_F = I_S\left(\exp\left(\frac{V}{V_T}\right) - 1\right)$$

where

$$I_S = Aq\left(\frac{D_p p_{no}}{L_p} + \frac{D_n n_{po}}{L_n}\right)$$

and represents the diffusion component of minority carrier flow across the junction. D and L stand for the diffusion coefficient and the diffusion length of the carriers in n- and p-type material. The p_{no} and n_{po} represent the hole and electron concentrations in the n- and p-type regions respectively.

For an n^+p junction such as the emitter–base, then $n_{po} \gg p_{no}$ and I_s is given by:

$$I_S \approx \frac{AqD_n n_{po}}{L_n}$$

For the base–collector junction $p_{no} \gg n_{po}$ and I_s is:

$$I_S \approx \frac{AqD_p p_{no}}{L_p}$$

Under reverse bias I_S is the leakage current and when evaluated from the above

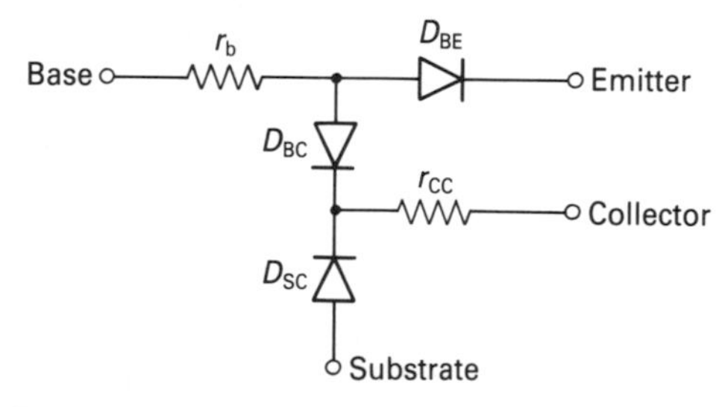

Figure 12.1 Schematic of the diodes associated with an npn transistor.

equations has a value of 10^{-14}–10^{-16} A. In practice the measured value of leakage current is several orders of magnitude greater than this. The discrepancy is due to the recombination and generation of hole–electron pairs in the depletion layer. This generation current is given by:

$$I_g = AqgX_m$$

where g is the generation rate of carriers and X_m is the width of the depletion layer. This current is typically 10^{-9}–10^{-12} A. The generation rate is given by[1]:

$$g = \frac{n_i}{\tau_e}$$

where τ_e is the effective lifetime of the carriers[2].

Thus for silicon diodes $I_s \approx I_g$. Since the generation rate (g) is inversely proportional to the effective lifetime (τ_e), then the saturation current is often larger in digital circuits where the lifetime may be deliberately reduced to improve the switching times. In the past, gold has been used to reduce lifetime, but it can affect the long-term reliability and with advanced technologies the improvement in speed can be achieved with reduced device geometries.

Under forward bias the current can depart from the ideal quite significantly and can be represented by:

$$I_F = I_S \exp\left(\frac{V}{\eta V_T}\right)$$

where the factor η accounts for the departure from the ideal diode. For low values of current when recombination–generation dominates, then $\eta \approx 2$. It equals 1 at intermediate currents when diffusion dominates, and it reverts to 2 again at high currents when the injected minority carrier density becomes comparable to the majority carrier density. At still higher currents the series resistance of the bulk material results in further departures from the ideal. The ideal and actual characteristics are shown in Figure 12.2.

There are six possible diode configurations, as shown in Table 12.1. The series resistance results from the resistors r_{bb} and r_{cc} shown in Figure 12.1. With the collector open-circuit as is shown in configuration (a) the series resistance is simply r_{bb}. For configuration (b), however, the transistor is active and based on the simple equivalent circuit shown in Figure 12.3. The input voltage v_i is given by:

$$v_i = i_b r_{bb} + (1 + \beta) r_e$$

and

$$R_i = \frac{v_i}{i_i}$$

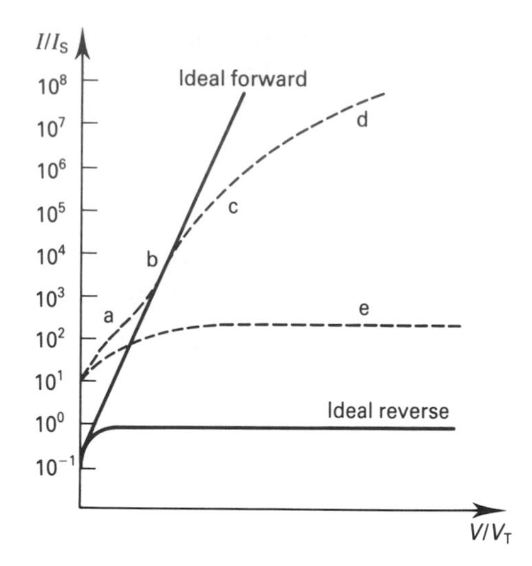

Figure 12.2 Ideal and actual current-voltage characteristics for a junction diode; (a) recombination current dominates, (b) diffusion current dominates, (c) high level injection, (d) bulk series resistance dominates, and (e) the reverse bias characteristics.

where $i_i = (1 + \beta)\, i_b$, which gives:

$$R_i = \frac{r_{bb}}{(1 + \beta)} + r_e$$

or

$$R_i = \frac{r_{bb}}{\beta} + r_e$$

that is, the series resistance is r_{bb}/β. For the other configurations the base and collector resistances are in series.

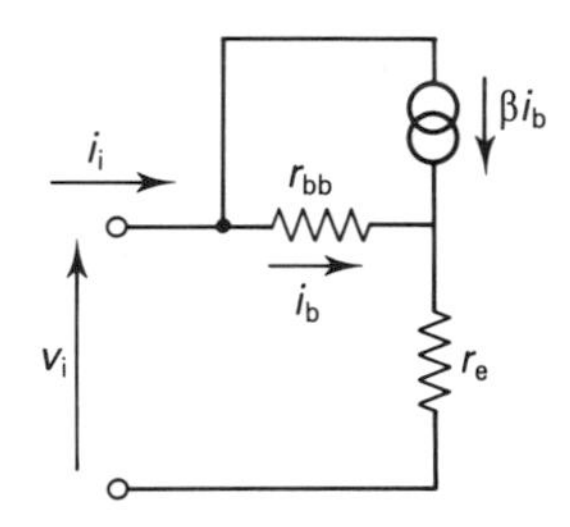

Figure 12.3 Equivalent circuit for the diode configuration shown in (b) of Table 12.1.

Table 12.1 Diodes based on the npn transistor

Diode connection	Series resistance	Reverse breakdown	Capacitance	Storage time
(a)	low $\approx r_{bb}$	low 5–7 V	C_{be} ~0.5 pF	high ≈70 ns
(b)	low $\approx r_{bb}/\beta$	low 5–7 V	C_{be} ~0.5 pF	low ≈5 ns
(c)	high $\approx r_{bb} + r_{cc}$	high >40 V	C_{bc} ~0.7 pF	high ≈120 ns
(d)	high $\approx r_{bb} + r_{cc}$	high >40 V	C_{bc} ~0.7 pF	high ≈90 ns
(e)	high $\approx r_{bb} + r_{cc}$	low 5–7 V	$C_{be} + C_{bc}$ ~1.2 pF	high ≈150 ns
(f)	high $\approx r_{bb} + r_{cc}$	high >40 V	C_{bc} ~0.7 pF	high ≈80 ns

To determine the breakdown voltage it is necessary to consider two pn junctions, the emitter–base and the collector–base. For voltages larger than the built-in voltage the junctions approximate to step-junctions and the graph of voltage breakdown versus resistivity of Figure 11.12 can be used to estimate the breakdown voltages. The emitter–base junction with the higher resistivity associated with the base region results in a breakdown voltage in the range of 5–7 V, while for the collector–base junction the resistivity of the collector is the determining factor and the breakdown voltage is in excess of 40 V.

The capacitance for a reverse-biased abrupt pn junction is:

$$C = A\sqrt{[q\varepsilon_o\varepsilon_{si}N/2(V + \phi_o)]}$$

where N is the impurity concentration of the more lightly doped region, ε_{si} is the dielectric constant of silicon, V is the applied voltage and ϕ_o is the built-in junction voltage. Assuming a base concentration of 1×10^{17} impurities/cm^3 and a collector concentration of 1×10^{15} impurities/cm^3 then the capacitance per unit area is 4×10^4 pF/cm^2 for the emitter–base and 4×10^3 pF/cm^2 for the base–collector for a reverse voltage of 5 V. For a typical small-signal transistor this would result in junction capacitance of 0.5 pF for C_{be} and 0.7 pF for the larger area base–collector C_{bc}.

There is an additional capacitance associated with the collector–substrate junction of perhaps 1–2 pF for the same transistor. This capacitance represents a stray capacitance to ground.

For digital applications the time required to respond to an input pulse is very important. When a pn junction is forward biased and is conducting current it cannot be immediately turned OFF when the applied bias is reversed. The injected minority carriers must first recombine and until this process is complete the diode will continue to conduct and will remain in a low-impedance state. The time required to change state is a function of the minority carrier lifetime and the number of minority carriers. The lifetime can be controlled by the fabrication process, for example the addition of small amounts of gold reduces the lifetime. However, this is no longer a recommended practice because of long-term reliability problems. The number of carriers varies with the diode configuration. The minority carrier distributions for the various configurations is shown in Figure 12.4. For the majority of the configurations, carriers exist in both the base and the collector after the diode has been forward biased. The exception is configuration (b) with the collector shorted to the base such that $V_{CB} = 0$. This configuration results in the smallest value of storage time.

The base–emitter junction with the collector shorted to the base has good features provided the low breakdown voltage is not a problem.

12.3 Schottky diodes

A Schottky barrier diode is formed when a metal is placed in contact with a semiconductor. These diodes are electrically similar to a pn junction diode where one side of the junction is heavily doped. However, there are a number of important differences which make the Schottky diode a useful circuit element.

First, the diode is a majority carrier device, unlike the pn junction, which is a

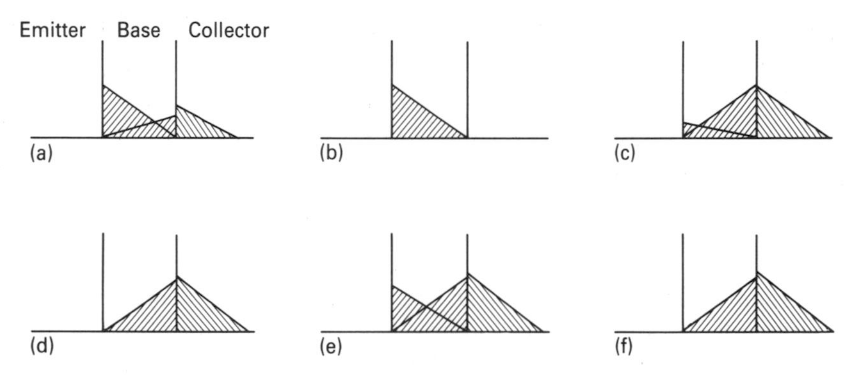

Figure 12.4 Minority carrier distributions for the six diode configurations. The shaded regions represent stored charge under forward bias.

minority carrier device. As a consequence the storage time associated with minority carriers is eliminated and the diodes are inherently fast switches.

Second, the turn-on voltage in the forward direction is smaller than for a silicon pn junction. This feature, together with the fast switching action, makes the diode very suitable for use as a clamp across the base–collector junction of npn transistors to prevent the transistor going into saturation.

The diode current is given by:

$$I_D = A^{**}T^2 \exp\left(\frac{-\phi_B}{V_T}\right)\left(\exp\left(\frac{V}{V_T}\right) - 1\right)$$

where ϕ_B is the barrier height of the metal–semiconductor contact and A^{**} is the Richardson constant for thermionic emission ($A^{**} = 120\ \mathrm{A/(cm^2\,K^2)}$). Some values of ϕ_B for various metals in contact with n- and p-type silicon are given in Table 12.2[3].

The last two entries in the table refer to metal silicides, which are formed when the metal and silicon are annealed at an elevated temperature. For platinum the forming temperature is 300 °C while for tungsten it is 650 °C. The metal silicide barriers are more reproducible than simple metal–silicon contacts.

Typical forward voltages for two Schottky diodes and a pn junction diode are listed in Table 12.3.

An important application for the Schottky diode is as a clamp across the base and collector junction of an npn transistor, as shown in Figure 12.5. The Schottky diode is formed by extending the base contact from the p-type base region to the n-type

Table 12.2 Metal-to-semiconductor barrier heights

Metal	n-type	p-type
Al	0.72	0.58
Au	0.80	0.34
Cr	0.61	0.50
Mo	0.68	0.42
Pt	0.90	0.25
PtSi	0.84	–
WSi	0.65	–

Table 12.3 Turn-on voltages for Schottky and pn diodes at 1 mA

Al–nSi	0.40 V
PtSi–nSi	0.58 V
pn junction	0.80 V

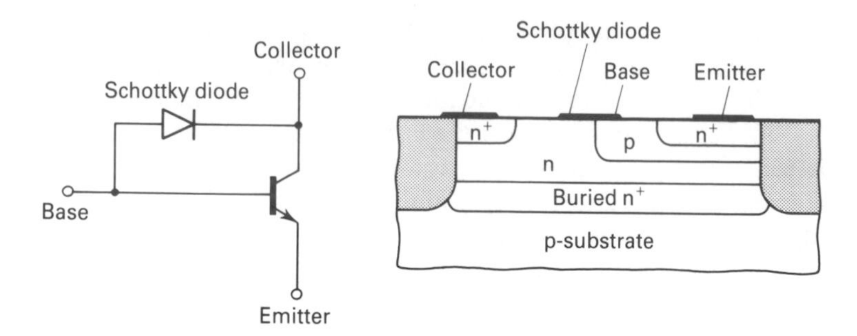

Figure 12.5 Schottky diode clamp. (a) The circuit diagram of the diode across the base-collector junction and (b) the cross-section showing the extended base contact which forms the Schottky diode.

epitaxial layer of the collector. The contact forms an ohmic contact to the base and a Schottky barrier to the n-type collector. The Schottky barrier conducts current at a lower voltage than the base–collector and thus when the transistor is turned ON and the collector voltage drops to its saturation level the Schottky barrier becomes forward biased rather than the base–collector. Stored charge in the transistor is reduced and, therefore, the switching time of the transistor is improved.

12.4 Integrated resistors

The value of the integrated resistor is controlled by the geometry and the material from which it is formed. The basic equation for resistance is:

$$R = \frac{\rho L}{A}$$

where ρ is the resistivity of the material and L is the length of a rectangular slab of cross-sectional area A. For a rectangle of width W and thickness t the area is Wt. For an integrated circuit the fabrication process is precisely defined and ρ and t are fixed by the process. It is therefore convenient to define the resistance as:

$$R = R_s \frac{L}{W}$$

where $R_s = \rho/t$ is known as the sheet resistance with the dimensions of ohm/square. A base diffusion may have a sheet resistance of between 200 and 600 ohm/square so that L/W for a 1000 ohm resistor would need to be 5 if $R_s = 200$ ohm/square, or 2 if $R_s = 500$ ohm/square.

The plan view and cross-section of a diffused resistor formed from the base diffusion is shown in Figure 12.6. Alternative contact layouts are shown where L/W for Figure 12.6(a) is less than L/W for Figure 12.6(b). Very small values of W can be used to

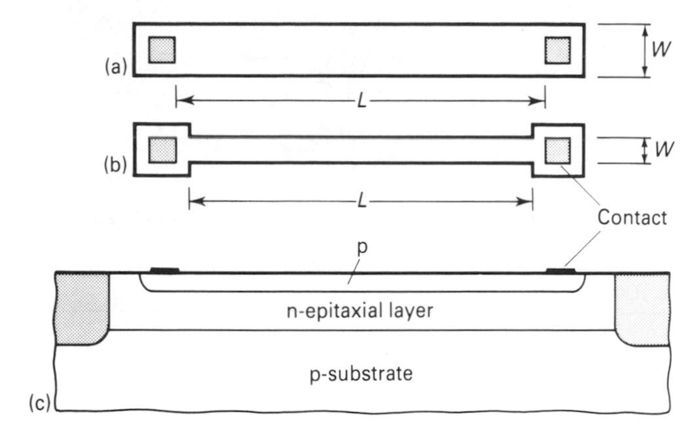

Figure 12.6 Diffused resistor with (a) and (b) representing plan views of alternative contacts and (c) the cross-section.

increase the value of resistance for a given length; however, if W is too small then geometrical errors introduced during photolithography and etching will reduce the accuracy of the resistor.

In practice, in order to save space it is probable that the resistors will be meandered, as shown in Figure 12.7(a). A correction can be made for the corners, as shown in Figure 12.7(b), where the effective length of a corner is a square.

Resistors are placed in one or more isolation pockets with the n-type epitaxial layer connected to the most positive circuit potential to ensure that the pn junction remains reverse-biased. The equivalent circuit for the resistor is shown in Figure 12.8. The reverse-biased pn junction results in a distributed capacitance along the length of the resistor.

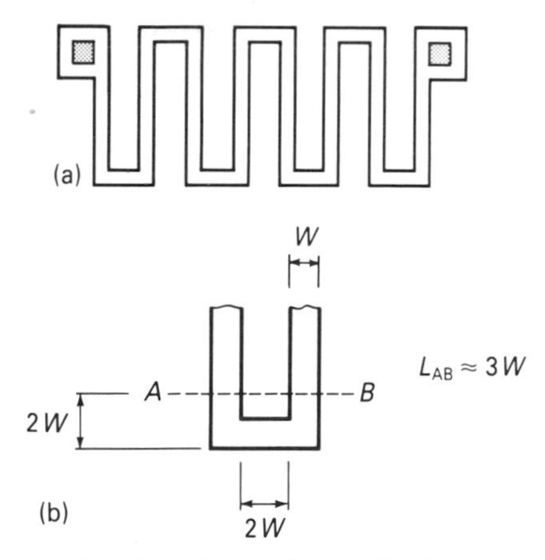

Figure 12.7 A meandered resistor shown in (a) plan view and (b) the correction for the corners.

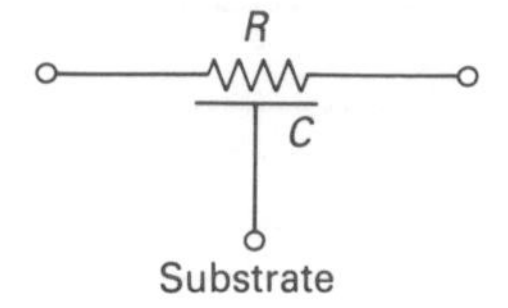

Figure 12.8 Equivalent circuit of a diffused resistor.

The capacitance is given by:

$$C = A\sqrt{[q\varepsilon_o\varepsilon_{si}N/2(V + \phi_o)]}$$

where A is the area of the total diffused region, which includes the area occupied by the contacts and the sidewalls.

The resistance and the capacitance act as a low-pass filter with a corner frequency given approximately by:

$$f_c = \frac{1}{3RC}$$

The maximum value of a resistor formed with the base diffusion is limited to a few tens of thousands of ohms.

Much higher values can be produced with a pinch resistor, as shown in Figure 12.9. Here an emitter diffusion is formed across the p-type region so that the p-type region is pinched between the n^+ emitter diffusion and the n-type epitaxial layer. The sheet resistance of the region beneath the n^+ diffusion may be 4000 ohm/square or more so that very large value resistors can be obtained in a very small area.

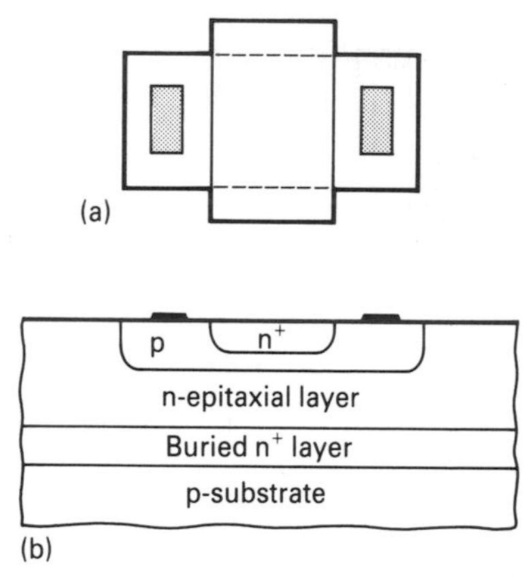

Figure 12.9 Plan and cross-section of a pinch resistor which makes use of the high-resistance region beneath the emitter region, that is, the base region of the npn transistor.

Very high value resistors can also be produced with shallow ion implanted layers which result in sheet resistance values similar to those obtained for pinch resistors.

The temperature coefficient of diffused resistors is positive and ranges from 1000 ppm for relatively high impurity concentrations ($\sim$50 ohm/square) to 5000 ppm for lightly doped layers ($>$ 300 ohm/square).

The absolute tolerance of the resistor value is dependent on a number of factors, including photolithography, etching and diffusion, and is typically $>5\%$. However, the ratio of two resistors can be better than 1 or 2%. For critical circuit functions requiring very close tolerance resistors thin metal films can be used, but at greatly increased complexity and cost.

12.5 Capacitors

The main problem with integrated circuit capacitors is surface area, with the capacitance being given by:

$$C = C_o \cdot \text{Area}$$

where C_o is typically in the range of 1×10^{-4}–10×10^{-4} pF/μm^2. A capacitor of 1 pF would require an area of between 10^4 and 10^3 μm^2. For a silicon integrated circuit this represents a large area and as a result capacitance values are limited to a few picofarads. There are two basic types – junction and metal-oxide–silicon.

The junction capacitor is simply a large-area diode based on one or more of the p- and n-type regions formed during the fabrication of the npn transistors. The largest value is obtained by using both the base–emitter and the base–collector junctions in parallel, as shown in Figure 12.10. The junction must remain reverse biased at all times and the value of the capacitance is inversely proportional to the square root of the applied bias.

A cross-section of an MOS capacitor is shown in Figure 12.11. The top electrode is a layer of aluminium and the lower electrode is the n$^+$ diffusion used for the emitter. The dielectric is the silicon dioxide grown over the emitter during the drive-in diffusion. The value of the capacitance is given by:

$$C = \frac{A\varepsilon_o\varepsilon_{ox}}{t_{ox}}$$

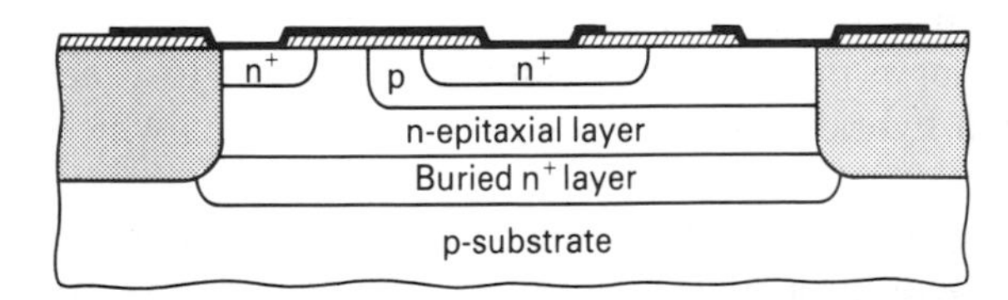

Figure 12.10 Junction capacitor which uses both the base–emitter and the base–collector junctions.

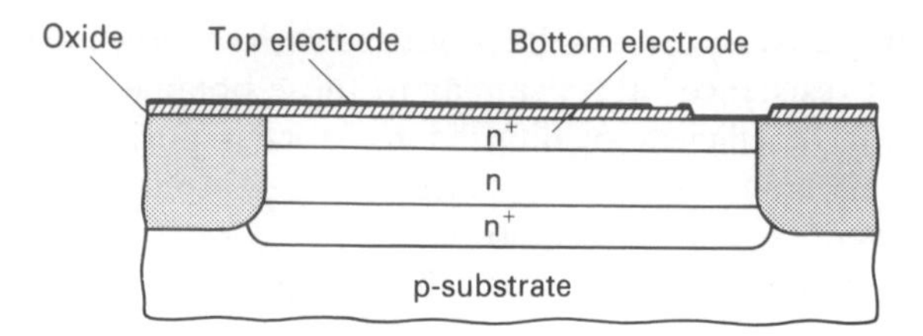

Figure 12.11 Cross-section of a MOS capacitor.

where ε_{ox} is the dielectric constant of the oxide (3.9 for SiO_2), A is the area of the active region and t_{ox} is the oxide thickness. For a thickness of 50 nm the capacitance per unit area is $\sim 7 \times 10^{-4}$ pF/μm^2. The advantage of the MOS capacitor is that it does not require a dc bias.

While the actual capacitance of a pn junction or MOS capacitor is very small, circuit design methods can be used to increase its effective value, for example the Miller effect can be used to simulate a capacitor which is equivalent to the product of a small capacitor and the gain of an associated amplifier. With this method, relatively large values of capacitance (1–10 nF) can be simulated for the purpose of frequency compensation in operational amplifiers, or for capacitor-based frequency-dependent filters.

12.6 Limitations

The range of integrated circuit components is limited in variety, range of values, power dissipation and tolerance and yet very many circuit functions which may have originally been developed with discrete components can be reproduced in integrated form, often with improved performance. Wherever possible, standard production methods are used to manufacture all components to keep cost low. However, non-standard methods can be used if the additional cost can be justified, for example thin metal film resistors for improved tolerance and temperature coefficient, special oxides for higher-value capacitors, vertical pnp high-gain transistors, thin film inductors for microwave applications, etc. Many manufacturers are already mixing analog and digital functions on the same chip either as an all bipolar process, or more often as a combination of MOS and bipolar processes.

The alternative to using special processes for the integrated circuit is to use a standard process for as much of the circuit as possible and to attach these circuits to a thin or thick film hybrid together with those components which cannot readily be integrated.

Power dissipation is limited by the need to remove large amounts of heat from very small areas and, generally, high-power devices are separated from low-power control or signal conditioning circuitry. High-voltage requirements are beginning to be met by special processing. New device designs which are optimized for high-voltage operation are becoming available for use in domestic appliances to operate directly from the mains supply.

Problems

1. Show graphically by plotting $I_F : V$ between 0.1 V and 0.7 V that if the saturation current (I_s) for a pn junction diode is 1×10^{-14} A, then the turn-on voltage for forward bias is approximately 0.55 V.

2. Show graphically by plotting $I_F : V$ between 0.1 V and 0.3 V that the turn-on voltage for a Schottky diode with a platinum–silicide contact ($\phi_B = 0.84$ V) is between 0.2 V and 0.25 V, given that the Richardson constant is 120 A/(cm^2 K^2).

3. Two different manufacturing processes have base sheet resistance values of 200 ohm/square and 600 ohm/square. If the n-type collector region has an impurity concentration of 5×10^{15}/cm^3 estimate the cut-off frequency for a 10 kohm resistor produced by each process assuming a minimum width of 5 μm and a simple rectangular shape.
[75 MHz, 225 MHz.]

4. The ratio of two resistors is generally considered to be of greater importance for circuit design than the absolute values. Consider the effect of over-etching the oxide window used to produce the resistors by 0.1 μm. Assume a nominal design width of 2 μm and consider the percentage change to the absolute value and the resistor ratio for a 10 kohm and a 25 kohm resistor assuming that the sheet resistance is 400 ohm/square.
[absolute: ~5%, ratio: 0%.]

5. What would happen to the resistors in Problem 4 if the lengths were constrained to be the same while the minimum width was maintained at 2 μm.
[absolute: ~2%, ~5%, ratio: ~3%.]

6. If the zero voltage capacitance for the base–emitter junction is 10×10^{-4} pF/μm^2 and for the base–collector it is 1×10^{-4} pF/μm^2, determine the area required for a capacitor of 10 pF which utilizes both junctions.
[~10^4 μm^2.]

7. Determine the area required for a MOS capacitor of 5 pF if the oxide thickness is 20 nm. Assume a dielectric constant of 4 for the oxide.
[~2.8×10^3 μm^2.]

References

[1] S. M. Sze (1981), *Physics of Semiconductor Devices* (2nd edn), p. 90, Wiley.
[2] C. T. Sah, R. N. Noyce and W. Schockley (1957), 'Carrier generation and recombination in p–n junction and p–n junction characteristics', *Proc. IRE*, **45**, 1228.
[3] S. M. Sze (1981), *Physics of Semiconductor Devices*, p. 291, ibid.

CHAPTER 13

Bipolar circuit elements

13.1 Introduction

Bipolar transistors are used in a wide range of both digital and analog circuits. Probably the most widely used class of circuit is the Transistor Transistor Logic (TTL) range of bipolar-based digital circuits. These circuits represent small- to medium-scale integration and a large number of logic functions are available from many different manufacturers. These circuits are used extensively in computer-based products. The designs are now well established and the main developments which are taking place are size reductions and design changes aimed at reducing power consumption.

The most important analog circuit which uses bipolar transistors is the operational amplifier. The basic circuit which can provide a gain of 10^5 or more is now available in a variety of configurations to suit many different requirements. The main improvements which are taking place are the development of new designs for additional analog functions which previously may have only been possible using discrete components. The new designs often extend the frequency range of existing products. Size reductions are not so important for analog circuits as they are for digital circuits, since the number of components is likely to be much less than one hundred.

Oxide isolation and ion implantation have greatly increased the options available to circuit designers: oxide isolation reduces stray capacitance and, therefore, improves the frequency response, while ion implantation allows much greater control over junction depth and sheet resistance values. Improved junction control has allowed base widths to be reduced to less than 0.2 μm with a consequent increase in the cut-off frequency of the transistors; the improved control of sheet resistance, and in particular the ability to produce much higher values than the 200 ohm/square, which was common with an all diffused process, has provided the designer with larger value resistors, which also have improved tolerances.

The number of components which are available to the integrated circuit designer, and their range of values, are limited and this affects the design. The major shortcomings of integrated circuit components are as follows:

- poor component value tolerance;
- poor temperature coefficient;
- limited range of component values;
- lack of large-value capacitors.

On the other hand, there are a number of advantages, as follows:

- large number of active devices;
- good matching and tracking;
- close thermal coupling;
- good control of device geometry.

Circuits have been designed which make use of these advantages. Designs have been developed which use constant current sources rather than voltage sources; gain stages are based on differential amplifiers which do not require decoupling capacitors, the coupling between stages is achieved with voltage level shifters to avoid the need for coupling capacitors and complementary output stages provide large undistorted output signals.

This section is not intended to be an authoritative account of bipolar design, but rather an introduction to some of the basic circuit configurations that are most widely used for the design of digital and analog circuits.

13.2 Digital circuits

A digital circuit is only required to respond to two signal conditions, one which corresponds to the Boolean condition 'true' and the other to the condition 'false'. This can be achieved with a switch which is either ON or OFF. Alternatively, it can be achieved with two different voltage (or current) levels. These two approaches, a switch and a change in a signal level, form the basis for all bipolar digital circuits.

A switch-based digital circuit uses the transistor as a switch. It is either OFF when no current flows, or ON when the maximum current permitted by the resistance in the circuit flows. In the OFF condition the voltage across the transistor is high, which may be taken to represent a Boolean logic variable '1' ('true'). In the ON case the voltage across the transistor is low, which corresponds to the Boolean '0' ('false'). These two conditions are obtained with a circuit in which a transistor is connected in series with a resistor. When the transistor is turned ON current flows and there is a large voltage drop across the resistor and a corresponding reduction in the voltage across the transistor. When the transistor is turned OFF there is no voltage drop across the resistor and a corresponding increase in the voltage across the transistor.

The alternative approach, of using a change in signal level, also uses a transistor with a resistive load, but the transistor is not allowed to be driven either to cut-off or into saturation. Instead, the signal levels are such that the transistor remains in its linear operating region. The change in voltage level is much less than for the switched approach, but the frequency at which the changes can be made is much greater because the transistor is not driven into saturation (see Section 11.10).

13.2.1 Transistor transistor logic

TTL digital circuits are probably the most widely used range of small- to medium-scale circuits. They were developed during the 1960s and are still being developed to improve speed and to reduce power consumption. The important feature of these circuits is the multi-emitter input transistor. The single-input transistor has a separate emitter for each input; there is one base connection and one collector. The basic circuit for a two-input TTL gate is shown in Figure 13.1.

In addition to the special input transistor Q1, there is a phase-splitter Q2 and a 'totem-pole' output comprising Q3, D and Q4. With the base of Q1 connected to the positive rail of the power supply, the input transistor will conduct if one or more of the emitters is connected to ground. In practice, the inputs are connected to the outputs of other gates and the condition which is equivalent to connecting one of the emitters of Q1 to ground is for the output voltage of the preceding gate to be at a low value (0.2 V). When Q1 conducts, the base will be approximately 0.9 V ($\approx$0.2 V from the preceding gate plus 0.7 V of the emitter–base voltage of Q1). This value is insufficient to forward bias the base–emitter junctions of Q2 and Q4. Thus Q2 and Q4 are OFF and the collector voltage of Q2 is at a high value ($\approx$4.5 V). This will cause the pull-up transistor Q3 to conduct and to pull the output up to a high value.

When all the inputs are high, which corresponds to the outputs of the preceding gates all being at a high voltage, then Q1 is turned OFF and the base voltage rises. This causes the base–collector junction of Q1 to become forward biased and for current to flow into Q2, which turns Q2 ON. The collector voltage of Q2 falls, which turns Q3 and D OFF. With Q3 and D both OFF and with Q4 ON, the voltage at the output falls to a low value (0.2 V).

Thus with one or more of the inputs low ('0') the output is high ('1'), while if all the inputs are high ('1') then the output is low ('0'). These Boolean conditions can be described using a table, which is known as a truth table. It is shown below:

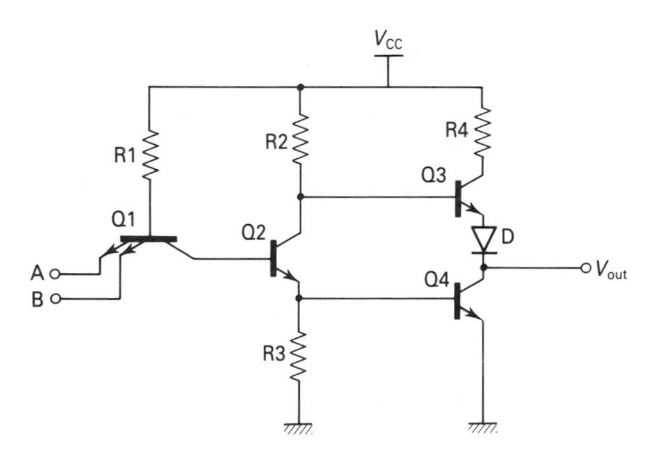

Figure 13.1 A two-input TTL NAND gate.

A	B	Output
0	0	1
0	1	1
1	0	1
1	1	0

This truth table describes the conditions for a two-input NAND gate.

The diode D in the output stage is important to ensure that only one of the output transistors is ON at any time. If it were absent, then when Q2 conducts, the voltage at its collector would be about 0.9 V (0.7 V from the emitter–base of Q4 plus the 0.2 V drop across the collector–emitter of Q2), which is sufficient to turn Q3 ON, that is, both Q3 and Q4 would conduct. With the diode present the turn-on voltage for Q3 and D is about 1 V (0.5 V for each junction). Thus the diode ensures that the totem-pole operates correctly with Q3 and D pulling the output up, and Q4 pulling the output down.

An important requirement of a logic gate is its ability to drive a number of other gates from a single output. This is described by its fan-out. Thus a number of inputs can be connected to a single output. When an input is low the emitter of the input transistor (Q1) conducts. The current which flows is determined by the value of R1 as:

$$I_{LO} = \frac{V_{CC} - 0.2 - 0.7}{R1}$$

With $V_{CC} = 5$ V and $R1$ typically 4 kohm then $I_{LO} \approx 1$ mA. This current must flow into the output of the preceding gate, that is, the equivalent of Q4. Because of the collector series resistance (r_{cc}) there is a voltage drop which raises the output voltage from the saturation value for an open-circuited output gate of 0.2 V to some higher value. A limit to the maximum voltage which can be accepted at the output is usually about 0.4 V. The value of the series resistance, together with the value of the input current (I_{LO}), determines how many inputs can be connected and hence the fan-out.

There are two problems with the basic TTL gate. First, the transistors Q1, Q2 and Q4 are driven into saturation when they are forward biased, which increases the time required to turn the transistors OFF. Second, each gate dissipates about 10 mW. With computer circuits requiring many thousands of gates, the combined power consumption can be excessive. With developments in technology a number of improvements have been possible which have largely overcome these problems. Most of the advanced TTL gates now use Schottky transistors, oxide isolation and sub-micron geometries in the transverse direction through a silicon wafer.

13.2.2 Schottky TTL

A significant improvement in switching speed is obtained by preventing the transistors from going into saturation by clamping the collector–emitter voltage above the saturation level. This is achieved with a Schottky diode connected across the base and collector (Section 12.3). Thus in Figure 13.1 transistors Q1, Q2 and Q4 are replaced by Schottky transistors. A further improvement is obtained by replacing Q3 with two transistors, as shown in Figure 13.2. The additional transistor increases the effective gain of Q3 and improves the ability of the output stage to drive capacitive loads. The presence of Q3 removes the need for the diode.

The power consumption can be decreased by increasing the resistor values, although this has the adverse effect of reducing the switching speed.

Further developments have resulted in the advanced low-power Schottky TTL (the ALS series of gates). This series of gates uses oxide isolation, sub-micron geometries and improved circuit designs to improve speed and to reduce power. A figure of merit for logic gates is the speed–power product. For the original 74#### series of TTL gates the speed–power product is 90 pJ; for the original Schottky TTL it is 60 pJ; and for the advanced ALS series it is 20 pJ.

13.2.3 The current switch

The transistors in the TTL range of circuits act as switches. An alternative approach is for the transistor to remain in its linear operating region, which allows it to operate at its maximum speed. The basic circuit is the current switch shown in Figure 13.3. The circuit is an emitter-coupled differential amplifier which is extensively used in analog operational amplifiers. The input is applied to Q1 while Q2 is biased at a fixed voltage V_{ref} . If V_{in} is equal to V_{ref} and the transistors are assumed to be identical, then the current I divides equally between the two transistors. If V_{in} is taken below V_{ref} then the current is diverted from Q1 to Q2. The collector voltage on Q1 rises and that on Q2

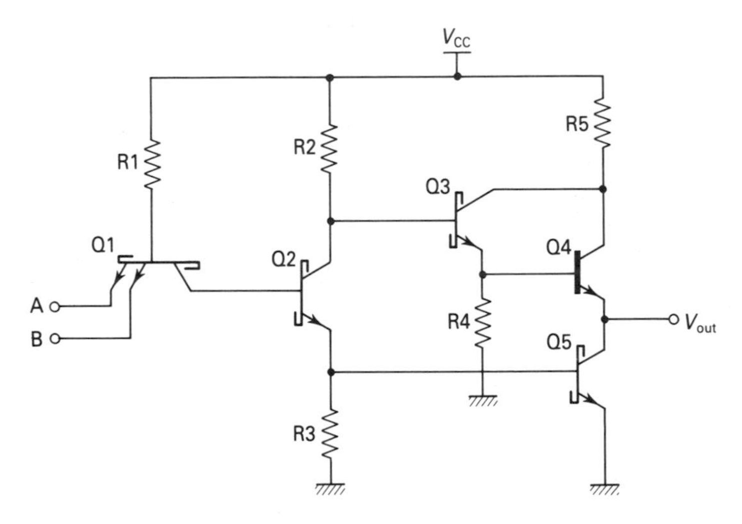

Figure 13.2 Schottky two-input TTL NAND gate.

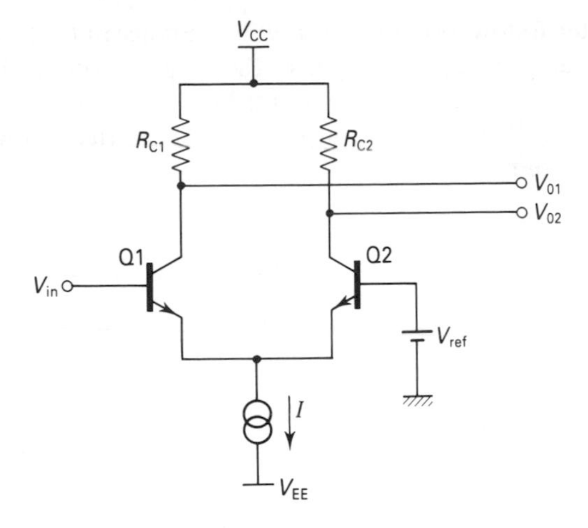

Figure 13.3 Non-saturating current switch.

falls. If V_{in} moves above V_{ref} then the current switches from Q2 to Q1 and the collector voltages swing in the opposite direction. The values of R_{c1}, R_{c2} and V_{ref} are selected to ensure that the transistors do not saturate for the given changes in V_{in}. The current switch forms the basis of the Emmitter-Coupled Logic (ECL) series of digital circuits. An early version of a two-input ECL gate is shown in Figure 13.4.

To cater for multiple inputs, additional transistors are placed in parallel with Q1 of the differential stage, that is, Q11, Q12, Q13, etc., as shown in Figure 13.4, where there are two inputs. The base of Q2 is connected to a reference voltage which is provided by additional circuitry. The collectors of the differential input stage are

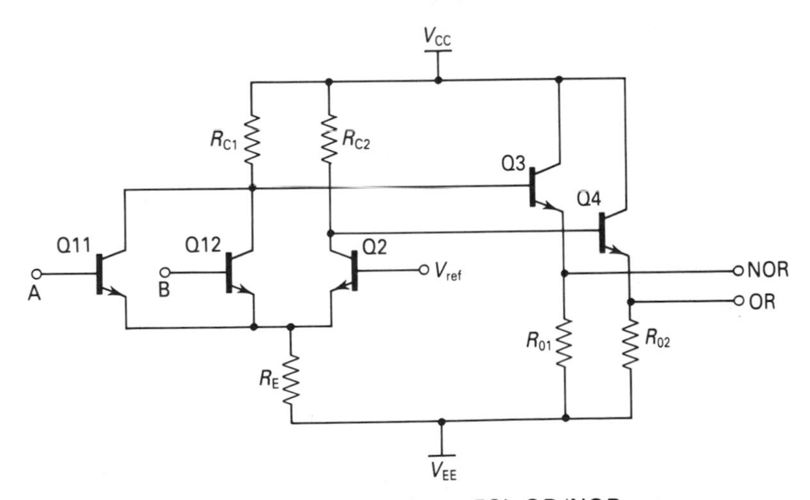

Figure 13.4 Basic two-input ECL OR/NOR gate.

connected to emitter-followers to provide low-impedance outputs. Because the outputs of the differential stage operate in anti-phase, the outputs from the ECL gate satisfy two logic functions. With one or more of the inputs high the common collector of the input transistors is low and, therefore, the output from its emitter-follower is also low. The truth table for this configuration is:

A	B	Output1	Output2
0	0	1	0
0	1	0	1
1	0	0	1
1	1	0	1

Output1 represents the NOR function while Output2 represents the OR function. The low output impedance of the emitter-followers, together with the high input impedance of the inputs, ensures a large fan-out capability. High speed is achieved by making the transistors very small and ECL circuits are capable of operating at frequencies well in excess of 500 MHz. The speed–power product is typically 25 pJ.

13.3 Analog circuits

Analog circuits, unlike digital circuits, must respond to signals which are continuously variable. A signal must be processed without losing any of the original content in terms of amplitude, frequency and phase. It is important for the transistors to operate in the linear region of their operating current–voltage range. This is achieved by establishing the correct steady-state values of direct current and voltage. The signal to be processed is superimposed upon these steady-state values and it is important that as the signal varies the transistor is not driven out of its linear operating region. When the amplitude of the input signal is increased it must not be distorted. This process of amplification, without distortion, must apply to as wide a range of frequency components as possible.

Some of the more important design considerations are dc biasing, dc level shifting and gain.

13.3.1 Bias circuit

The bipolar transistor is a low-impedance current operated device, and one approach to obtaining the correct dc conditions is to control the dc current. This can be achieved with a bias circuit which is based on the current mirror shown in Figure 13.5. The collector and base of Q1 are connected together in order that they act as a diode.

The current flowing in Q1 establishes a base–emitter voltage V_{BE}. If Q2 is an identical transistor then this value of V_{BE} will result in a collector current in Q2 which is the same as the current in Q1. The current I_{ref} is given by:

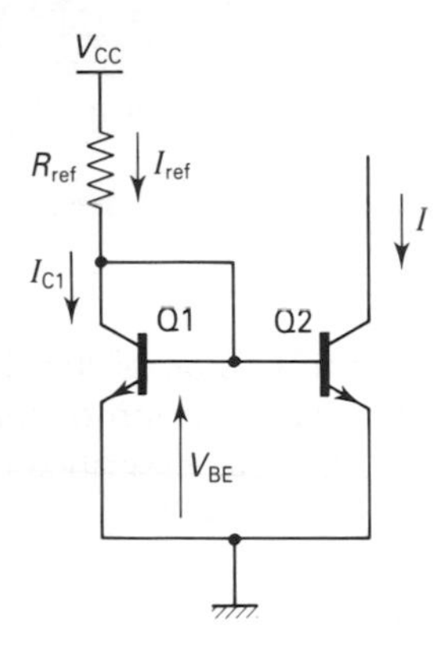

Figure 13.5 Schematic of a simple current mirror.

$$I_{ref} = \frac{V_{CC} - V_{BE}}{R}$$

If $\beta > 100$ then the base current is small and $I_{C1} \approx I_{ref}$ and thus $I = I_{ref}$. The circuit will act as a constant current source as long as Q2 remains in its active region, that is, so long as $V_{CE2} > V_{BE}$. The collector current is proportional to the area of the emitter and if the area of Q2 is twice that of Q1 then $I = 2I_{ref}$, that is,

$$I = \frac{I_{ref} A_2}{A_1}$$

The effect of a finite value of β is to make I smaller than I_{ref}. The analysis can be obtained from Figure 13.6, where the transistors Q1 and Q2 are assumed to be identical, in which case, for a given value of V_{BE},the emitter currents are equal. Then

$$I = \beta I_B$$

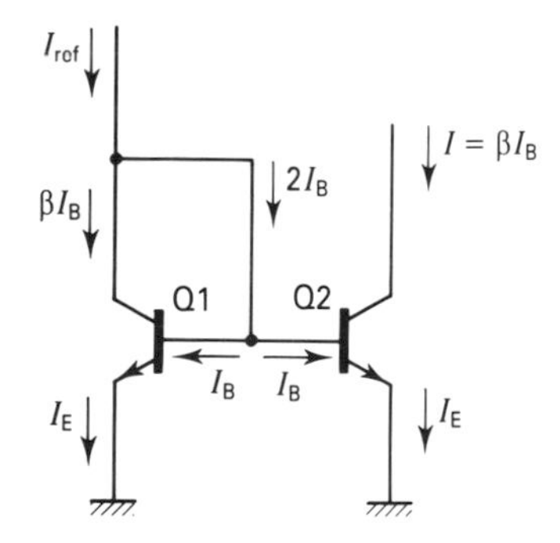

Figure 13.6 Current division for the analysis of the current mirror.

and

$$I_{ref} = 2I_B + \beta I_B$$

$$\frac{I}{I_{ref}} = \frac{\beta}{2 + \beta} = \frac{1}{1 + 2/\beta}$$

Provided that $\beta \gg 1$ then $I = I_{ref}$. For a value of $\beta = 100$ the error is 2%.

Another factor that affects I is the variation of collector current with collector voltage. This is caused by the Early effect, as described in Section 11.4. The current I will only equal I_{ref} when the collector voltage of Q2 is equal to V_{BE} as determined by I_{ref} in Q1. As the collector voltage of Q2 increases then I will increase by an amount that is determined by the output resistance r_o which is related to the Early voltage as:

$$r_o = \frac{V_A}{I}$$

For an ideal current source the output resistance is infinite, so that the simple current mirror described above is not ideal. However, for many applications it is perfectly adequate.

The current mirror can be used to provide multiple sources, as shown in Figure 13.7. The effect of β becomes more of a problem as the number of sources is increased since the base current of each additional source must be supplied by I_{ref}. By varying the areas of the multiple sources it is possible to scale the currents with respect to I_{ref}. For example, the currents may be scaled I, $2I$, $4I$, $8I$, etc., for some form of analog to digital converter.

13.3.2 Improved current source

A serious limitation of the simple current source is the effect of a finite β, particularly if multiple sources are required. The circuit in Figure 13.8 attempts to reduce this

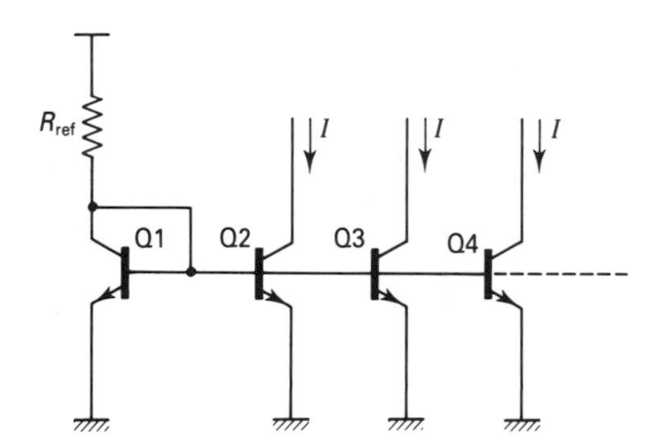

Figure 13.7 Current mirror with multiple sources.

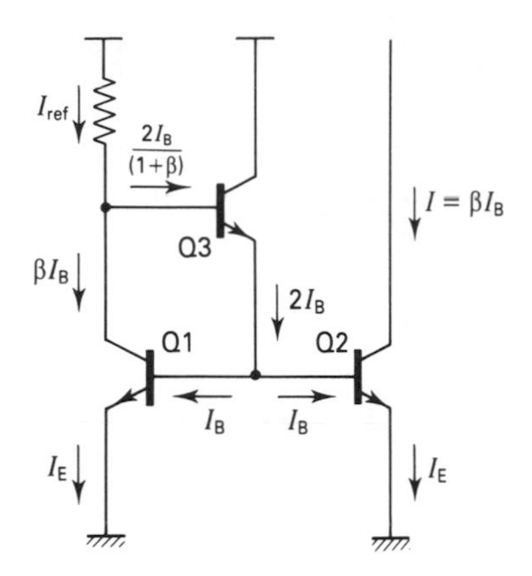

Figure 13.8 Current mirror with separate base current supply.

problem with an additional transistor to supply the base current. The emitter currents are again assumed to be equal if Q1 and Q2 are identical; then from the circuit it can be seen that:

$$I = \beta I_B$$

and

$$I_{ref} = \beta I_B + \frac{2I_B}{1+\beta}$$

Thus

$$\frac{I}{I_{ref}} = \frac{\beta}{\beta + 2/(1+\beta)}$$

$$\approx \frac{1}{1 + 2/\beta^2}$$

which means that the error introduced by a finite β is reduced from $2/\beta$ to $2/\beta^2$.

13.3.3 Wilson mirror

The Wilson mirror shown in Figure 13.9 has the same current gain compensation as the circuit of Figure 13.8, but with the added benefit of a greatly increased output resistance. It is left to the reader to show that for equal values of β then:

$$\frac{I}{I_{ref}} = \frac{1}{1 + 2/(2+\beta)^2}$$

$$\approx \frac{1}{1 + 2/\beta^2}$$

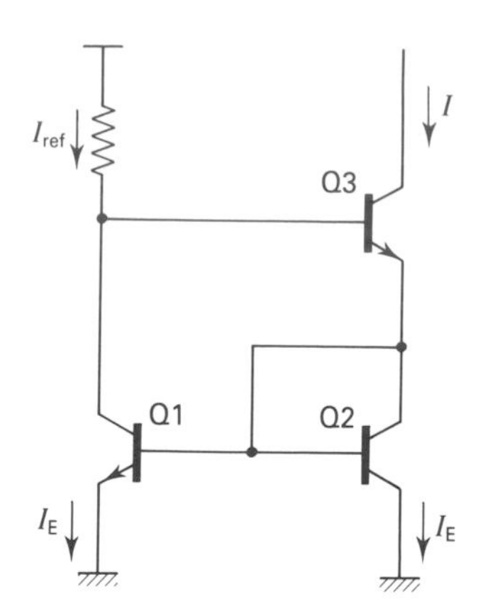

Figure 13.9 Schematic of the Wilson current mirror.

The Wilson mirror is a current feedback circuit. When the current I_{ref} is applied, Q3 responds with a current of I, but its emitter current is sensed by Q2 and establishes the V_{BE} of Q_2. This value of V_{BE} is mirrored back to the Q_1, where a balance is established between I_{C1} and I_{B3} . In effect, I_{B3} is the error signal of the feedback loop. One of the results of negative feedback is to increase the output resistance, and it can be shown that it is approximately $\beta r_o/2$ or $\beta/2$ times greater than the simple current source.

The quality of the Wilson source is very dependent on the matching of the βs. If they are not matched then the Wilson source is worse than the emitter-follower source of Figure 13.8. Mismatch can occur if the transistors have a very small emitter area, because then errors introduced by over- or under-etching become significant. Quality is also affected by variations in the current gain. This is a particular problem for current mirrors formed from lateral pnp transistors where in addition to mismatch, the βs may also be less than 100, in which case the base current can degrade the performance of the circuit.

13.3.4 Widlar mirror

For many applications involving current mirrors the value of the current which is required is a few tens of microamps. However, because of the limited range of resistor values the reference current may typically be several hundred microamps. The ratio between the reference current and the controlled current can be achieved by varying the area of the transistors. However, this becomes impractical for ratios much greater than ten because of the size of the transistors and the errors in the photolithography and etching. An alternative approach is to use equal area transistors but to include a resistor in the emitter of Q2, as shown in Figure 13.10. The Widlar mirror is analyzed as follows:

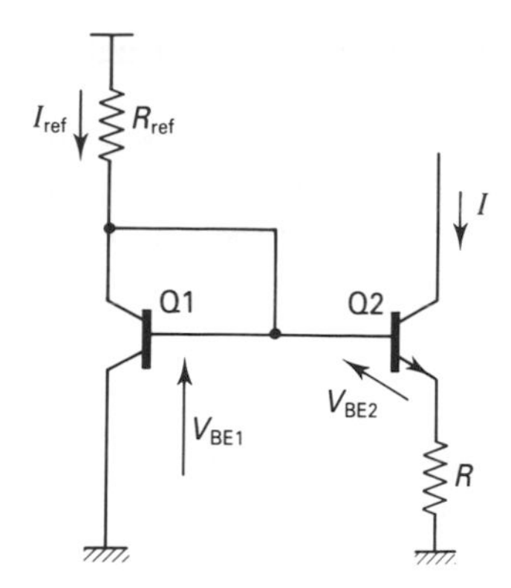

Figure 13.10 Schematic of the Widlar current mirror.

$$V_{BE1} = V_T \ln\left(\frac{I_{ref}}{I_s}\right)$$

and

$$V_{BE2} = V_T \ln\left(\frac{I}{I_s}\right)$$

where I_s is the saturation current and is assumed to be the same for both transistors, and V_T is the thermal voltage (kT/q).

Combining these two equations by taking the difference gives:

$$V_{BE1} - V_{BE2} = V_T \ln\left(\frac{I_{ref}}{I}\right)$$

From the circuit in Figure 13.10 it can be seen that:

$$V_{BE1} \approx V_{BE2} + IR$$

Thus

$$IR = V_T \ln\left(\frac{I_{ref}}{I}\right)$$

This equation is transcendental for I but can easily be solved for R given I and I_{ref}, or for I_{ref} given R and I. However, when I_{ref} and R are given, then an approximate solution for I can be obtained by solving the right-hand side of the equation for an initial guess of the current I. This calculation provides a value of I which will probably differ from the initial guess. This new value is substituted into the right-hand side of the equation and the calculation is repeated. With a judicious choice of I for the initial guess, a sufficiently accurate value of I is usually obtained after three or four substitutions.

The output resistance of the Widlar source is increased because of the negative feedback introduced by R. It is not as large as for the Wilson mirror but it is typically an order of magnitude greater than that of the simple mirror.

13.3.5 pnp mirror

The basic circuit for the pnp current mirror is shown in Figure 13.11(a). The operation is the same as for the npn version, with the exception that the β of pnp transistors is generally less than 100 and, therefore, the approximations which are made for the npn version regarding the base current are even less accurate for the pnp mirror. The structure of the lateral pnp is such that separate collectors can be formed around a single emitter (see Section 11.12 for details of the structure). Thus Q1 and Q2 of Figure 13.11(a) can be combined into a single transistor, as shown in Figure 13.11(b). The collector current is proportional to the collector periphery adjacent to the emitter and, therefore, the current ratio I/I_{ref} can be determined by the ratio of the periphery of the two collectors.

13.3.6 Supply voltage variations

For all the current sources the reference current (I_{ref}) is proportional to the supply voltage, namely:

$$I_{ref} = \frac{V_{CC} - V_{BE}}{R_{ref}}$$

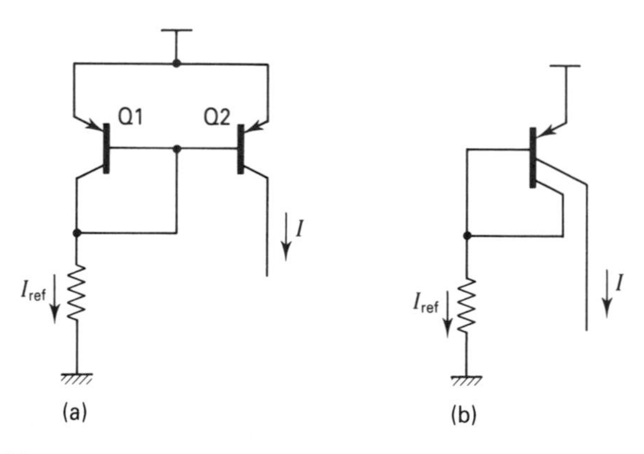

Figure 13.11 Schematic of a pnp current mirror with (a) separate transistor and (b) with a lateral pnp with two collectors.

or

$$I_{\text{ref}} = \frac{V_{CC} + V_{EE} - V_{BE}}{R_{\text{ref}}}$$

if both positive and negative supplies are used. Thus although the controlled current I is constant provided the supply voltage remains constant, it will change for different values of V_{CC} and V_{EE}.

The variation of I with supply voltage is reduced using the Widlar source because of the logarithmic dependence of I upon I_{ref}. Thus a 10% change in the supply voltage which produces a corresponding 10% variation in I_{ref} only results in a 2.7% change in the controlled current.

For the current to be independent of supply voltage variations it would be necessary to obtain the reference current from a constant voltage source, for example a Zener diode. The emitter–base junction of an npn transistor breaks down at between 5 and 7 V and could be used to establish a constant voltage source. However, for the majority of applications the absolute value of current is of less importance than the ratio of two or more currents.

13.4 dc level shifting

The lack of coupling capacitors means that when npn transistors are being used the voltage at the collector is positive with respect to ground and if more than one stage is cascaded then the voltage increases progressively until it approaches the supply

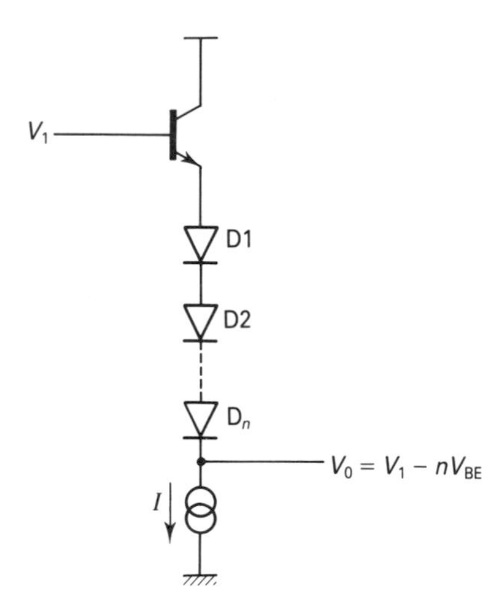

Figure 13.12 Level shifter employing forward-biased diodes and an emitter follower.

voltage. For an operational amplifier the output is required to be at zero volts and should be capable of both positive and negative voltage excursions with respect to ground. Thus it becomes necessary to reverse the upward voltage trend as the signal progresses from the input to the output. This can be achieved with a voltage level shifter.

One of the simpler circuits is shown in Figure 13.12, where the npn transistor acts as an emitter-follower with little loss of gain. The dc level of the output voltage is lowered with respect to the input by means of a number of forward-biased diodes. Each diode introduces a voltage drop of approximately 0.7 V. A single Zener diode could be used in place of the forward-biased diodes, but Zener diodes produce electrical noise as a result of random generation of carriers in the junction depletion layer during the reverse breakdown process.

A more versatile circuit is the V_{BE} multiplier shown in Figure 13.13. The voltage across the transistor is:

$$V_{CE} = I_R(R1 + R2)\ldots \text{ ignoring the base current}$$

$$= V_{BE2}\left(1 + \frac{R1}{R2}\right)$$

Thus the circuit multiplies V_{BE} by the factor $(1 + R1/R2)$. It is a simple matter to control the ratio of the resistors to obtain the required multiplication factor. The value of V_{BE} is obtained from:

$$V_{BE2} = V_T \ln\left(\frac{I_{C2}}{I_s}\right)$$

where I_s is the saturation current for the transistor (1×10^{-12} A) and $I_{C2} = I - I_R$.

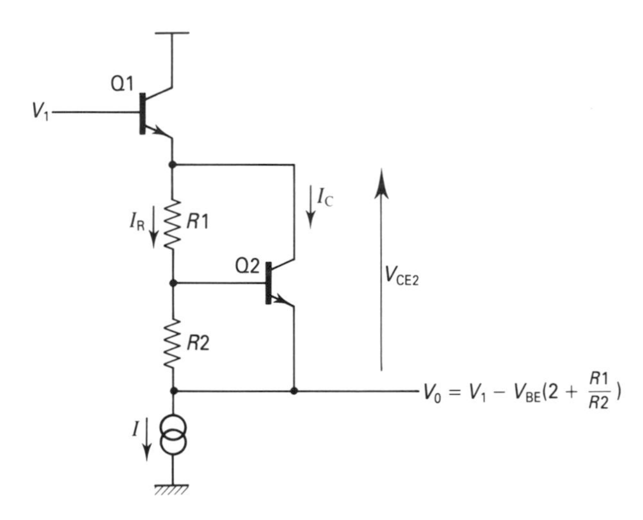

Figure 13.13 Schematic of the VBE multiplier level shifter.

The value of I_R must be selected to ensure that there is adequate current for Q2 to prevent it being cut off. Thus:

$$R1 + R2 = \frac{V_{CE2}}{I_R}$$

For a given value of V_{BE2} as determined from the equation above the value of $R2$ is:

$$R2 = \frac{V_{BE2}}{I_R}$$

A pnp transistor can provide both level shifting and gain, as shown in Figure 13.14. The output voltage is $I_C R$ above the negative supply voltage.

13.5 Temperature-independent voltage reference

The simplest form of voltage reference is a potential divider formed from two resistors and placed across the supply. The value of the voltage is determined by the resistor ratio of the potential divider. As a voltage reference it is not ideal because of the large internal resistance resulting from the resistors in the potential divider. In addition, the temperature coefficient of diffused resistors is positive, rather large and variable and, therefore, the voltage is dependent on the temperature, the supply voltage and any variation of the load resistance.

Two voltage sources which are independent of the supply and which have well-defined temperature coefficients are:

- forward-biased diode where V_{BE} has a negative temperature coefficient of $-2\,\text{mV/°C}$;
- avalanche breakdown of the emitter–base diode with a positive temperature coefficient of 200–500 ppm/°C, which at 7 V results in a voltage temperature coefficient of about $+3\,\text{mV/°C}$.

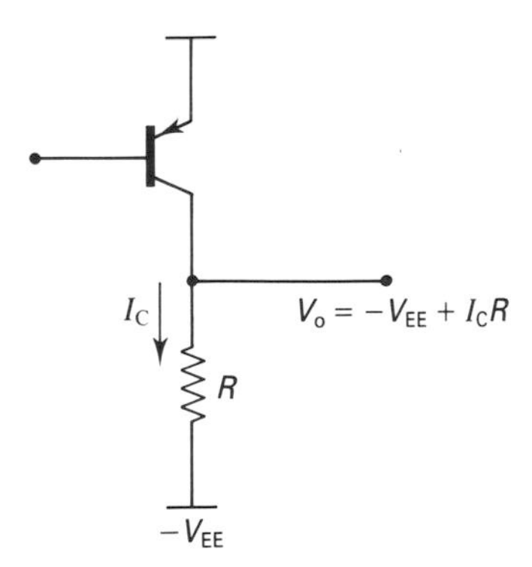

Figure 13.14 Level shifting with a pnp transistor.

Both sources provide a simple means of defining a voltage which is independent of the supply, except that in the case of the Zener diode the supply must be greater than the Zener voltage (>7 V).

A voltage source which is independent of temperature can be produced from two sources which have opposite temperature coefficients. The negative temperature coefficient of a forward-biased diode could be combined with the positive temperature coefficient of avalanche breakdown. However, a preferred source with a positive temperature coefficient is the thermal voltage kT/q. A schematic of such a voltage reference which uses these two effects is shown in Figure 13.15.

The forward-biased emitter–base junction of an npn transistor provides V_{BE} with a temperature of -2 mV/°C. The thermal voltage V_T can be generated from a Widlar current source where the current is equal to $(V_T/R)\ln(I_{ref}/I)$. A voltage proportional to V_T is obtained by taking the voltage drop across a resistor and at the same time scaling the voltage by a factor K.

To determine the factor K, note that the thermal voltage is equal to kT/q. At 300 K it has a value of 25.875 mV and at 301 K it has a value of 25.9612 mV, that is, V_T has a positive temperature coefficient of 86.25 μV/°C. This must be scaled up to 2 mV by the factor K, that is, K is equal to 23.2. The final output voltage is the sum of V_{BE} and KV_T. With V_T approximately equal to 26 mV at room temperature, the output voltage is:

$$V_o = 0.6 + (23.2)(26\,\text{mV}) = 1.2\,\text{V}$$

This voltage is close to the band-gap of silicon and the voltage source is often known as a band-gap reference.

A simple implementation of the schematic in Figure 13.15 is shown in Figure 13.16. Q1 and Q2 form a Widlar current source so that the current flowing in $R3$ is:

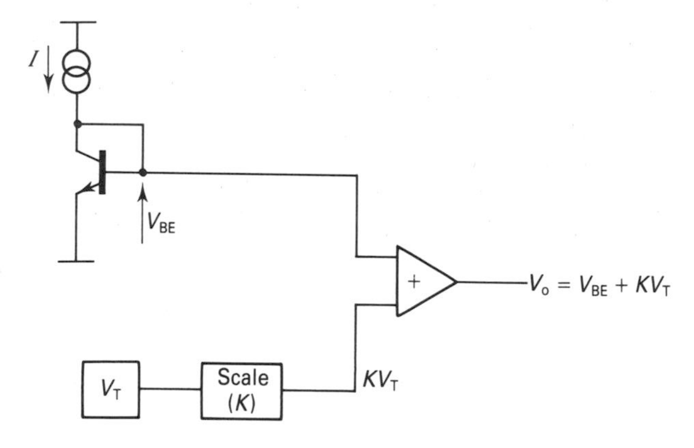

Figure 13.15 Schematic diagram of a temperature-independent voltage source.

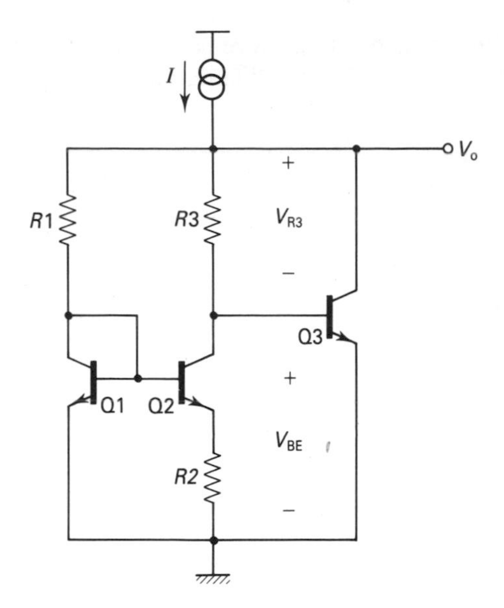

Figure 13.16 Circuit of a band-gap voltage reference source.

$$I_{R3} = \frac{V_T}{R2}\ln\left(\frac{I_1}{I_2}\right)$$

and

$$V_{R3} = V_T\frac{R3}{R2}\ln\left(\frac{I_1}{I_2}\right)$$

and

$$V_o = V_{BE} + KV_T$$

where

$$K = \frac{R3}{R2}\ln\left(\frac{I_1}{I_2}\right)$$

Practical versions of this circuit have temperature coefficients of 30–60 ppm/°C.

13.6 Gain stages

The basic gain stage for the input of an integrated circuit operational amplifier is the differential amplifier. An important feature of the differential amplifier is that it does not require a decoupling capacitor to achieve large values of gain. A schematic of the amplifier is shown in Figure 13.17. The differential gain (A_{vd}) is:

$$A_{vd} = g_m R_C$$

where

$$g_m = \frac{I/2}{0.026}$$

The common mode gain is:

$$A_{vc} = \frac{R_C}{2r_o}$$

where r_o is the output resistance of the current source.

The common mode rejection ratio ($CMRR$) is defined as the ratio of the difference mode gain to the common mode gain (A_{vd}/A_{vc}) and is given by:

$$CMRR = 2g_m r_o$$

For the simple current mirror which is used to establish the current source the output resistance is:

$$r_o = \frac{V_A}{I} \quad (V_A \text{ is the Early voltage})$$

and since

$$g_m = \frac{I/2}{V_T}$$

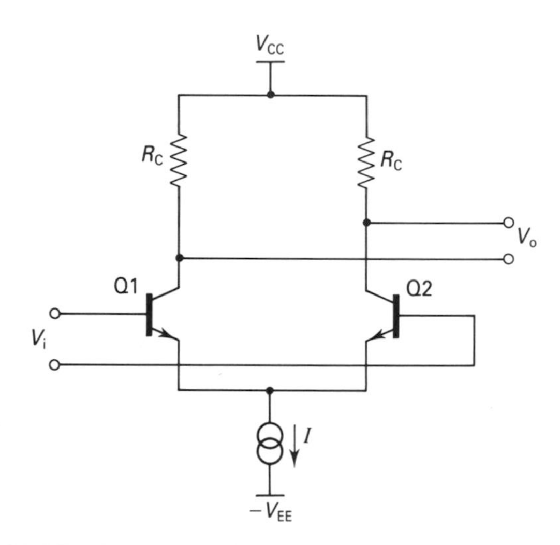

Figure 13.17 Schematic diagram of a simple differential amplifier.

then

$$CMRR = 2\left(\frac{I}{2V_T}\right)\left(\frac{V_A}{I}\right) = \frac{V_A}{V_T}$$

For an npn transistor with $V_A = 100\,\text{V}$ then

$$CMRR = \frac{100}{26\,\text{mV}} = 4000 \quad \text{or} \quad 73\,\text{dB}$$

In practice a $CMRR$ of 80–90 dB is required, so that the simple current mirror is not ideal. The Widlar current source has a larger output resistance and is more suitable when a large CMRR is required.

The input resistance for a single-stage common-emitter amplifier is:

$$R_{in} = \beta \frac{V_T}{I_E}$$

where I_E is the emitter current. For a differential pair the emitters are joined together, and, as far as the small-signal current is concerned, they are regarded as being at ground potential. The input resistance for a common-emitter stage with the emitter at ground is:

$$R_{in} = 2\beta \frac{V_T}{I/2} = 4\beta \frac{V_T}{I}$$

If $\beta = 250$ and $I = 25\,\mu\text{A}$ then $R_{in} = 1\,\text{M}\Omega$.

Larger values of input resistance can be achieved with larger values of β or with smaller values of current. Reducing the current reduces g_m and hence the gain (A_{vd}) so the usual approach to increase R_{in} is to increase β. This can be achieved with the Darlington circuit shown in Figure 13.18.

The extra current source is usually 10% of the main source (I). The effective β of the transistors Q1, Q11 is β^2 assuming that the βs of the two transistors are equal.

The alternative to the Darlington circuit is the field effect transistor and there are now a wide selection of operational amplifiers which use FETs for the input stage and bipolar devices for the remainder of the circuit. The FET, whether it is a JFET or a MOSFET, has an input resistance (R_{in}) which is much greater than $10\,\text{M}\Omega$.

13.6.1 Active loads

The gain of a differential stage with resistive loads is limited by the maximum value of diffused resistors and by the dc voltage drop which occurs if large values are used (for example, pinch resistors – see Section 12.4). A more effective solution is to use transistors connected as current sources which have a very large output resistance. The

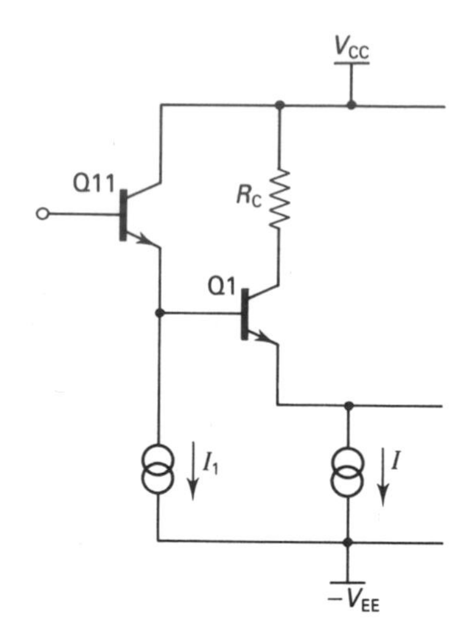

Figure 13.18 Darlington connected transistors at one of the inputs of the differential stage to increase the input resistance.

circuit of an amplifier with active loads and a single-ended output is shown in Figure 13.19.

The transistors Q3 and Q4 form a pnp current mirror which results in:

$$I_{C4} = I_{C1} = I/2$$

The stage can be analyzed by considering the effect of applying a small signal symmetrically to the inputs such that the base voltage increases on Q1 and decreases by an equal amount on Q2, that is,

$$v_i/2 \text{ causes } I_{C1} \text{ to increase to } I_{C1} + i = I/2 + i$$

$$v_i/2 \text{ causes } I_{C2} \text{ to decrease to } I_{C2} - i = I/2 - i$$

Thus at the output:

$$\begin{aligned} i_{out} &= I_{C4} - I_{C2} \\ &= I_{C1} - I_{C2} \\ &= (I/2 + i) - (I/2 - i) \\ &= 2i \end{aligned}$$

The transconductance (g_m) is defined as:

$$g_m = \frac{\delta_i}{\delta v_i} = \frac{i}{v_i/2}$$

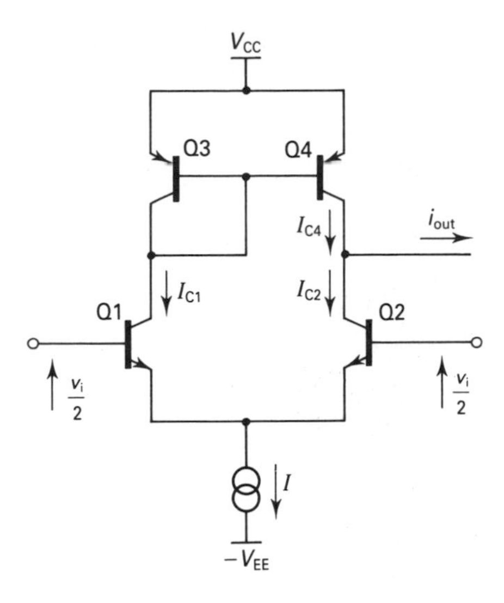

Figure 13.19 Differential amplifier with active loads.

or

$$i = g_m \frac{v_i}{2}$$

and

$$i_{out} = g_m v_i$$

or, in terms of output voltage:

$$v_{out} = i_{out} R_o$$

that is,

$$v_{out} = g_m v_i R_o$$

where R_o is the effective output resistance. Q2 is in effect a common-emitter stage and its output resistance is r_{o2}. The transistor Q4 is a current source with a resistance r_{o4}. Thus

$$R_o = r_{o2} \| r_{o4}$$

The output resistances of the transistors can be expressed in terms of their Early voltages as

$$r_{o2} = \frac{V_{A2}}{I_{C2}} = \frac{V_{A2}}{V_T g_m}, \quad \text{where} \quad g_m = \frac{I_C}{V_T}$$

$$r_{o4} = \frac{V_{A4}}{V_T g_m}$$

and

$$A_v = g_m R_o$$

$$= \frac{1}{V_T}\left(\frac{V_{A2} \cdot V_{A4}}{V_{A2} + V_{A4}}\right)$$

Assuming that $V_{A2} = V_{A4} = 50\,V$ then

$$A_v = \frac{1}{25\,mV}\left(\frac{50 \cdot 50}{100}\right)$$

$$= 1000 \text{ or } 60\,dB$$

More elaborate current mirror circuits can be utilized to achieve even higher gains.

13.6.2 Single-ended gain stage

Current mirror active loads are also used in the design of single-ended gain stages and one example of such a stage is shown in Figure 13.20. Q1 is a class A amplifier with a high-impedance current source acting as a load. The current source is a pnp current mirror with Q3 providing the reference current which is mirrored into Q2.

The load seen by Q1 is the small-signal output resistance of Q2 given by:

$$r_{o2} = \frac{V_{A2}}{I_{C2}}$$

where V_{A2} is the Early voltage and I_{C2} the collector current, for example with $V_{A2} = 50\,V$ and $I_{C2} = 0.5\,mA$ then $r_{o2} = 100\,K$. The output resistance of Q1 is in parallel with that of Q2 and the gain is

$$A_V = g_m r_{o2} || r_{o1}$$

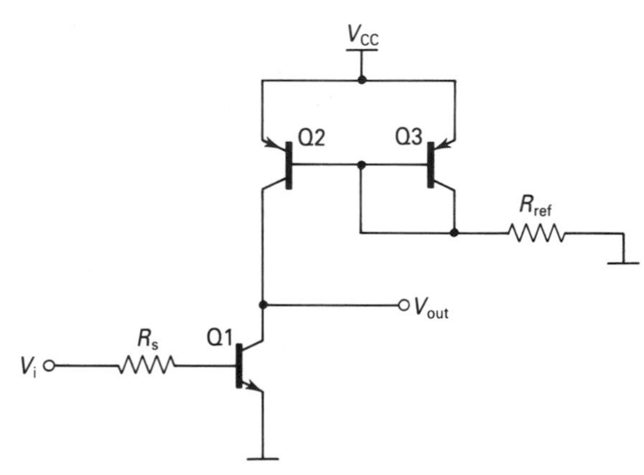

Figure 13.20 Single-ended gain stage with current mirror active load.

However, in order to achieve high values of gain, the input impedance of the next stage must be high, for example an emitter-follower with a gain of one. A single-ended stage such as this would be used as an intermediate stage between the differential input stage and the output stage.

13.7 Output stages

The output stage of an integrated circuit amplifier must deliver the signal to a load, which may have a low impedance, without loss of gain. Because the output stage is the last stage in the amplifier it must handle large signals without introducing distortion. Finally, it must transfer the power to the load efficiently without any significant loss of power in the output transistors.

The most commonly used output stage is the complementary emitter-follower class B stage, as shown in Figure 13.21(a) where Q2 is a substrate pnp. With this stage only one transistor conducts at a time. Thus for a positive input signal Q1 conducts and Q2 is OFF, while for a negative signal Q2 conducts and Q1 is OFF. However, silicon transistors require at least 0.5 V of forward bias before they start to conduct and as a result the output waveform suffers from cross-over distortion, as shown in Figure 13.21(b).

Cross-over distortion is removed with class AB operation in which each transistor is forward-biased to the point of conduction. The modified circuit is shown in Figure 13.22. The forward-biased diodes provide approximately 1.2 V of bias which is sufficient to produce a quiescent current in Q1 and Q2. As the input voltage increases, the current in one of the transistors increases and that in the other decreases, but since both are conducting current for small values of v_i the transition from Q1 conducting and Q2 OFF to Q2 conducting and Q1 OFF is smooth and cross-over distortion is

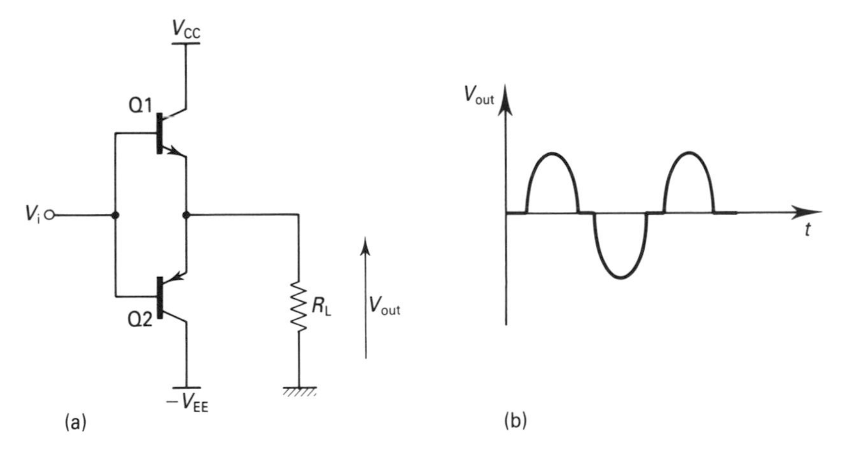

Figure 13.21 (a) Simplified diagram of the class B output stage and (b) the output waveform with cross-over distortion.

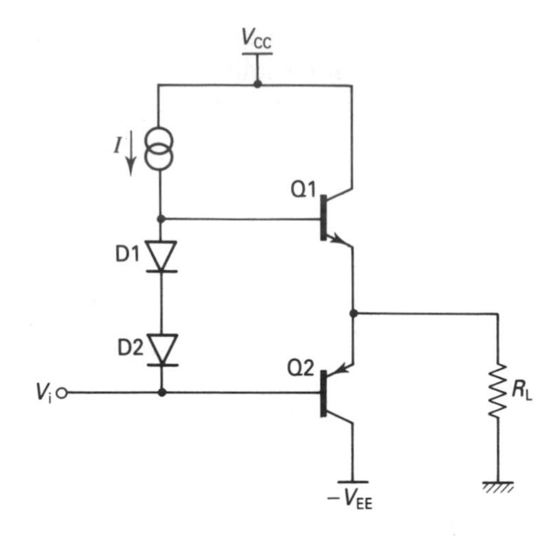

Figure 13.22 Class AB output stage.

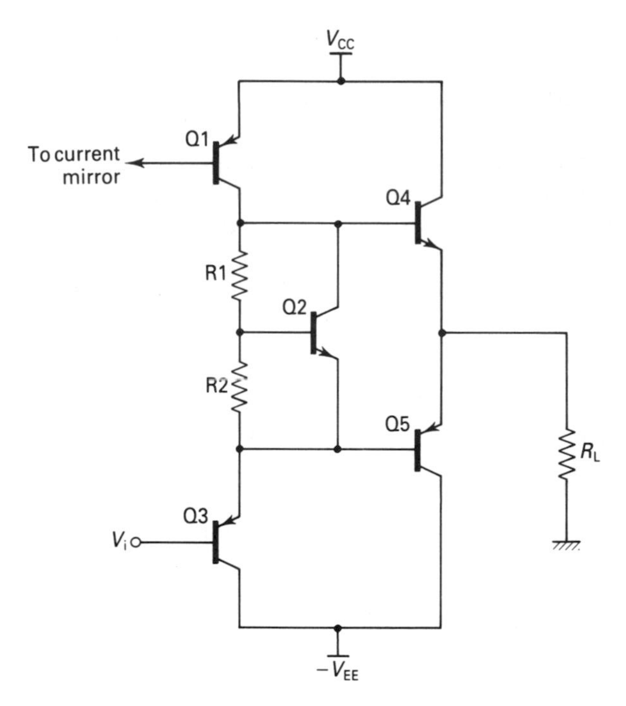

Figure 13.23 Circuit schematic of a class AB output stage with V_{BE} multiplier.

almost totally eliminated. The power relationships for class AB are almost the same as those for class B with a theoretical efficiency of 78%.

An alternative to the two diodes, which offers the designer much greater flexibility, is the V_{BE} multiplier shown in Figure 13.23. The resistors R1 and R2 are selected to provide the two V_{BE} drops required for the output transistors, with the current being provided by the pnp current mirror Q1. The transistor Q3 is a substrate pnp acting as an emitter-follower.

An important requirement for an output stage is that it should be protected against accidental or deliberate short-circuiting of the output to ground. Under short-circuit conditions a large current flows in the output transistors and there is the possibility of permanent damage. A suitable circuit is shown in Figure 13.24. In the event of a short-circuit a large current will flow through Q4. This current develops a voltage across R_{E4} which causes Q6 to turn ON. The collector current of Q6 flows at the expense of the base current to Q4. The current through Q4 is, therefore, reduced to a safe level. The value of R_{E4} is selected to provide a voltage of about 550 mV at the safe operating current of the output transistor.

This form of protection is effective but the emitter resistors result in some loss of signal. On the other hand, the resistors do provide some protection against thermal runaway.

13.8 Operational amplifier example

An example of a simple operational amplifier circuit is shown in Figure 13.25. The amplifier contains many of the basic circuit elements described above. There is a

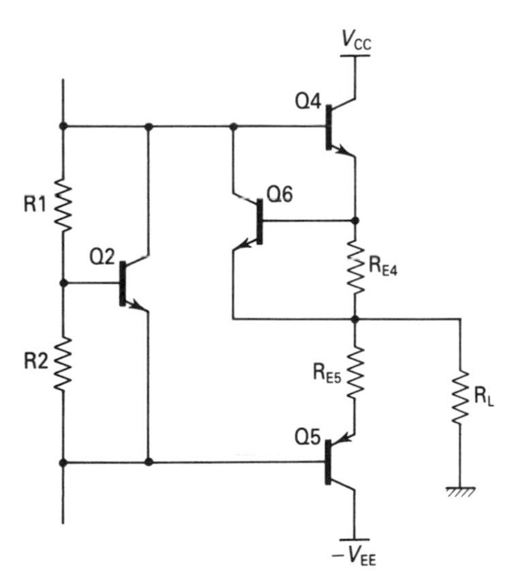

Figure 13.24 Output stage with short-circuit protection.

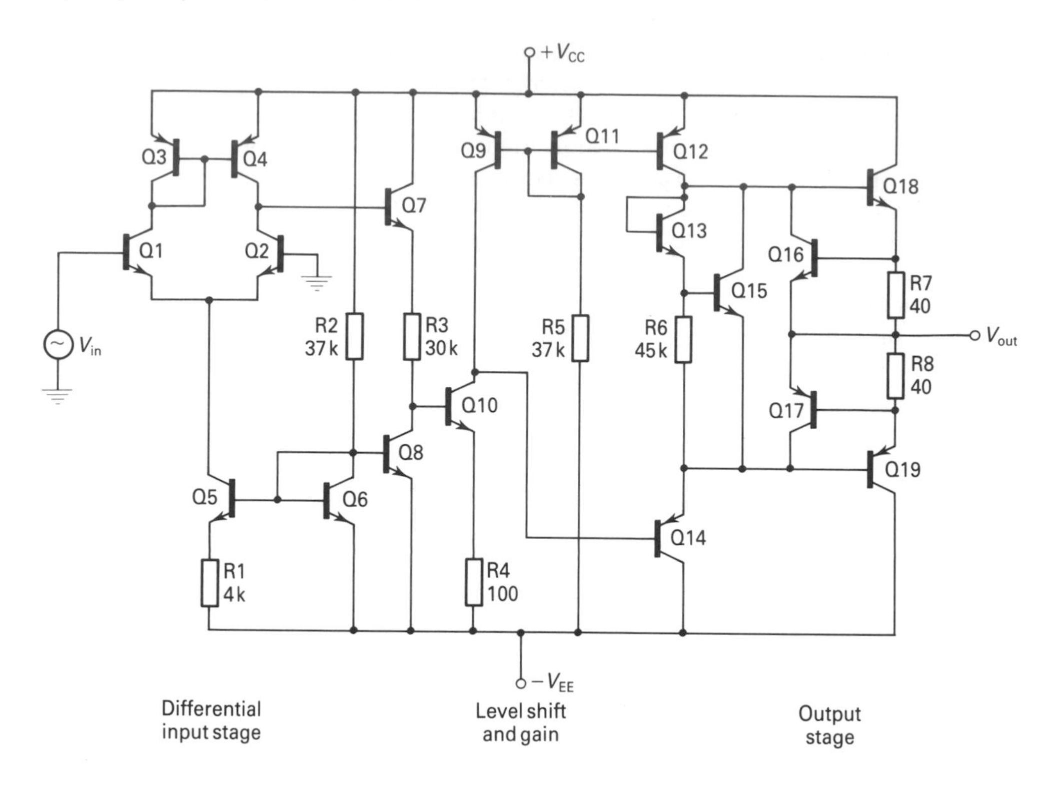

Figure 13.25 Circuit schematic of a simple operational amplifier.

differential input stage (Q1, Q2) with active loads (Q3, Q4) supplied from a Widlar constant current source (Q5, Q6, R1, R2). This is followed by an emitter-follower level shifter (Q7, R3) with a current source (Q8) and a common-emitter gain stage (Q10) with an active load (Q9). The current for this load is provided by a current mirror (Q11, R5). The class B output stage comprises Q18 and Q19 with a V_{BE} multiplier (Q13, R6, Q15). The output stage has short-circuit protection provided by Q16, Q17, R7 and R8. The link between the level shifter and gain stage is provided by an emitter-follower (Q14), which presents a high impedance to Q10. The most efficient means of analyzing a circuit of this complexity is by means of a circuit simulator (see Problem 11), but the simplified equations presented above provide the means for ensuring that the simulator is used sensibly in a design situation where it may be necessary to optimize the performance to satisfy a design specification.

13.9 Summary

This chapter has dealt with some of the basic circuit configurations which are used in both digital and analog bipolar circuits. In practice, the circuits of commercial integrated devices are much more complex but many of the basic features can be observed. Many of the more interesting analog designs are based on the current-mode approach[1] which is much more appropriate for current controlled bipolar transistors, and a very wide range of analog circuit functions are now available.

Problems

1. If in the circuit of Figure 13.1 R1 is 2.7 kohm and one of the emitters is low at 0.2 V estimate the current in the emitter which is low. Assume that V_{CC} is 5 V.
[1.52 mA.]

2. If the pull-down transistor in the circuit described in Problem 1 is capable of sinking 20 mA, estimate the fan-out.
[13.]

3. For the circuit described in Problem 1, R2 is 750 ohm. The static power consumed by the gate may be estimated as the average of the power ($P_{diss} = IV_{CC}$) consumed when the inputs are all high and when at least one input is low. With the inputs high the current I is given approximately as the sum of I_{R1} and I_{R2}, while when an input is low the current I is I_{R1}. Estimate the average static power for the gate.
[20 mW.]

4. Determine the value of R_{ref} to provide a constant current of 10 μA if the area of Q2 is one-tenth of Q1.
[93 kΩ.]

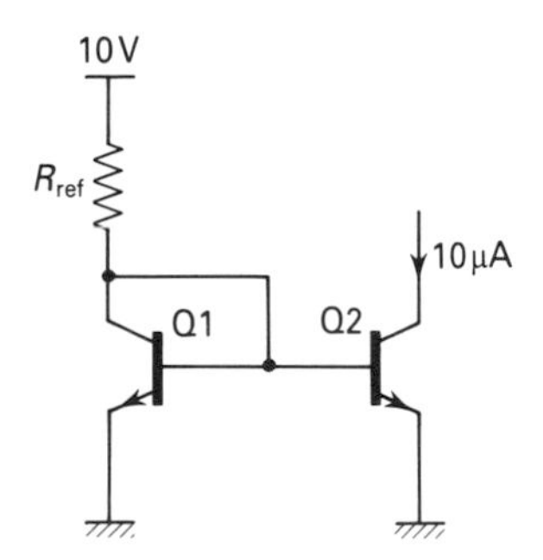

5. If the two transistors have the same area and I_{ref} is 1 mA, determine suitable values for R_{ref} and R_E.
[9.3 kΩ, 12 kΩ.]

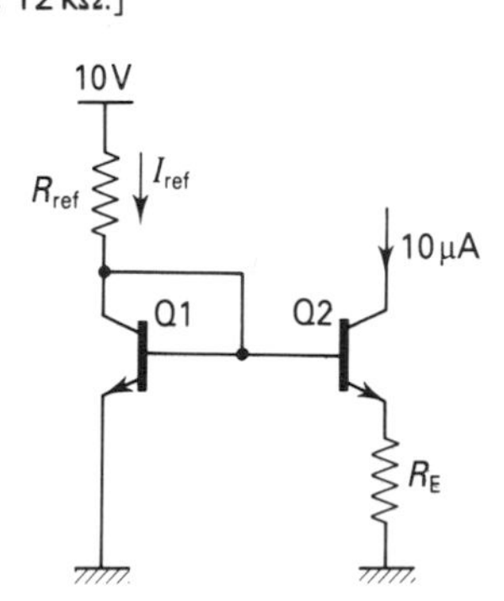

6. For the circuit shown determine the constant current I.
[11 A.]

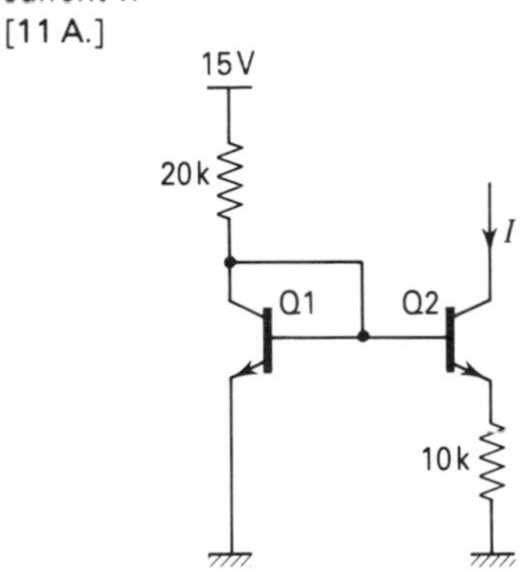

7. For the circuit shown in Problem 6 determine the percentage change in I if V_{CC} changes from 15 V to 10 V.
[8%.]

8. Estimate the voltage gain V_{out}/V_{in} for a differential gain stage with 30 kΩ resistive loads as shown.
[170.]

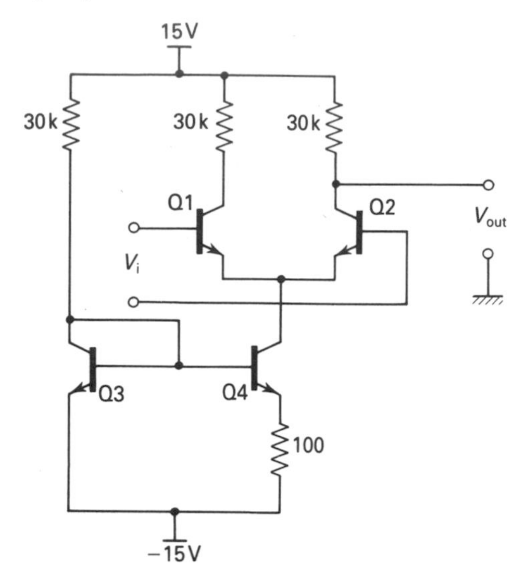

9. Estimate the voltage gain for the differential stage with pnp active loads if the Early voltage for the npn is 50 V and for the pnp 30 V.
[720.]

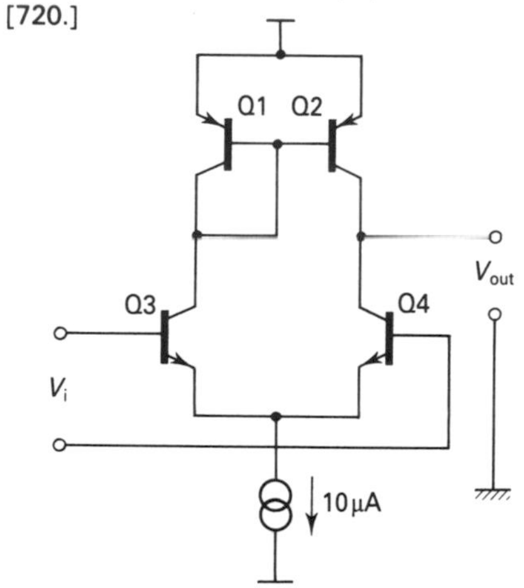

10. Estimate the voltage gain for the common-emitter stage with a pnp active load if the Early voltage for the two transistors is 50 V, and if the collector periphery C1 of transistor Q1 is five times smaller than the periphery C2.
[960.]

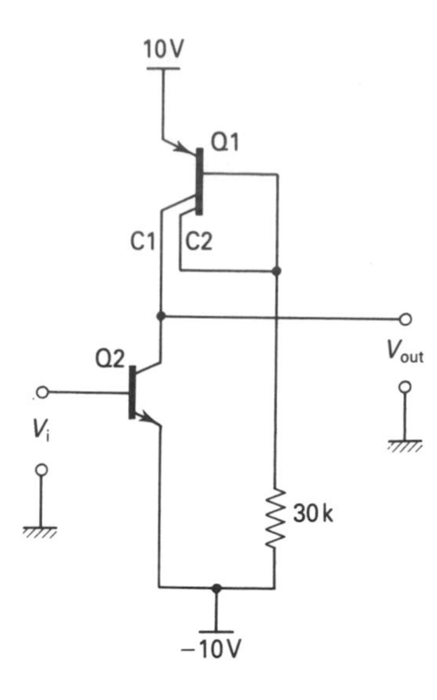

11. Generate source code or a circuit schematic for an analog circuit simulator (e.g. SPICE/PSPICE/HSPICE) for the input gain stage of the amplifier shown in Figure 13.25. The circuit should contain the differential stage, the Widlar current mirror and a signal source. The signal source should have an $AC = 1$ value for small-signal analysis, plus a 1 kHz sine generator with an amplitude of about 1 mV. Identify the transistors as 'nmod' for the npn devices and 'pmod' for the pnp devices and include the 'model' information given below to describe the devices. Assume that the power supply is **±15 V.**

```
.model nmod npn(bf=200 tf=4e-10 rb=1k
cjc=1p cje=1.2p ccs=1.8p rc=100 vaf=100)
.model pmod pnp(bf=60 tf=5e−8 rb=1k
cjc=1p cje=1.2p ccs=1.8p rc=100 vaf=60)
```

Run the simulation and observe the small-signal frequency response, mid-band gain, dc voltages and the sinewave output at the collectors of Q4 and Q2.

From the results confirm that the collector current for Q5 is given approximately by:

$$I_{C5} = \frac{0.026}{4k} \ln \frac{I_{R2}}{I_{C5}}$$

Confirm also that the gain of the stage is given

approximately by:

$$A_v = 20 \log \left\{ \frac{1}{V_T} \left(\frac{V_{A2} \times V_{A4}}{V_{A2} + V_{A4}} \right) \right\}$$

where V_{A2} and V_{A4} are the Early voltages (the term 'vaf' in the SPICE parameter list for the transistor models) for Q2 and Q4 respectively.

12. Add the remainder of the circuit, as a text file or a circuit schematic, and simulate the complete response. (NB: it will be necessary to include a small dc voltage offset in the sinewave input source generator to force the dc voltage level at the output to 0 V – a value of about 5 mV should be sufficient). Note the increase in gain and the overall frequency response.

13. Investigate the effect of applying negative feedback to the circuit described in Q12 by adding a 1 kohm resistor between the input signal source and the base of Q1, and a 10 kohm resistor from the output to the base of Q1. This creates an inverting amplifier with a gain of 10. Perform a small-signal analysis to obtain the frequency response. Confirm that the gain is 10, but also confirm that there is a 'peak' in the frequency response curve. This peak indicates instability at the frequency where the gain reaches a maximum.

The peak, and the instability, can be removed by applying frequency compensation to the integrated circuit in the form of a small-value (10 pF) capacitor connected from the collector of Q10 to the base of Q7. Verify that the frequency response with the compensating capacitor in place is improved at the expense of bandwidth. (Refer to an appropriate textbook on analog circuits to obtain more information on frequency compensation.) The value of 10 pF is far too large for inclusion in an integrated circuit (refer to Chapter 12 for typical values), but the Miller effect can be used to simulate a larger value. Suggest how this may be achieved. (See the Solution Guide p. 270 for more details.)

References

[1] C. Toumazou, F. J. Lidgey and D. G. Haigh (eds) (1990), *Analog IC Design: The Current-mode Approach*, London: Peregrims.

CHAPTER 14

The MOS capacitor

14.1 Introduction

The properties of the silicon–oxide interface are important for the Metal-Oxide–Semiconductor (MOS) class of device, with the MOS transistor being the most widely used of these devices. Unlike bipolar devices where the active regions are within the body of the silicon, for the MOS device the active region is at (or very near) the surface.

To understand the operation of these devices it is necessary to consider the surface physics associated with the MOS capacitor, as shown in Figure 14.1. The MOS capacitor was first used by Terman[1] to study the thermally oxidized silicon surface and has since been used by many authors to study oxide charge and its effect on the performance of devices, as a manufacturing process monitor and for studying device reliability.

14.2 The MOS capacitor

The metal-oxide–semiconductor capacitor as shown in Figure 14.1 is formed by evaporating a layer of metal on to the surface of a semiconductor which is covered with

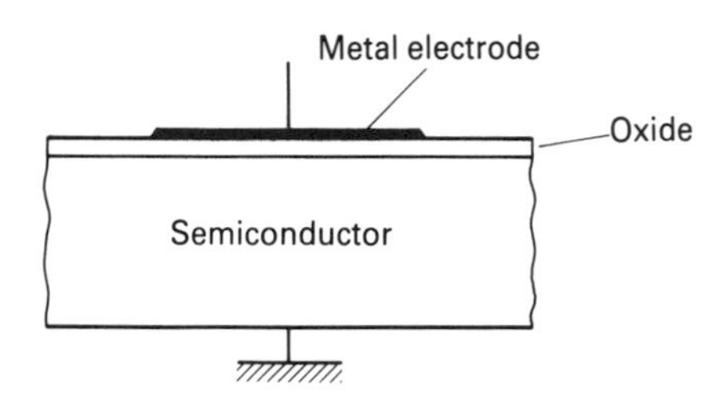

Figure 14.1 Metal-oxide–semiconductor capacitor.

a layer of oxide. For silicon the oxide is formed by thermal oxidation and the metal is usually aluminium, which is deposited by evaporation. Photolithography is used to produce electrodes which may be 0.5–3 mm in diameter.

When a voltage is applied to the metal electrode with respect to the substrate the mobile electrons or holes in the silicon are either driven from the surface or attracted to it depending on the polarity of the applied voltage. The redistribution of charge within the silicon affects the Fermi potential and causes the energy bands near the surface to bend. The Fermi potential is given by:

$$\phi_f = V_T \ln\left(\frac{N_a}{n_i}\right)$$

where V_T is the thermal voltage (kT/q) and N_a is the acceptor impurity concentration of the p-type substrate (N_d for an n-type substrate).

The band diagram for four bias conditions for a p-type substrate is shown in Figure 14.2. For an ideal capacitor with no voltage applied, the energy bands are flat, as shown in Figure 14.2(a). If a negative voltage is applied to the metal electrode, then positively charged holes from the bulk of the silicon are attracted to the surface. The increased concentration of holes at the surface is shown in the band diagram as an upward bending of the bands with respect to the Fermi level.

The Fermi level acts as a reference level and is shown as a horizontal line in the diagram. The accumulation of charge is shown in Figure 14.2(b). Applying a positive voltage repels holes from the surface to create a depletion region. At the same time electrons are attracted to the surface. The depletion of holes and the increase in electrons is shown in the band diagram as a downward bending of the bands, as shown in Figure 14.2(c).

It is useful to introduce the concept of a surface potential (ϕ_s). For the flat-band condition $\phi_s = 0$; for the depletion condition $\phi_s > 0$. As the positive voltage applied to the metal electrode is increased further, ϕ_s continues to increase as more carriers are displaced from the surface until $\phi_s = \phi_f$. Further increase of the voltage causes the intrinsic energy level (E_i) to bend below the Fermi level. This position of E_i with respect to E_f corresponds to the position adopted for the energy bands in an n-type semiconductor. The surface has in fact been inverted and the number of minority carrier electrons exceeds the number of majority carrier holes at the surface. This formation of an n-type region does not involve a metallurgical junction. An arbitrary definition of inversion is that $\phi_s = 2\phi_f$ because for this value of surface potential the number of electrons in the conduction band at the surface equals the number of holes in the valence band in the bulk. Further increase of the external bias does not greatly change the width of the depletion layer but instead the increased voltage is balanced by an increase of the negative charge in the inversion layer.

The distribution of charge for the conditions of accumulation, depletion and inversion is shown diagrammatically in Figure 14.3.

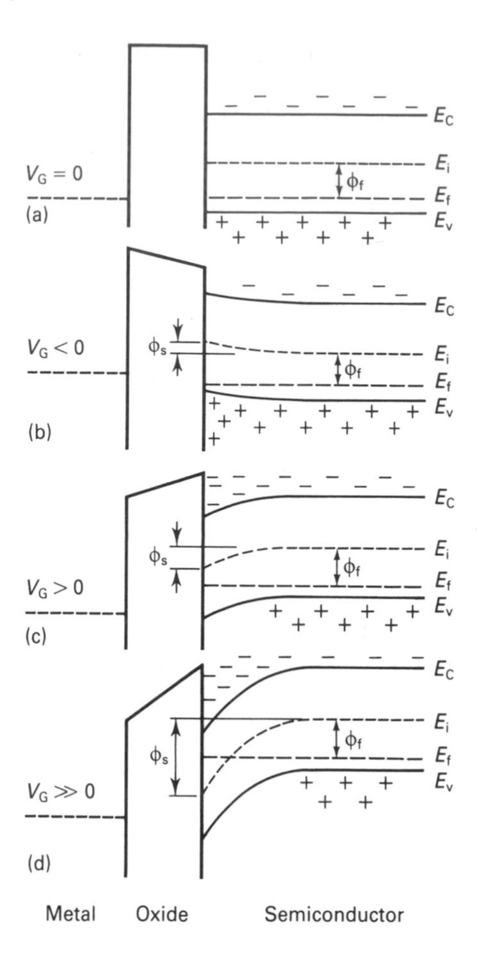

Figure 14.2 Energy-band diagram for an ideal MOS capacitor on p-type silicon for (a) flat-band, (b) accumulation, (c) depletion and (d) inversion.

14.3 The ideal surface

If a negative voltage is applied to the metal electrode the majority holes accumulate as a thin sheet of charge at the silicon surface, as shown in Figure 14.3(a). With a positive voltage holes are driven from the surface to expose the negatively charged acceptor impurity atoms. Unlike the holes, however, the atoms are fixed in the silicon lattice and the charge cannot be concentrated into a thin sheet at the surface. The number of exposed atoms is a function of the applied voltage and the impurity concentration. The charge in the depleted region is given by:

$$Q_s = Q_B = -qN_a x_d$$

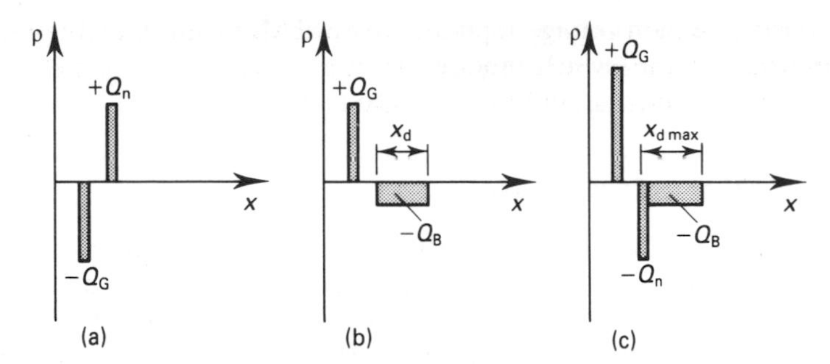

Figure 14.3 Charge distribution for (a) accumulation, (b) depletion and (c) inversion.

where x_d is the width of the depletion layer, as shown in Figure 14.3(b) and N_a is the acceptor impurity concentration. With further increase of the positive voltage the surface becomes inverted and minority carriers (electrons) collect at the surface. Like the holes under accumulation, these electrons are able to form a thin sheet of charge at the surface, as shown in Figure 14.3(c). The surface charge is now given by:

$$Q_s = -Q_n - qN_a x_{d\max}$$

where $x_{d\max}$ is the maximum width of the depletion layer. Further increase of the applied external voltage will not greatly increase the width of the depletion layer since the charge from the minority carriers effectively shields the bulk region of the silicon from the electric field at the surface.

The depletion region at the surface is similar to that associated with a step pn junction for which the width is given by:

$$x_d = \left(\frac{2\varepsilon_o \varepsilon_{si} \phi_s}{qN_a}\right)^{1/2}$$

where ϕ_s is the surface potential. The depletion width is a maximum when $\phi_s = 2\phi_f$.

The voltage applied to the metal electrode with respect to the substrate is divided between the voltage drop across the oxide and the surface potential:

$$V_G = V_{ox} + \phi_s$$

From the simple relationship between charge, capacitance and voltage ($Q = CV$) the voltage across the oxide can be expressed in terms of the surface charge and oxide capacitance:

$$V_G = \frac{Q_s}{C_{ox}} + \phi_s$$

At the onset of inversion $\phi_s = 2\phi_f$ so that the voltage for the onset of inversion is:

$$V_{G\,inv} = \left(\frac{2\varepsilon_o \varepsilon_{si} q N_a 2\phi_f}{C_{ox}}\right)^{1/2} + 2\phi_f$$

This value for the inversion voltage applies to an ideal MOS interface. In practice, there are two important factors which modify the above expression, namely the metal–semiconductor work function and the interface charge.

14.3.1 The Work function

The energy-band diagrams for an ideal MOS capacitor for n- and p-type silicon are shown in Figure 14.4. The diagram includes an energy level which corresponds to vacuum. If an electron is raised to the vacuum level then it is free to leave the metal or the semiconductor. This could be achieved by applying heat. In Figure 14.4 the vacuum level only serves as a reference against which the energies in two different materials can be compared. For the metal the energy difference between the Fermi level and the vacuum level is $q\phi_m$ where ϕ_m is the work function for the metal. For the semiconductor the appropriate energy level is based on the energy difference between the vacuum level and the Fermi level. The energy difference between the vacuum level and the top of the conduction band is known as the electron affinity χ.

The work function is:

$$\phi_s = \chi + \frac{E_g}{2} - \phi_f \quad \text{for n-type}$$

where

$$\phi_f = V_T \cdot \ln(N_d/n_i)$$

$$\phi_s = \chi + \frac{E_g}{2} + \phi_f \quad \text{for p-type}$$

where

$$\phi_f = V_T \cdot \ln(N_a/n_i)$$

For aluminium $\phi_m = 3.2$ V. The electron affinity for silicon is 3.25 V and for a

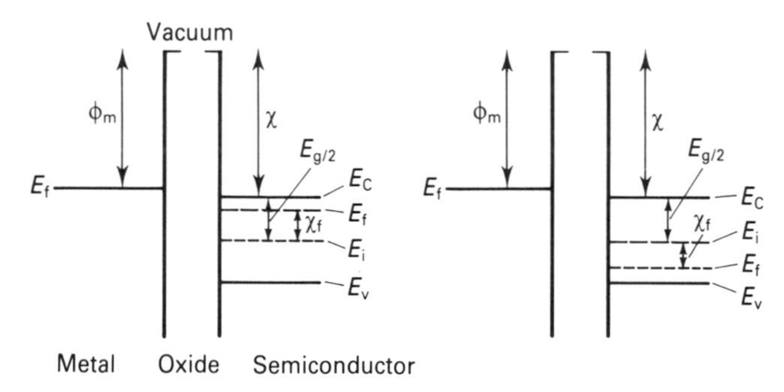

Figure 14.4 Energy-band diagram for a metal electrode and (a) an n-type semiconductor and (b) a p-type semiconductor.

carrier concentration of 10^{15} atoms/cm^3, $\phi_f = 0.29$ V at 300 K. With $E_g = 1.1$ V the silicon work function is:

$$\phi_s = 3.25 + 0.55 - 0.29 = 3.51 \text{ V for n-type}$$

and

$$\phi_s = 3.25 + 0.55 + 0.29 = 4.09 \text{ V for p-type}$$

The work function between aluminium and silicon is:

$$\begin{aligned}\phi_{ms} &= \phi_m - \phi_s \\ &= 3.20 - 3.51 = -0.3 \text{ V for n-type} \\ &= 3.20 - 4.09 = -0.9 \text{ V for p-type}\end{aligned}$$

Thus with no external voltage applied to the metal electrode but with the metal connected to the semiconductor there will exist a voltage equal to ϕ_{ms} which will cause the bands to bend.

Another commonly used electrode material, particularly for MOS transistors, is polycrystalline silicon. The energy-band diagram for this structure is shown in Figure 14.5. It is usual for the polysilicon to be heavily doped so that the Fermi level is very close to the valence band for p-type doping, as shown in Figure 14.5(a), or the conduction band for n-type doping, as shown in Figure 14.5(b). The work function difference for n-type silicon with an impurity concentration of 10^{15} atoms/cm^3 and p-type polysilicon is:

$$\begin{aligned}\phi_{ms} &= \chi + E_g + \left[\chi + \left(\frac{E_g}{2} - \phi_f\right)\right] \\ &= 3.25 + 1.10 - [3.25 + (0.55 - 0.29)] \\ &= +0.84 \text{ V}\end{aligned}$$

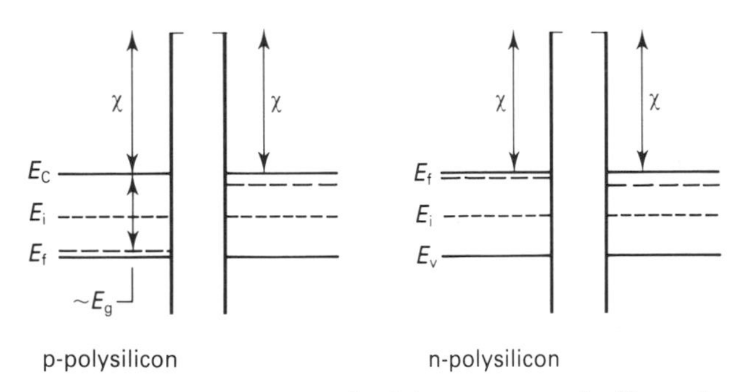

Figure 14.5 Energy-band diagram for (a) an n-type polysilicon electrode and (b) a p-type polysilicon electrode with an n-type substrate.

and for n-type polysilicon

$$\phi_{ms} = \chi - \left[\chi + \left(\frac{E_g}{2} - \phi_f\right)\right]$$
$$= 3.25 - 3.51$$
$$= -0.26\ \mathrm{V}$$

Similar calculations can be made for other electrode materials and for other impurity concentrations.

An additional non-ideal property of the MOS capacitor is the presence of charge at the silicon–silicon dioxide interface and in the oxide itself, as shown in Figure 2.4.

14.3.2 Oxide charge

As described in Chapter 2, charge can exist at the interface and in the oxide itself. The most important component is the fixed oxide charge located at the interface (Q_f) but in addition there may be interface-trapped charge (Q_{it}), trapped-oxide charge (Q_{ot}) and mobile charge (Q_m). All of these charges can be lumped together to form a surface charge (Q_{ss}) which represents the charge per unit area. The number of charges per unit area is $N_{ss} = Q_{ss}/q$.

The presence of oxide charge influences the bending of the energy bands when an external voltage is applied to the ideal model of a MOS capacitor. The sum total of oxide charge usually results in a positive charge which attracts negatively charged electrons to the interface from the bulk of the silicon. Thus with no external voltage applied, n-type silicon is in a state of accumulation and p-type silicon is in a state of depletion.

The voltage which results from this charge is $-Q_{ss}/C_{ox}$.

14.4 Flat-band voltage

The flat-band voltage represents an external voltage which must be applied to the MOS capacitor to make the energy bands flat from the interface to the bulk. It is a measure of the non-ideal features of the MOS capacitor, that is, the work function difference and the oxide charge.

If the oxide charge is assumed to be located at the interface then the effective voltage produced by this charge is Q_{ss}/C_{ox} and the flat-band voltage is:

$$V_{fb} = \phi_{ms} - \frac{Q_{ss}}{C_{ox}}$$

14.5 Inversion voltage

In considering an ideal MOS capacitor an expression was obtained for the external voltage required to invert the surface defined as $\phi_s = 2\phi_f$. If the work function and the surface charge are both taken into account then the voltage required to produce inversion is:

Aluminium electrode

$$\text{p-type} \quad V_{inv} = -0.6 - \phi_f - \frac{Q_b}{C_{ox}} - \frac{Q_{ss}}{C_{ox}}$$

$$\text{n-type} \quad V_{inv} = -0.6 - \phi_f + \frac{Q_b}{C_{ox}} - \frac{Q_{ss}}{C_{ox}}$$

n-type silicon electrode

$$\text{p-type} \quad V_{inv} = +0.55 - \phi_f - \frac{Q_b}{C_{ox}} - \frac{Q_{ss}}{C_{ox}}$$

$$\text{n-type} \quad V_{inv} = -0.55 + \phi_f + \frac{Q_b}{C_{ox}} - \frac{Q_b}{C_{ox}}$$

14.6 *C–V* curves

It has been shown that the variation of the external voltage applied to the metal electrode with respect to the substrate produces variations in the charged carriers at the surface of the semiconductor which can be described by means of the bending of the energy bands. The variation of the voltage can also be described in terms of variation of capacitance.

When majority carriers are driven from the surface to create a depletion region the effect is similar to that of a reverse-biased diode, and, like the diode, the variation of the depletion layer can be observed as a variation of capacitance. For the MOS capacitor there are in fact two capacitors, the oxide capacitance in series with the capacitance of the depletion layer, as shown in Figure 14.6. The total capacitance is:

$$C = \frac{C_{ox} C_s}{C_{ox} + C_s}$$

where C_{ox} is the oxide capacitance and C_s is the capacitance of the space charge layer. For a given oxide thickness the value of C_{ox} is constant and represents the maximum capacitance of the MOS capacitor. The capacitance C_s is voltage dependent and for a p-type semiconductor the variation with voltage is as shown in Figure 14.7. The explanation of the curve is as follows.

Starting with a negative voltage the surface is in accumulation and there is a thin sheet of positive charge from the accumulated holes. The lines of the electric field in the oxide terminate in this layer of charge and the only capacitance is that of the oxide through which the electric field passes. As the negative voltage decreases the accumulated holes disperse and a depletion layer is formed. Some of the lines of the electric field now terminate on the exposed impurity atoms and the capacitor dielectric

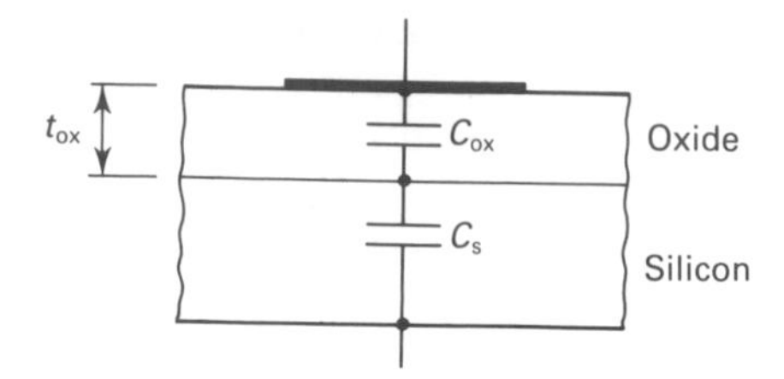

Figure 14.6 Equivalent circuit for the MOS capacitor.

now includes silicon as well as the oxide, that is, there is now a component of C_s in series with C_{ox}, and the total capacitance decreases.

At a certain value of the applied voltage the surface inverts and minority electrons start to collect at the surface. The lines of the electric field now start to terminate on the electrons rather than the acceptor atoms. Because the electric field now extends over a smaller distance the capacitance starts to increase. Thus the capacitance goes through a minimum and then increases as the inversion layer is established. Finally with a sufficiently large positive voltage a continuous layer of minority carriers exists at the surface and the value of the capacitance is again that of C_{ox}, as shown in curve (a) of Figure 14.7. However, this ability for the capacitance to increase after passing through a minimum depends on the minority carriers being able to follow the alternating signal which is used to measure the capacitance. This variation of capacitance is only obtained at low frequencies (<100 Hz). At very high frequencies (1 MHz) the minority carriers are unable to follow the ac signal and as a consequence the capacitance does not show an increase but remains at a low level for positive voltages, as shown in curve (b) of Figure 14.7.

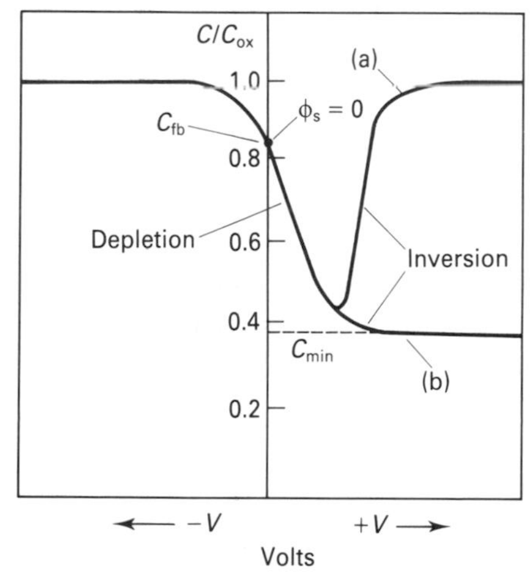

Figure 14.7 Capacitance–voltage variation for a MOS capacitor on a p-type semiconductor at (a) low frequency and (b) high frequency.

The variation of capacitance with voltage provides a useful means for measuring certain properties of the surface. The maximum capacitance provides a measure of the oxide thickness if the dielectric constant and the area of the electrode are both known. The minimum capacitance measured at 1 MHz enables the impurity concentration in the substrate to be determined as:

$$N_{sub} = \frac{4\phi_f}{q\varepsilon_o\varepsilon_{si}}\left(\frac{C_{s\,min}}{A}\right)^{1/2}$$

$$\phi_f = \frac{kT}{q}\ln\left(\frac{N_{sub}}{n_i}\right)$$

where $C_{s\,min}$ is the minimum value of space charge capacitance and A is the area of the metal electrode. The space charge capacitance is given by:

$$C_{s\,min} = \frac{C_{ox}C_{min}}{C_{ox} + C_{min}}$$

where C_{min} is the minimum value of the measured capacitance.

An important parameter is the voltage required to produce the flat-band condition. The capacitance under this condition is given by:

$$C_{fb} = \frac{C_{ox}C_{sfb}}{C_{ox} + C_{sfb}}$$

where C_{sfb} is the capacitance associated with the space charge region when the energy bands are flat up to the silicon–oxide interface. The capacitance under this condition can be defined in terms of the Debye length L_D, that is:

$$\frac{1}{C_{fb}} = \frac{1}{C_{ox}} + \frac{1}{C_{sfb}}$$

$$= \frac{t_{ox}}{A\varepsilon_o\varepsilon_{ox}} + \frac{L_D}{A\varepsilon_o\varepsilon_{si}}$$

$$C_{fb} = \frac{A\varepsilon_o\varepsilon_{ox}}{t_{ox} + (\varepsilon_{ox}/\varepsilon_{si})L_D}$$

where

$$L_D = \sqrt{(kT\varepsilon_o\varepsilon_{si})/(q^2N_{sub})}$$

The flat-band voltage is that bias voltage required to provide a measured value of capacitance equal to the calculated value of flat-band capacitance as given above.

The oxide charge can be determined as follows:

$$Q_{ss} = \frac{C_{ox}}{A}(\phi_{ms} - V_{fb})$$

where A is the area of the metal electrode and ϕ_{ms} the work function of the metal with respect to the silicon.

The presence of mobile charge can be detected by subjecting the MOS capacitor to temperature-bias stress and comparing the $C-V$ curves to those obtained at room temperature. The shift in the flat-band voltage between the two curves provides a measure of the mobile charge.

The variation of the capacitance with voltage for the MOS capacitor together with the above relationships for $C_{s\,min}$, C_{fb} and C_{ox} form the basis of microprocessor controlled measurement equipment for evaluating the oxide–semiconductor interface. $C-V$ plots are important for monitoring the manufacturing process for MOS integrated circuits.

14.7 Summary

An understanding of the conditions which exist at the interface of silicon and its native oxide are important for a complete understanding of the operation of MOS devices. The theoretical variation of the capacitance of the MOS capacitor provides a useful means for observing the effects that different processes may have on the surface conditions, for example the effects of impurities in the oxide or the effects of radiation.

Problems

1. Estimate the inversion voltage for an ideal silicon surface for which $\phi_{ms} = 0$ and $Q_{ss} = 0$, if the impurity in the silicon substrate is p-type with a concentration of 5×10^{16} atoms/cm^3 and if the substrate is covered with 80 nm of silicon oxide. Assume that $\varepsilon_{ox} = 3.9$ and $\varepsilon_{si} = 11.7$.
[3.34 V.]

2. If for the silicon substrate in Problem 1 the voltage on the metal electrode is twice the inversion voltage, estimate the number of charges which collect in the inversion layer.
[$\sim 9 \times 10^{11}$/cm^2.]

3. An n-type silicon substrate has an impurity concentration of 5×10^{15} atoms/cm^3 and a surface-state charge density (Q_{ss}/q) of 3×10^{11}/cm^2. If the metal electrode is aluminium with an aluminium to silicon work function given by $\phi_{ms} = -0.6 + \phi_f$, determine the inversion voltage for:
(a) a thin oxide of 80 nm;
(b) a thick oxide of 1000 nm.
[-2.8 V, -24 V.]

4. A p-type silicon substrate has an impurity concentration of 1×10^{16} atoms/ cm^3. For ⟨111⟩ silicon the surface-state charge density at the silicon–silicon oxide interface is 5×10^{11}/cm^2 and for ⟨100⟩ silicon it is 9×10^{10}/cm^2. Determine the inversion voltages for both types of silicon for the following oxides:
(a) 100 nm,
(b) 1200 nm,
where the work function between aluminium and silicon is given by $\phi_{ms} = -0.6 - \phi_f$.
[-1.13 V, $+0.72$ V.]
[-10.86 V, $+11.38$ V.]

References

[1] L. M. Terman (1962), 'An investigation of surface states at a silicon/silicon dioxide interface employing metal-oxide–silicon diodes', *Solid State Electronics*, **5**, 285.

CHAPTER 15

The MOS transistor

15.1 Introduction

The Metal-Oxide-Semiconductor-Transistor (MOST) is the most important device for very large-scale integrated circuits. It relies on the surface effects which are described for the MOS capacitor in Chapter 14 and in particular the formation of an inversion layer at the interface between silicon and silicon-oxide. The inversion layer is created by a voltage which is applied to a gate electrode placed over the oxide. This creates a conducting channel through which current can flow between source and drain regions which are formed at either end of the channel. The MOST is a high-impedance voltage operated device, unlike the bipolar transistor which is a low-impedance current operated device. A particular advantage of the MOST for VLSI circuits is that the transistors do not have to be isolated from each other in the way that bipolar devices require isolation. As result the number of devices per unit area can be very large.

The analysis of the device is greatly simplified if the length of the channel is assumed to be much larger than the sum of the depletion layer widths of the source and drain. For modern day devices, however, this assumption is no longer valid, but the long-channel model still provides a useful starting point for the analysis. The effect of short channels may then be considered with reference to the long-channel model. A number of complex mathematical models which take account of short-channel effects have been developed for use in computer simulation.

The channel length has been reduced from about 10 μm for early devices produced during 1960 to 1970 to less than 1 μm in 1990. There has been a corresponding increase in the number of components per integrated circuit from a few thousand in 1970 to more than a million in 1990.

The fact that the gate electrode is separated from the channel region by a very good insulator means that the capacitor formed by the gate and the channel region is a very low loss device which is capable of retaining charge for long periods of time. This feature lends itself to the formation of a memory device and has resulted in the development of very large dynamic random access memories (DRAMs). The single transistor DRAM cell uses a transistor as a switch and a MOS capacitor as the storage element. However, a further development of this charge-retaining property of the MOS capacitor is the development of the electronically programmable read-only memory (EPROM), and, more recently, the electronically erasable and programmable read-only memory (EEPROM). While the charge on the DRAM cell must be refreshed periodically, the memory cell is volatile; the EPROMs are non-volatile and retain charge for very long periods. These devices could lead to the solid-state version of the computer hard disk as a read/write mass storage medium.

15.2 Basic device characteristics

The basic structure of the MOS transistor is illustrated in Figure 15.1. It is a four-terminal device with a source, gate, drain and substrate, although in the initial analysis to follow, the substrate is not used as an active electrode. For an n-channel device the substrate is a p-type semiconductor and the source and drain are n-type regions formed by diffusion or ion implantation. The opposite polarities occur in an n-channel device. The substrate between the source and drain is covered with a thin layer of oxide and the gate electrode is deposited on this oxide.

The gate is either metal (aluminium), or a heavily doped polysilicon or a combination of polysilicon and a silicide. The important device parameters are the channel length (L), which is the distance between the edges of the metallurgical junctions of the source and drain, the channel width (W), the oxide thickness (t_{ox}) and the substrate impurity concentration (N_a or N_d).

The gate is equivalent to the metal electrode of a MOS capacitor described in Chapter 14. When a positive voltage is applied to the gate with respect to the source for an n-channel device, an inversion layer is formed as electrons are attracted to the surface. The n-type inversion layer connects the n-type source to the n-type drain and application of a positive voltage to the drain causes a current to flow between source and drain. For a given drain voltage the magnitude of the current is determined by the

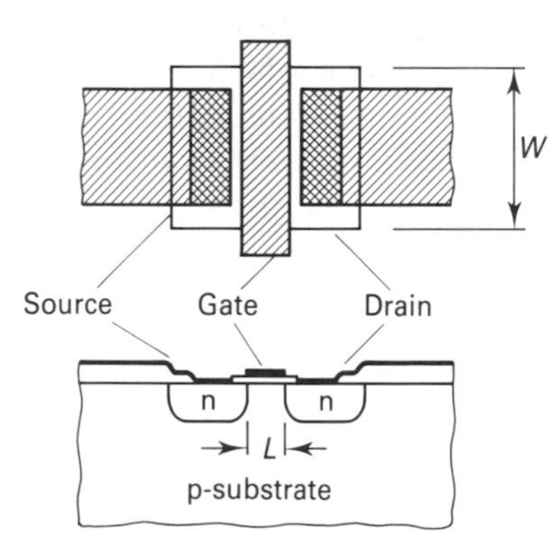

Figure 15.1 Plan view and cross-section of an n-channel MOS transistor.

gate voltage. The voltage applied to the substrate with respect to the source also influences the channel current. The source is often connected to the substrate, but for many circuit configurations this is not possible and account must be taken of the effect of substrate bias on the electrical characteristics.

The current–voltage characteristics of the device are shown in Figure 15.2. For $V_{GS} = 0$ the drain current is very small and is equal to the leakage current of the reverse-biased pn junction of the drain. When a sufficiently large voltage has been applied to the gate an inversion layer forms and current flows between the source and the drain.

With increasing drain voltage, for a fixed gate voltage, the drain current increases linearly but then saturates and remains constant for further increase of V_{DS}. A family of curves is obtained for different values of V_{GS}. This device is known as an enhancement MOST to describe the creation or enhancement of the channel beneath the gate.

There is an alternative mode of operation, which is illustrated in Figure 15.3. For this device the n-type channel is formed during the fabrication process and is shown in Figure 15.3(a). Current flows when $V_{GS} = 0$ since the conducting channel already exists. The application of a negative voltage to the gate drives electrons out of the channel and the current decreases, as shown by the $I-V$ curves in Figure 15.3(b). This device is known as a depletion device since the channel is depleted of carriers by the application of the gate voltage.

If the p-type substrate is replaced with an n-type substrate then p-channel enhancement and depletion MOST devices can be fabricated.

15.3 Long-channel approximation

It can be seen from the $I-V$ curves shown in Figure 15.2 that for low values of V_{DS} the drain current I_D is proportional to V_{DS} but beyond a certain level the current saturates. These relationships can be explained qualitatively with reference to the diagrams in

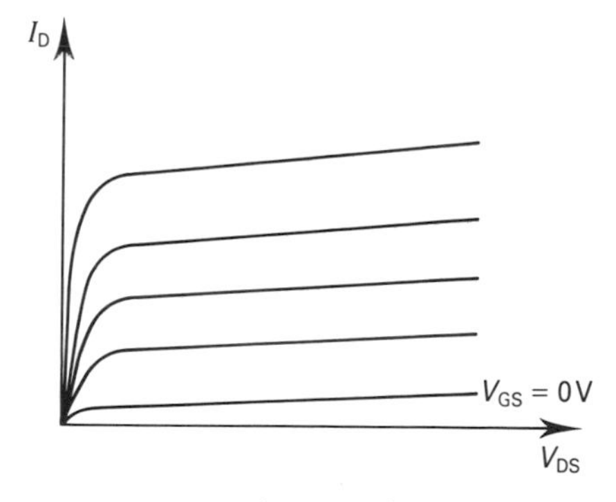

Figure 15.2 $I-V$ characteristics for an enhancement-mode MOS transistor.

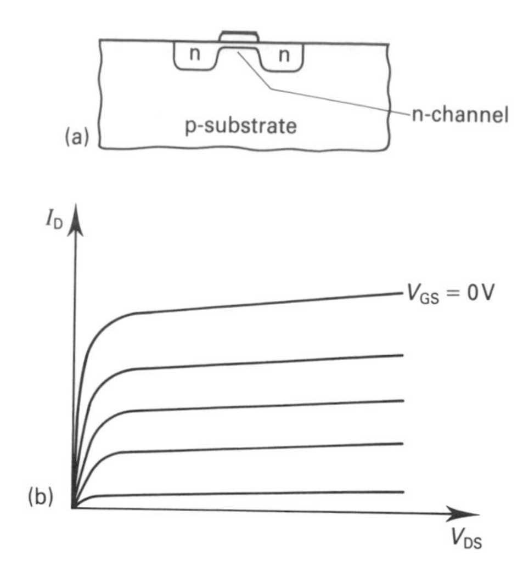

Figure 15.3 (a) Cross-section of a depletion-mode MOS transistor and (b) the $I-V$ curves.

Figure 15.4. With the gate voltage greater than the value required to create the inversion layer ($V_{GS} > V_T$), for low values of V_{DS}, shown in Figure 15.4(a), the channel extends from source to drain and the current path is resistive. For this region I_D is proportional to V_{DS} and there is a voltage drop along the channel, from 0 V at the source end to V_{DS} at the drain end. The channel voltage at the silicon–silicon-oxide interface is the difference between the voltage on the gate and that in the silicon substrate. Because of the change of voltage along the length of the channel, the voltage between the gate and substrate at the drain end is less than at the source end. As the drain voltage increases, the gate-to-substrate voltage at the drain end diminishes until eventually it is less than V_T. At this point the channel disappears or is 'pinched off'. This condition is shown in 15.4(b), where the value of V_{DS} is defined as V_{Dsat}. Intuitively, this value of voltage is defined by:

$$V_{GS} - V_{Dsat} = V_T$$

Although the channel does not exist at the drain for $V_{DS} > V_{Dsat}$ current still continues to flow because carriers arriving at the drain end of the channel exit into the depletion region of the drain–substrate junction and are swept by the electric field into the drain.

The current, however, is now controlled by Ohm's law in the region between the source and the drain end of the channel and not by the drain voltage. As the drain voltage increases further the additional voltage appears across the high-impedance of the drain-junction depletion layer rather than across the low resistance of the channel. There will, however, be a slight increase in I_D caused by the reduction of the effective channel length from L to L' as the depletion layer extends towards the source. This is

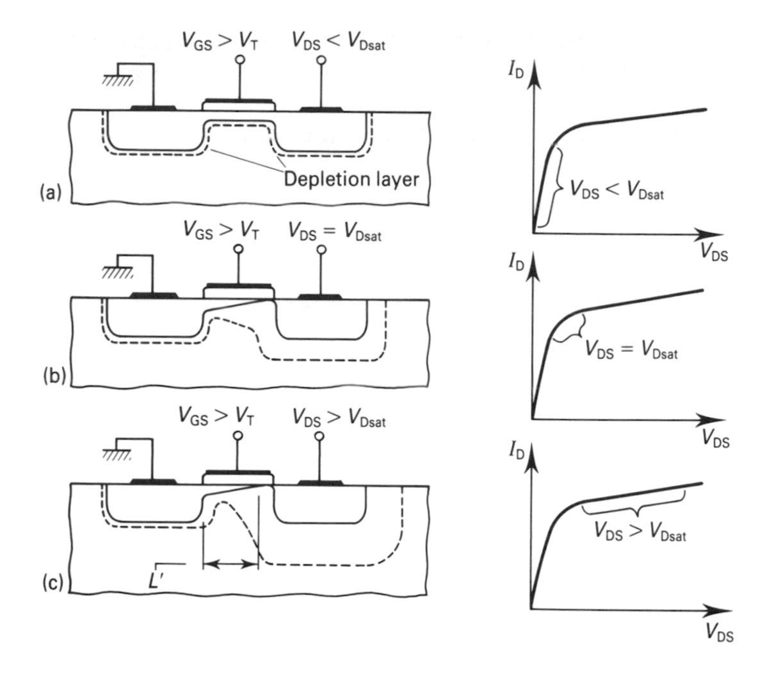

Figure 15.4 (a) MOST operating below pinch-off in the linear region, (b) on the verge of saturation with pinch-off occurring at the drain end of the channel and (c) in saturation with the effective channel length less than the geometrical length.

shown in Figure 15.4(c), where the length of the channel is reduced. This small increase in the drain current accounts for the finite slope of the $I-V$ characteristics in the saturation region.

An analytical expression can be obtained for the $I-V$ characteristics with the simplifying assumption that the charge in the channel is uniform along the length of the channel. This assumption holds if the channel length is long with respect to the depletion layer associated with the source and drain junctions. Under these conditions the charge of the carriers in the channel is related to the gate voltage and the gate capacitance as:

$$Q_C = C_{ox} V_G$$

where V_G is the effective gate voltage which supports the charge in the channel and for $V_{DS} = 0$ is given by:

$$V_G = V_{GS} - V_T$$

where V_T is the threshold voltage required to create the inversion layer which forms the channel.

When V_{DS} is applied there is a voltage drop along the channel, from 0 V at the source to V_{DS} at the drain. As a consequence the effective gate voltage V_G varies along the length of the channel. The relationship between charge and capacitance can be described in terms of an average value of source-drain voltage as:

$$Q_C = C_{ox}\left[(V_{GS} - V_T) - \frac{V_{DS}}{2}\right]$$

where $V_{DS}/2$ represents an average value of the effect of the drain voltage.

Current is defined as the rate of flow of charge as:

$$I_D = \frac{\delta Q_C}{\delta t}$$

The time to traverse the channel is:

$$t = \frac{length}{velocity}$$

where the velocity (v) is related to the mobility and the electric field ($\mathbf{E}_{DS}$) as:

$$v = \mu \mathbf{E}_{DS}$$

where $E_{DS} = V_{DS}/\text{length}$. Thus the transit time is:

$$\tau_{SD} = \frac{L^2}{\mu V_{DS}}$$

and the drain current is:

$$I_D = \frac{C_{ox}\mu}{L^2}\left[(V_{GS} - V_T)V_{DS} - \frac{V_{DS}^2}{2}\right]$$

The capacitance is expressed as:

$$C_{ox} = \frac{\varepsilon_o \varepsilon_{ox} WL}{t_{ox}}$$

and the drain current is rewritten as:

$$I_D = \frac{KW}{L}\left[(V_{GS} - V_T)V_{DS} - \frac{V_{DS}^2}{2}\right]$$

where $K = \varepsilon_o \varepsilon_{ox}/t_{ox}$ and represents a process-dependent gain factor for the MOST. The gain for the device is given by $\beta = KW/L$.

This expression for the drain current only applies for small values of V_{DS}, that is, below pinch-off. As V_{DS} approaches V_{Dsat} the slope $\delta I_D/\delta V_{DS}$ decreases until beyond V_{Dsat} the ideal slope is zero, that is:

$$\frac{\delta}{\delta V_{DS}} \frac{KW}{L}\left[(V_{GS} - V_T)V_{DS} - \frac{V_{DS}^2}{2}\right] = 0$$

The solution of this equation is:

$$V_{DS} = V_{GS} - V_T$$

Substituting this value of V_{DS} into the equation for I_D gives:

$$I_{Dsat} = \frac{KW}{2L}(V_{GS} - V_T)^2$$

This equation is independent of V_{DS} which means that I_D remains constant beyond the point at which $V_{DS} = V_{Dsat}$. In practice this is not the case, as is illustrated in Figure 15.4(c). As was described qualitatively above, the increase in I_D can be attributed to the reduction in the effective length of the channel by the encroachment of the drain depletion layer. From simple pn junction theory the width of the depletion layer is:

$$x_d = \sqrt{(2\varepsilon_o \varepsilon_{si} V_{DS})/(qN_a)}$$

where N_a is the impurity concentration of the p-type substrate. Thus the effective length of the channel is:

$$L' = L - x_d$$

and the drain current in the saturation region is:

$$I_{Dsat} = \frac{KW}{2L'}(V_{GS} - V_T)^2$$

The equations hold for both enhancement and depletion devices but with a change in the value of the threshold voltage to account for the different characteristics of the two devices. They are sufficiently accurate for long-channel devices to provide reasonable agreement between theory and experiment, but with channel lengths of less than 3 μm the simplified approach adopted above no longer holds and a much more detailed analysis is required. The starting point for such analysis would be Sze, Chapter 8[1].

15.4 Threshold voltage

When a voltage is applied to the gate with respect to the source, the surface potential (ϕ_s) in the semiconductor is altered. The threshold voltage is defined as the voltage required to make $\phi_s = 2\phi_f$, that is, to create an inversion layer. For an n-channel device the threshold voltage is (Section 14.5):

$$V_T = \phi_{ms} - \frac{Q_m}{C_{ox}} + 2\phi_f + \sqrt{\frac{2\varepsilon_o \varepsilon_{si} q N_a 2\phi_f}{C_{ox}}}$$

For an enhancement device $V_T \approx 1$ V, while for a depletion device $V_T \approx -4$ V. The polarities change for p-channel devices.

The above equation takes no account of the substrate bias. When this is applied, as shown in Figure 15.5, the width of the depletion layer is increased beyond that which would have occurred with gate bias alone.

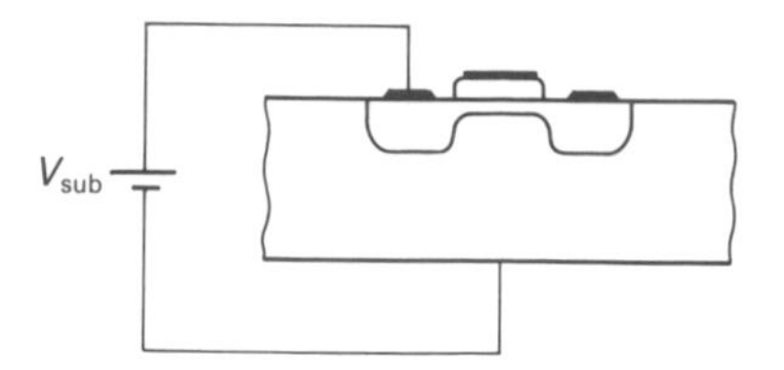

Figure 15.5 Application of substrate bias.

The number of exposed acceptor impurity atoms within the depletion layer, and hence the charge, is increased when V_{sub} is applied. The voltage across the depletion layer is given by:

$$\frac{Q_b}{C_{ox}} = \sqrt{\frac{2\varepsilon_o \varepsilon_{si} q N_a (2\phi_f + V_{sub})}{C_{ox}}}$$

The threshold voltage with substrate bias may be expressed as:

$$V_T = V_{T(0)} + \gamma \sqrt{V_{sub}}$$

where $V_{T(0)}$ is the threshold voltage with $V_{sub} = 0$ and

$$\gamma = \frac{t_{ox}}{\varepsilon_o \varepsilon_{ox}} \sqrt{2\varepsilon_o \varepsilon_{si} q N_a}$$

Gamma (γ) is a process-dependent variable, for example if $t_{ox} = 80\,\text{nm}$, $N_a = 5 \times 10^{15}$ atoms/cm^3, $\varepsilon_{ox} = 4$ and $\varepsilon_{si} = 11.7$, then $\gamma = 0.9$, and a 5 V substrate bias would result in a 2 V shift in the threshold voltage.

It is common practice to use ion implantation to adjust the threshold voltage to satisfy the particular circuit requirements. For example, for NMOS circuits there is an enhancement-mode driver and a depletion-mode load. The standard manufacturing process for n-channel transistors produces enhancement devices with a threshold voltage of about 1 V. Ion implantation is used to change the threshold voltage to the negative value required for a depletion device. It may also be necessary to change the value required for an enhancement device if required. The possible changes are illustrated in Figure 15.6.

An estimate of the number of ions required can be obtained from the expression for the threshold voltage with implanted ions:

$$V_T = V_{T(0)} \pm \frac{q N_I}{C_{ox}}$$

where $V_{T(0)}$ is the threshold voltage with no implant and N_I is the number of implanted ions per cm^2.

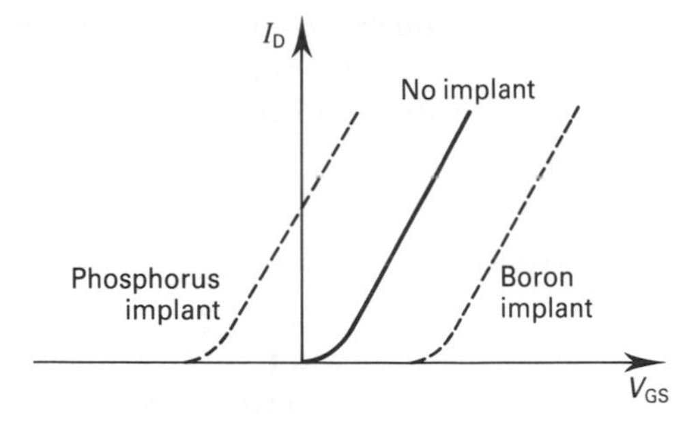

Figure 15.6 Threshold voltage adjustment using ion implant for n-channel devices.

15.5 Transconductance

The transconductance (g_m) is expressed as:

$$g_m = \left.\frac{\delta I_D}{\delta V_{GS}}\right|_{V_{DS} = \text{constant}}$$

The active region for the transistor is the saturation region for which the current is:

$$I_D = \frac{KW}{2L}(V_{GS} - V_T)^2$$

and the transconductance is:

$$g_m = K\frac{W}{L}(V_{GS} - V_T)$$

or

$$g_m = \frac{\mu\varepsilon_o\varepsilon_{ox}}{t_{ox}}\frac{W}{L}(V_{GS} - V_T)$$

It can be seen that g_m is proportional to the mobility and inversely proportional to the oxide thickness. Thus for high values of g_m the gate oxide should be as thin as possible. The limitation is the breakdown voltage of the oxide, but typical values for t_{ox} are 20–50 nm. For silicon the surface mobility for electrons is about three times larger than the hole mobility and, therefore, n-channel devices will have the larger g_m for the same dimensions as a p-channel device. The g_m is also proportional to the aspect ratio W/L and there are obvious advantages in making L as small as possible. The channel length is usually determined by the smallest feature which can be reproduced with the current photolithographic process. For advanced technology devices this may be less than 0.5 μm.

The g_m for a MOS device is much less than that for a bipolar device of similar area. At 1 mA the g_m for a bipolar transistor is I_E/V_T and has a value of 38 000 siemen. For an n-channel MOS transistor with 20 nm of gate oxide, and $W/L = 1$, the g_m would be approximately 50 siemen.

15.6 Figure of merit

The input impedance of the MOS transistor between the gate and source can be represented by a capacitor (C_g). The output current is $g_m v_g$, where v_g is the small-signal input voltage. A simplified equivalent circuit for the transistor which is based on these two components is shown in Figure 15.7. Under normal conditions the small-signal output current is much greater than the input current. However, as the frequency increases, a point will be reached when the input current equals the output current. The frequency at which this occurs can provide a figure of merit for the transistor. Thus:

$$i_i = g_m v_g$$

or

$$v_g(2\pi f_T C_g) = g_m v_g$$

and

$$f_T = \frac{g_m}{2\pi C_g}$$

$$f_T = \frac{\mu}{2\pi L^2}(V_{GS} - V_T)$$

This equation shows that the highest frequency is obtained with very short n-channel devices. Since switching speed is related to maximum frequency, then the same conditions apply if high switching speeds are required.

15.7 Short-channel effects

When the channel length is reduced to improve the switching speed and increase the packing density (by reducing the overall size of the device), the depletion layers

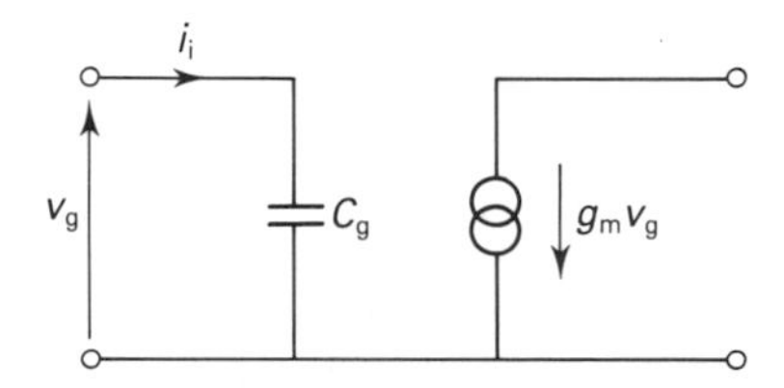

Figure 15.7 Simplified equivalent circuit of MOST.

associated with the source and drain junctions cannot be ignored. The potential distribution along the channel is no longer one-dimensional with a single component parallel to the current flow. Instead, there are two components, a lateral component and a transverse component. This two-dimensional field degrades the behaviour of the device below the threshold voltage (sub-threshold) and introduces a dependence of the threshold voltage on channel length and the biasing voltages.

The mobility becomes field-dependent, with the transverse component producing a reduction of the surface mobility and a resultant lowering of the velocity saturation. Velocity saturation limits the value of the drain saturation current.

If a reduction in channel length is not accompanied by a corresponding reduction in the drain voltage (for example, if V_{DD} is maintained at 5 V), then the increased electric field can result in increased carrier multiplication as carriers are accelerated through the depletion layer at the drain end of the channel. This carrier generation results in a substrate current which flows through the bulk to the substrate contact. The product of this current and the bulk resistance can forward bias the source–substrate junction of a parasitic npn transistor formed by the source, substrate and drain of the n-channel MOST which can lower the breakdown voltage of the drain.

Electrons which are accelerated by the electric field may acquire sufficient energy to be injected into the gate oxide. These electrons produce oxide charge and also interface traps. Both affect the threshold voltage and the transconductance. The electron charge may leak away when the field is removed, but the interface traps remain. As a consequence hot-electrons generated by high electric fields result in progressive deterioration with an increase in the threshold voltage and a reduction in the transconductance.

One approach to minimizing hot-electron effects is to reduce the electric field in the vicinity of the drain. This can be achieved by reducing the impurity concentration of the drain so that the depletion-layer extends both into the channel and into the drain, as illustrated in Figure 15.8. The depletion layer width for a step-junction is given by:

$$x_d = \sqrt{\frac{2\varepsilon_o \varepsilon_{si} V}{q}\left(\frac{N_a + N_d}{N_a N_d}\right)}$$

For a one-sided junction with $N_a \gg N_d$

$$x_d = \sqrt{\frac{2\varepsilon_o \varepsilon_{si} V}{q N_d}}$$

and where $N_a = N_d$

$$x_d = \sqrt{\frac{4\varepsilon_o \varepsilon_{si} V}{q N_d}}$$

For the same impurity concentration and voltage the width of the depletion layer is increased when the impurity concentration of the drain is similar to that of the substrate. With an increased width the electric field is reduced.

A lightly doped drain, however, would result in an unacceptably high series

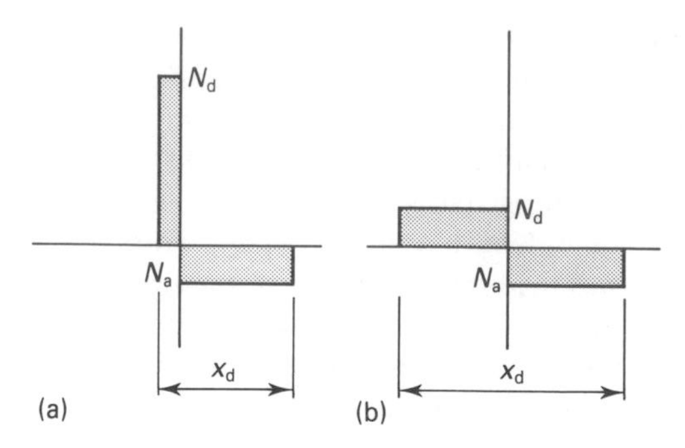

Figure 15.8 Depletion layers associated with (a) a one-sided step-junction and (b) a step-junction with approximately equal concentrations on either side of the junction.

resistance. This can be overcome by having a lightly doped region adjacent to the channel and a heavily doped region for the contact, as illustrated in Figure 15.9. This and other process developments have been used to minimize the problems associated with reduced dimensions.

15.8 Non-volatile MOS memory devices

The most widely used memory element is the RAM cell, with a single transistor used as a switch to select the cell, and a MOS capacitor to store charge to represent either a '1' or a '0'. However, the capacitor forms part of the circuit forming the memory element and is attached to the transistor, and as such is not completely isolated. When the select transistor is OFF, the impedance is high, but not so high as to prevent loss of charge. The memory is *volatile* and charge is lost after several tens of milliseconds and must be replaced by means of special circuits on the memory chips. When power is removed the memory (charge) is lost. The *non-volatile* memory retains the charge, and thus the contents of the memory cells, when the power is removed. There are two basic types of non-volatile memory, the EPROM and the EEPROM, but there are many variations of each type which have been developed by different manufacturers.

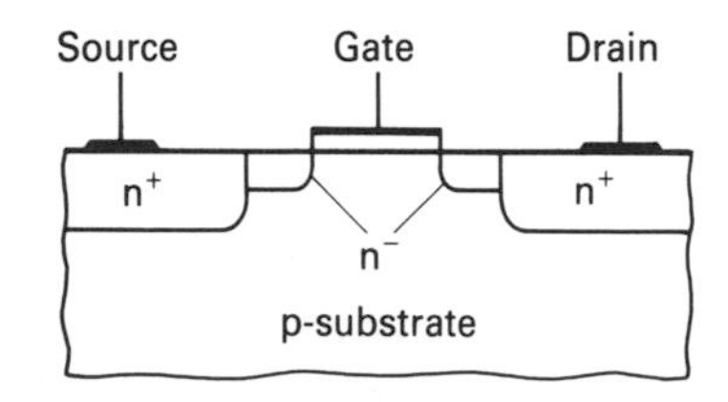

Figure 15.9 Cross-section of a MOST with lightly doped drain (LDD).

15.8.1 EPROM

In the EPROM the gate electrode is completely isolated from any other circuit elements, and any charge deposited on the gate is retained for very long periods of time – tens of years. The basic device is still a MOS transistor, as shown in Figure 15.10, but with an electrically floating gate plus a second control gate. The charge on the floating gate affects the band structure of the underlying silicon and thus the threshold voltage, so that with no charge, V_T may be small and the device behaves as a normal n-channel enhancement device. When negative charge is present the threshold voltage is shifted so that when a sensing voltage which is less than the threshold voltage is applied to the drain, the device does not conduct. The change in the threshold voltage, which is illustrated in Figure 15.11, provides the means for storing a '1' or a '0', with V_{T1} being the zero charge threshold and V_{T2} representing the charged condition. The most important distinguishing feature of the EPROM is that the floating gate is electrically isolated from any other part of the circuit and the charge is retained even when the power to the circuit is removed.

Charge is placed on the floating gate by avalanche injection of carriers from the

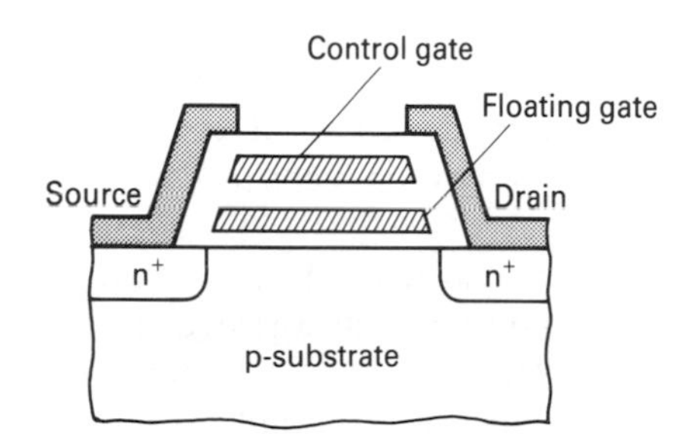

Figure 15.10 Cross-section of EPROM memory cell (FAMOS EPROM).

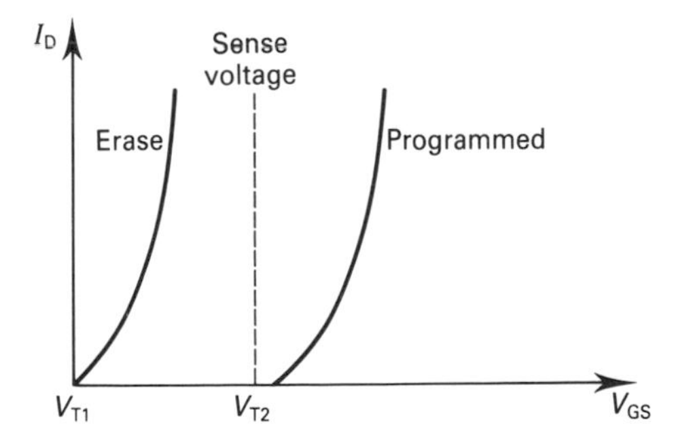

Figure 15.11 Change in the I_D: V_{GS} curves after programming.

depletion layer and into the oxide. In the floating gate avalanche injection MOS device (FAMOS) a polysilicon gate is placed over a short n-channel MOS transistor. When a positive voltage of suitable magnitude is placed on the drain and the control gate, the large electric field at the drain end of the channel generates hot-electrons by impact ionization – the carriers are accelerated in the depletion layer, strike silicon atoms and release other electrons. Some of these secondary electrons have sufficient energy (are hot-electrons) to overcome the barrier presented by the oxide and because of the field generated by the positive voltage on the control gate are attracted towards the control gate. These electrons and the associated negative charge come to rest on the floating gate. When the voltage is removed from the control gate this charge produces its own electric field which drives electrons from the silicon surface, thus making the formation of a conducting channel between the source and drain more difficult; the threshold voltage is increased.

The charge is removed, or erased, by the application of high-intensity ultra-violet light which excites the electrons in the oxide surrounding the floating gate so that it conducts and the charge on the gate leaks away to the substrate. The removal of charge is not selective and the ultra-violet is directed on to the complete memory chip through a window in the packaging material.

After programming, when a voltage of perhaps 20 V is applied to the control gate and 10 V to the drain, the state of the device can be read by applying, say, 5 V to the drain. This voltage is insufficient to generate hot-electrons and will not influence the charge on the gate. An erased device conducts, equivalent to a '1', but after programming the device continues to conduct if a '1' is present, but will not conduct if a '0' has been programmed.

During the ultra-violet erase all of the devices return to the low-threshold voltage state. This is the erase state and represents a stored '1'. During programming the '1' state may be retained or it may be changed to a '0' state. If the '1' is to be retained then no charge is required on the floating gate and during programming no voltage is applied to the drain. To program a '0' it is necessary to transfer charge to the floating gate which is achieved by applying the programming voltage to the drain and control gate.

The relationship between the voltage on the control gate (V_G), and the charge (Q) on the floating gate can be obtained by the application of Gauss's law. If the thickness of the oxide between the silicon and the floating gate is t_{ox1} and between the two gates is t_{ox2}, then by Gauss's law:

$$\varepsilon_o \varepsilon_{ox1} E_1 = \varepsilon_o \varepsilon_{ox2} E_2 + Q$$

and

$$V_G = V_1 + V_2 = t_{ox1} E_1 + t_{ox2} E_2$$

By substitution:

$$V_G = t_{ox1} E_1 - \frac{t_{ox2} Q}{\varepsilon_o \varepsilon_{ox2}} + \frac{t_{ox2} \varepsilon_{ox1} E_1}{\varepsilon_{ox2}}$$

and

$$E_1 = \frac{V_G}{\left(t_{ox1} + t_{ox2}\frac{\varepsilon_{ox1}}{\varepsilon_{ox2}}\right)} + \frac{Q}{\left(\varepsilon_o \varepsilon_{ox1} + \varepsilon_o \varepsilon_{ox2}\frac{t_{ox1}}{t_{ox2}}\right)}$$

or if $\varepsilon_{ox1} = \varepsilon_{ox2} = \varepsilon_{ox}$, then the field in the thin oxide is:

$$E_1 = \frac{V_G}{t_{ox1} + t_{ox2}} + \frac{Q}{\varepsilon_o \varepsilon_{ox}\left(1 + \frac{t_{ox1}}{t_{ox2}}\right)} \text{ V/cm}$$

The amount of charge which accumulates on the floating gate is simply the product of current density and time. Typically, the programming voltage may have to be maintained for several tens of microseconds to allow sufficient charge to accumulate on the gate. If current flow is confined entirely to the thin oxide, then the charge is:

$$Q(t) = \int_0^t J_1(E_1)\,dt\; C/\text{cm}^2$$

where the current density J_1 is produced by the hot-electron current.

The threshold voltage which is observed when the voltage is removed from the control gate is:

$$V_T = V_{T(0)} + \frac{Q}{C_{ox1}}$$

where $V_{T(0)}$ is the zero-charge threshold, as given in Section 15.4 above. The change in the threshold is:

$$\Delta V_T = Q\frac{t_{ox1}}{\varepsilon_o \varepsilon_{ox1}}$$

The change in V_T is a function of the time of the programming step, the voltage applied to the control gate and the current which flows through the oxide.

15.8.2 EEPROM

The disadvantage of the EPROM is the need for ultra-violet light to erase the contents of the memory cells. This process is not selective and applies to all the cells. It is a read-only memory. To provide the ability to be selective for both 'erase' and 'write' it is necessary to provide a means of erasing the contents of individual cells electrically instead of optically. In the conventional EPROM the oxide between the floating gate and the silicon substrate is typically 30–50 nm thick and very high energy electrons (hot-electrons) are required to overcome this barrier. These electrons are generated within the depletion layer of the drain–substrate pn junction. This effect can be used to drive electrons from the silicon on to the floating gate, but it offers no solution for the reverse process of transferring electrons from the gate to the silicon. Another process is required.

In one version of the EEPROM a small section of the oxide over the drain diffusion is reduced in thickness to 10 nm. The floating gate thin oxide EEPROM or FLOTOX EEPROM is shown in Figure 15.12.

For oxides of the order of 10 nm in thickness it is possible to generate electric fields in the oxide, which are sufficient to create a finite probability for electrons in the conduction band of either the floating gate or the silicon substrate to tunnel through the oxide barrier. This current is described by the Fowler–Nordheim equation:

$$\mathbf{J} = A\mathbf{E}^2 \exp\left(\frac{-B}{\mathbf{E}}\right)$$

where A and B are constants, and $\mathbf{E}$ is the electric field.

Possible values for the two constants are[2]:

$$A \approx 1.8 \times 10^{-7}\ \mathrm{A/V^2}$$

$$B \approx 2.1 \times 10^{8}\ \mathrm{V/cm}$$

The band diagram through the thin oxide region is shown in Figure 15.13. Programming and erasing are now achieved by application of the appropriate voltages to the control gate and the drain.

Programming is accomplished by application of a positive voltage to the control gate which is sufficient to generate a current through the tunnel oxide, but not through the oxide which separates the control gate from the floating gate, and also a positive voltage to the drain. The charge accumulates on the floating gate, as illustrated in Figure 15.13(a), by a combination of both tunnel and hot-electron currents. In the programmed state the charge on the floating gate repels electrons from the surface of the silicon and this increases the threshold voltage for the MOS transistor as for the EPROM. This is illustrated in Figure 15.13(b) by the upward bending of the bands at the silicon–silicon-oxide interface.

In the erase mode a negative voltage is applied to the control gate and the drain is

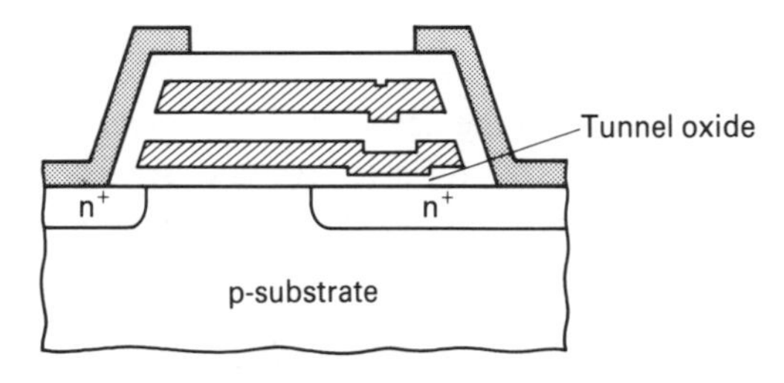

Figure 15.12 Cross-section of an EEPROM with tunnel oxide over the drain diffusion (FLOTOX EEPROM).

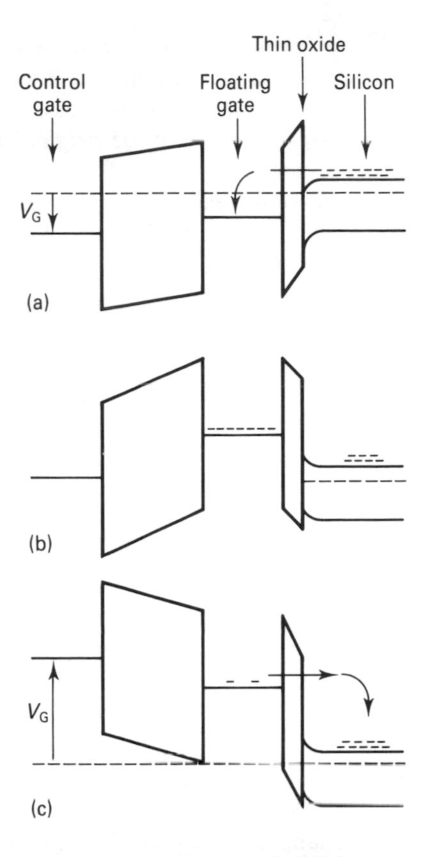

Figure 15.13 Energy-band diagram of an EEPROM with (a) showing the programming mode, (b) showing the charge on the floating gate and (c) showing the erase mode.

grounded, as illustrated in Figure 15.13(c) (or alternatively the control gate is grounded and a positive voltage is applied to the drain with the source floating). The electric field that is developed across the tunnel oxide over the drain region is now in the opposite direction to the programming mode and electrons flow from the gate to the drain.

The floating gate is the memory element for both the EPROM and the EEPROM, but an additional 'select' transistor is required for the EEPROM because during the erase cycle when a large voltage is applied to the drain the absence of a select transistor would result in the voltage being applied to the drains of other transistors connected to the same row or column of a two-dimensional array. The select transistor is used as an ON/OFF switch to select the memory cell to be erased. Thus the complete memory cell comprising the floating gate transistor and the select transistor occupies more area than the EPROM.

There are a number of variations of the EEPROM. In some the tunnel oxide is placed over the source diffusion rather than the drain, while in others the thin oxide region extends over the complete channel region and is similar in appearance to the EPROM, which is shown in Figure 15.10. The hot-electron current now flows through the thin oxide rather than the thicker oxide of the original FAMOS device shown in Figure 15.10 or the FLOTOX device shown in Figure 15.12. Because the oxide is thinner the current is larger and the increased current reduces the programming time. This type of device is used for the 'flash EEPROM'. However, the reliability of the device is currently less than the FLOTOX structure because of the damage produced by the high-energy hot-electrons which create traps (damage) in the oxide. Because the oxide is thin the effect of the traps is much greater than in the conventional EPROM, where the oxide is typically 30–50 nm thick. The traps are generated each time the device is programmed and after a few hundred cycles the effect of the trapped charge in the oxide is greater than the charge on the floating gate and the cell remains in the '0' state.

15.9 Future developments

Much of the present effort in developing the technology is directed towards reducing the length of the channel. Channel lengths of between 0.3 and 0.2 μm have been achieved in the research laboratories and will soon be used for production. Some of the first circuits to use such small geometries will be random access memory chips. The size of memory arrays keeps doubling and the next generation will be 64 Mbytes, with 256 Mbytes in the planning stage. Each reduction in size is accompanied by a revision of the complete process. Vertical dimensions are also affected when the lateral dimensions change and these changes usually involve changes to the impurity profiles and device properties.

Apart from the technological difficulties of reproducing such small dimensions accurately, there are also the problems of developing suitable electrical models for such short-channel lengths. Many of the parameters become dependent on channel length, for example mobility and threshold voltage, and developing accurate analytical models becomes increasingly difficult as the dimensions are reduced. Many workers are developing numerical methods to analyze these devices, and two such programs are MINIMOS and PISCES (or MEDICI). These programs, which make use of Poisson's equation and the continuity equation, allow workers to investigate the effects of changes to the geometry of the device before the device is fabricated. However, the numerical methods are computationally intensive and much work remains to be done before such programs can be used to simulate a complete circuit.

Problems

1. The following parameters apply to an n-channel MOS process:

gate oxide thickness (t_{ox}): 80 nm
substrate doping (N_a): $5 \times 10^{15}/\text{cm}^3$
work function (ϕ_{ms}): 0.6 V
surface-state density (Q_{ss}/q): $5 \times 10^{10}/\text{cm}^2$

$\varepsilon_{si} = 11.7$ $\varepsilon_{ox} = 3.9$
$\mu n = 580\,\text{cm}^2/(\text{V s})$ $\mu p = 230\,\text{cm}^2/(\text{V s})$

(a) determine the unimplanted threshold voltage for zero substrate bias.
(b) determine the number of implanted impurity ions required to create a threshold voltage of -4 V.
[0.66 V, $1.25 \times 10^{12}/\text{cm}^2$.]

2. If the saturation drain current of the device in Problem 1(a) is given by:

$$I_D = \frac{\beta}{2}(V_{GS} - V_{TE})^2$$

estimate I_D when $V_{GS} = 5$ V if the W/L aspect ratio for the channel region is unity.
[54 μA.]

3. A MOS transistor with the properties described in Problem 1 has a channel length of 3 μm and a width of 30 μm.

(a) determine the transconductance g_m for the enhancement device described in Problem 1(a) for $V_{GS} = 2.5$ V assuming that the drain current is given by:

$$I_D = \frac{KW}{L}\frac{(V_{GS} - V_{TE})^2}{2}$$

(b) estimate the maximum frequency of operation.
[4.6×10^{-4} μmho, 1.9 GHz.]

References

[1] S. M. Sze (1981), *Physics of Semiconductor Devices*, 2nd edn, Chichester: Wiley.
[2] S. Keeney, R. Bez, D. Cantarelli, F. Piccinini, A. Mathewson, L. Ravazzi and C. Lombardi (1992), 'Complete transient simulation of flash EEPROM devices', *IEEE Trans. Electron. Devices*, **39** (12), 2750–7.

CHAPTER 16

The MOS inverter

16.1 Introduction

The simplest form of MOS switching circuit is an inverter consisting of a transistor switch and a resistive load. The input is applied to the gate to turn the device ON and OFF and the output is taken from the drain. When the transistor is ON, current flows through the load, producing a voltage drop which reduces the output to a low value. That is, a logic 'high' or '1' applied to the input produces a logic 'low' or '0' at the output.

The first MOS transistors to be developed were p-channel enhancement devices with threshold voltages of between 3 V and 5 V which required power supplies in excess of 10 V for successful operation. A transistor was used as a load, with the gate connected to the positive supply, but with enhancement devices there is a problem in achieving a logic 'high' which should be equal to V_{DD}. In order for the load to conduct current, there must be a voltage difference between the source and gate equal to the threshold voltage (V_T). This requirement imposes a maximum limit on the value of the output voltage of $V_{DD} - V_T$. With V_T values of 3–5 V this represents a significant reduction in the voltage required to represent the logic '1'. It is possible to overcome the problem by connecting the gate of the load to a separate supply, but this requires two power supplies. Complicated multi-phase clocks were also developed to overcome this problem but at greatly increased circuit complexity. The introduction of ion implantation provided the solution by enabling enhancement and depletion devices to be fabricated on the same wafer. This development and improvements to the manufacturing processes in general provided the means for designing NMOS and CMOS circuits with a performance equivalent to that of the bipolar circuits. Both processes are capable of operating from 5 V supplies and are, therefore, compatible with bipolar TTL circuits.

Many of the present-day VLSI circuits are based on n-channel devices (NMOS) but many new circuits now use CMOS.

16.2 NMOS inverter

The schematic of an n-channel inverter with an enhancement-mode driver (Q1) and a depletion-mode driver (Q2) is shown in Figure 16.1(a).

The gate of the depletion transistor is connected to the source, and with a permanent channel between the source and the drain it acts as a high-value nonlinear resistor. With V_{in} at 0 V the driver is OFF and the output voltage is at V_{DD}. When V_{in} exceeds the threshold voltage, V_{TE}, the driver conducts and the output voltage falls to a low value. The complete switching characteristic is shown in Figure 16.1(b).

The transfer characteristics can be constructed manually by superimposing the static $I-V$ characteristics for the two devices, as shown in Figure 16.2. A family of curves for the enhancement transistor is shown which result from the different values of gate voltage, V_{GS1}, which are expressed as multiples of V_{DD}. A single curve is shown for the depletion transistor corresponding to $V_{GS2} = 0$ V. When $V_{GS1} = 0$ V the driver is OFF, no current is flowing and the voltage drop across the load is zero. When $V_{GS1} = V_{DD}$ the maximum current flows and the voltage across the load is approximately V_{DD}. The intermediate points of the transfer characteristic are obtained by noting the values of V_{DS2} at the intersections of the curve for $V_{GS2} = 0$ V with each of the curves for V_{GS1}.

16.2.1 Static analysis

The shape of the transfer characteristic is dependent on the ratio of the effective resistance of the channel regions of the two devices. If these are R_L and R_D for the load and driver, respectively, then, for increasing values of R_L/R_D, the switching characteristic moves towards the ideal rectangular curve, while a decreasing ratio degrades the characteristic, as illustrated in Figure 16.3. The ideal curve requires R_L to be very large, which means that the area occupied by the load is large but, more importantly, a large-value resistance increases the switching time (as will be shown in Section 16.2.3).

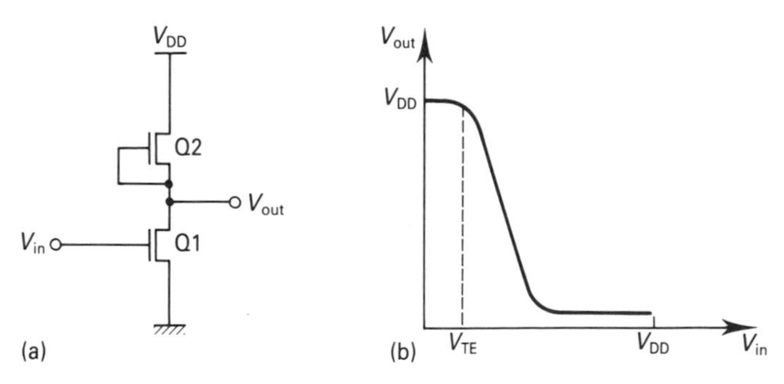

Figure 16.1 (a) N-channel inverter with an enhancement driver and depletion load and (b) the transfer characteristic.

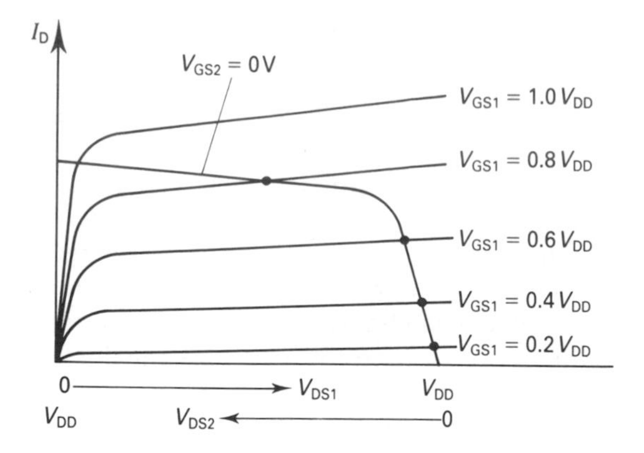

Figure 16.2 Static *I–V* curves for the enhancement and depletion n-channel devices.

A small value of R_L improves the switching time but, because of the gradual change from one logic level to another, as illustrated in the static characteristics, the inverter is more prone to random switching caused by electrical noise. Since the two devices are in series, a relationship between the device parameters can be obtained for a given input condition by equating the two drain currents. Consider a voltage of V_{DD} applied to the input of the driver. With $V_{GS1} = V_{DD}$ the driver is ON and since $V_{DS1} < V_{Dsat}$ it is in the linear region of the I–V characteristics, while the load with $V_{GS2} = 0$ is in saturation. Thus:

$$I_{D1} = I_{D2}$$

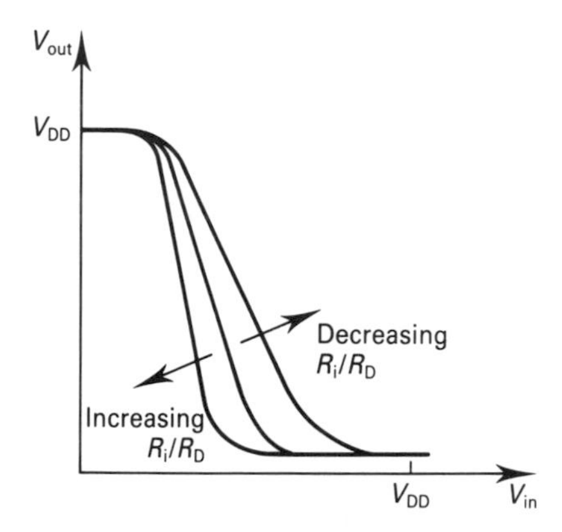

Figure 16.3 Transfer characteristics for different channel resistance ratios.

$$K_D \frac{W_D}{L_D}[(V_{GS1} - V_{TE})V_{DS1}] = \frac{K_L}{2}\frac{W_L}{L_L}(V_{GS2} - V_{TD})^2$$

{linear} {saturation}

where K_D, K_L are the gain factors of the driver and load, respectively; W_D/L_D, W_L/L_L are the aspect ratios of the driver and load.

For the load device $V_{GS2} = 0\,V$ and for the driver $V_{GS1} = V_{DD}$. The gain factors are dependent on the process and are the same for both devices. Thus:

$$\frac{W_D}{L_D}[(V_{DD} - V_{TE})V_{DS1}] = \frac{W_L}{2L_L}(-V_{TD})^2$$

and

$$\frac{W_D/L_D}{W_L/L_L} = \frac{V_{TD}^2}{2(V_{DD} - V_{TE})V_{DS1}}$$

This expression can be simplified if:

$$V_{TD} = -0.8\,V_{DD}$$

$$V_{TE} = 0.2\,V_{DD}$$

$$V_{DS1} = 0.1\,V_{DD}$$

Then

$$\frac{W_D/L_D}{W_L/L_L} = 4$$

This value of the aspect ratio is obviously dependent on the values chosen for the threshold voltages, but it is found in practice that a depletion threshold voltage of between -3 V and -4 V, an enhancement threshold of $+1$ V and an aspect ratio of 4:1 produces a transfer characteristic which has good discrimination between ON and OFF and which also has good transient performance.

For the linear region of the $I-V$ characteristics the slope is a measure of the resistance. It is given by:

$$\frac{1}{r_D} = \frac{\delta I_D}{\delta V_{DS}} = \frac{\delta}{\delta V_{DS}}\left\{\frac{K_D W_D}{L_D}[(V_{GSI} - V_{TE})V_{DS1}]\right\}$$

$$= \frac{K_D W_D}{L_D}(V_{GS1} - V_{TE})$$

or

$$r_D = \frac{L_D}{K_D W_D}\frac{1}{(V_{GS1} - V_{TE})}$$

or

$$r_D \propto \frac{L_D}{W_D}$$

and

$$r_L \propto \frac{L_L}{W_L}$$

Thus if the constant of proportionality is the same for both devices, then the aspect ratio can be defined in terms of the channel resistance as:

$$\frac{R_L}{R_D} = 4$$

The effects of this ratio on the transfer characteristics of the inverter are shown in Figure 16.3. Increasing the ratio by increasing the resistance of the load device improves the dc switching characteristics. The increase in resistance corresponds to either increasing the length of the channel for the load device or increasing the width, and hence reducing the resistance, of the driver device.

16.2.2 Pass transistor plus inverter

The very high input impedance and the low ON resistance between the source and drain make the MOS transistor ideally suited for use as a series connected switch, as shown in Figure 16.4.

The use of a pass transistor increases the number of logic options without having to resort to complex circuitry. However, when the pass transistor is turned ON ($V_{Gpass} = V_{DD}$) account must be taken of the voltage drop across the transistor. Consider the input V_{in} at 0 V, then the voltage at B is V_{DD} and with $V_{Gpass} = V_{DD}$ the voltage at C is $V_{DD} - V_{TEpass}$.

The threshold voltage of the pass transistor is greater than that of the driver transistor of the inverter because the source of the pass transistor cannot be connected to the substrate. Thus when node B is at V_{DD} there is a voltage difference between the source of the pass transistor and the substrate of V_{DD}. From Section 15.4 the change in

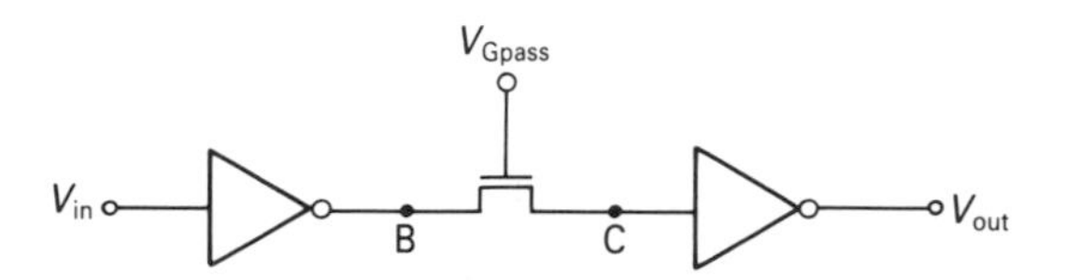

Figure 16.4 N-channel inverters with a pass transistor.

the threshold voltage is $\gamma(V_{sub})^{1/2}$. Typically $\gamma = 0.9$ so that if $V_{DD} = 5$ V the change is approximately 2 V.

Equating the currents for the driver and load of the inverter with the pass transistor in place gives:

$$\frac{W_D}{L_D}[(V_{DD} - V_{TEpass} - V_{TE})V_{DS1}] = \frac{W_L}{2L_L}(-V_{TD})^2$$

assuming the following conditions:

$$\begin{aligned} V_{TEpass} &= 0.4\, V_{DD} \\ V_{TE} &= 0.2\, V_{DD} \\ V_{TD} &= -0.8\, V_{DD} \\ V_{DS1} &= 0.1\, V_{DD} \end{aligned}$$

then

$$\frac{W_D/L_D}{W_L/L_L} = 8$$

To obtain a transfer characteristic with the same discrimination between ON and OFF as for the simple invertor, the inclusion of a pass transistor requires the resistance ratio to be increased from 4:1 to 8:1. As will be seen in Section 16.2.3, the transient response is degraded by the increase.

In summary, when an n-channel inverter is driven directly by another inverter the resistance ratio is 4:1. However, if an inverter is driven through one or more pass transistors then the resistance ratio must be 8:1.

16.2.3 Transient response

An important factor for MOS circuits is the transient response. This arises from the fact that the transistors are high-impedance, voltage-operated devices where the charging and discharging of capacitance can be a major limitation. The output of one inverter (or gate) is usually connected to one or more inputs of other gates. Each input is effectively a MOS capacitor, which, together with the stray capacitance of the interconnections, can be lumped together as a single capacitor. The gate capacitance is voltage-dependent (Section 14.6) but it is instructive to assume that it is constant and that it is added to the capacitance of the interconnection to form a single lumped capacitor (C_L). It is then necessary to consider two conditions, one when the capacitor charges from 0 V to V_{DD} and the other when it discharges from V_{DD} to 0 V.

The two conditions are shown in Figure 16.5, where in Figure 16.5(a) the capacitor is initially assumed to be charged to V_{DD}. The driver transistor is turned ON, the capacitor discharges through the driver (or pull-down transistor) and the voltage across C_L decays to 0 V, as shown in Figure 16.5(a). In Figure 16.5(b) the capacitor is initially assumed to be at 0 V. The driver transistor is turned OFF and the capacitor charges through the load (or pull-up transistor) towards V_{DD}.

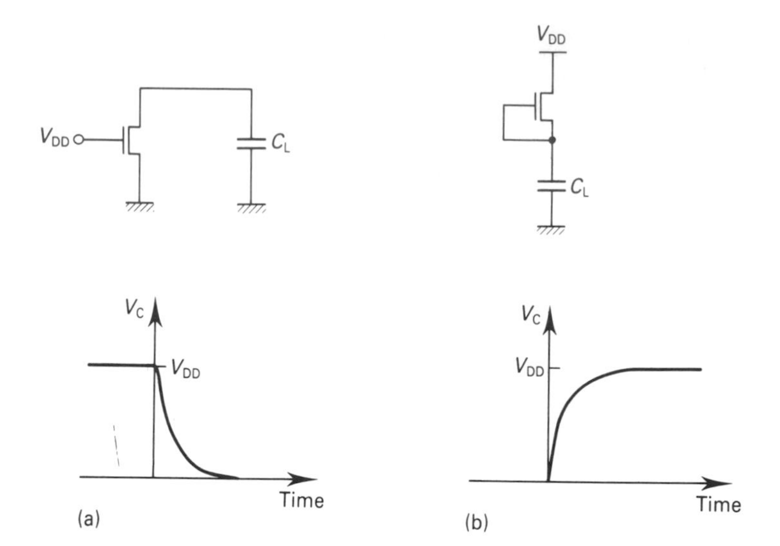

Figure 16.5 Circuit and graph of (a) the discharging of the load capacitance and (b) the charging of the capacitance.

With reference to Figure 16.6(a) the capacitor is initially charged to V_{DD} at point P_1 and in discharging follows the path $P_1P_2P_3$. Between P_1 and P_2 the discharge current is approximately constant and:

$$I = C_L \frac{\delta V}{\delta t}$$

or

$$t_1 = \frac{C_L(V_{P1} - V_{P2})}{I}$$

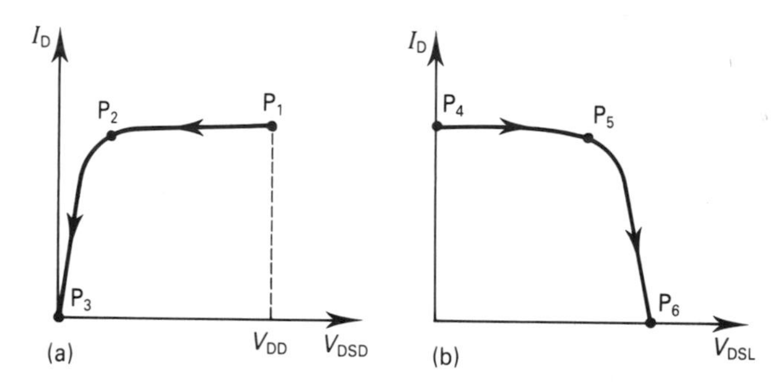

Figure 16.6 The current–voltage path for (a) the capacitor discharging from P_1 to P_3 and (b) charging from P_4 to P_6.

Between P_2 and P_3 the capacitor is effectively discharging through the resistance of the channel of the driver transistor, as shown in Figure 16.7. The resistance is given by:

$$\frac{1}{r_{DSD}} = \frac{\delta I_D}{\delta V_{DS}} = \frac{\delta}{\delta V_{DS}}\left[\beta_D(V_{DD} - V_{TE})V_{DS} - \frac{\beta_D V_{DS}^2}{2}\right]$$

where the input (V_{GS}) is assumed to be V_{DD} and β_D is the gain of the driver:

$$\frac{1}{r_{DSD}} = \beta(V_{DD} - V_{TE}) - \beta_D V_{DS}$$

and since V_{DS} is small below P_2:

$$r_{DS} \approx \frac{1}{\beta_D(V_{DD} - V_{TE})}$$

The discharge time for the simple RC circuit shown in Figure 16.7 between the 10% and 90% limits is:

$$t_2 = 2.2\, r_{DSD} C_L$$

In practice, $t_2 \gg t_1$ so that the discharge time with the driver transistor ON is:

$$t_{ON} \approx 2.2\, r_{DSD} C_L$$

When the driver is turned OFF the capacitor is initially at 0 V and charges up towards V_{DD} through the load. The current follows the $I-V$ curve shown in Figure 16.6(b) starting at P_4 and moving to P_5 and then to P_6. Between P_4 and P_5 the current is constant, while between P_5 and P_6 the capacitor charges through the channel resistance of load. The charging time with the driver OFF, and if the same simplifying assumptions as were used above for the discharge are applied, is given by:

$$t_{OFF} = 2.2\, r_{DSL} C_L$$

where

$$r_{DSL} \approx \frac{1}{\beta_L(-V_{TE})}$$

where β_L is the gain of the load transistor and V_{GS} of the load is 0 V.

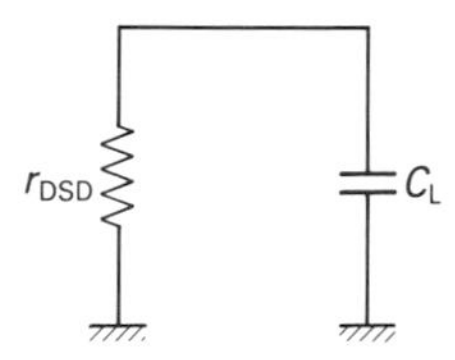

Figure 16.7 Simplified representation of the discharge of the load capacitance through the channel resistance of the driver transistor.

Thus the ratio of the OFF to ON times is:

$$\frac{t_{OFF}}{t_{ON}} = \frac{r_{DSL}}{r_{DSD}}$$

Assuming $V_{TD} = 0.2V_{DD}$, $V_{TE} = -0.8V_{DD}$ and substituting $\beta = K(W/L)$ gives:

$$\frac{t_{OFF}}{t_{ON}} = \frac{W_D/L_D}{W_L/L_L} = 4$$

The output waveform for an NMOS inverter is shown in Figure 16.8. This value of 4 applies to an inverter driven by another inverter. If there is a pass transistor between the two inverters then the resistance ratio of the second inverter is 8:1 and the ratio for the OFF and ON times is also 8:1.

16.3 CMOS inverter

The schematic of a CMOS inverter is shown in Figure 16.9. Both transistors are enhancement mode devices with Q1 being a p-channel device and Q2 an n-channel.

The source and substrate of the n-channel device are connected to ground, while the source and substrate of the p-channel device are connected to the positive supply. The technologies available to manufacture the inverter are n-well, p-well or n- and

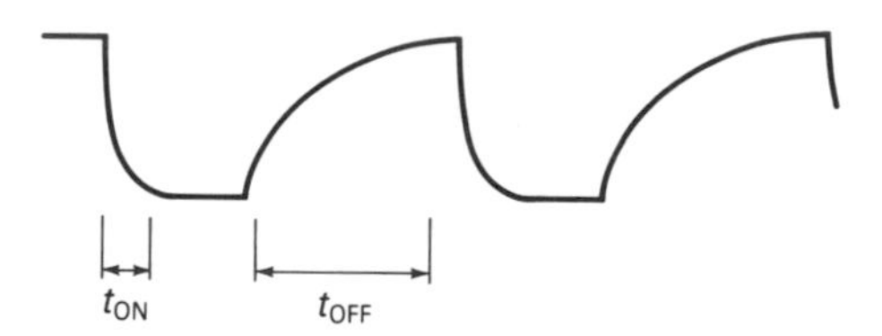

Figure 16.8 Output waveform for the NMOS inverter.

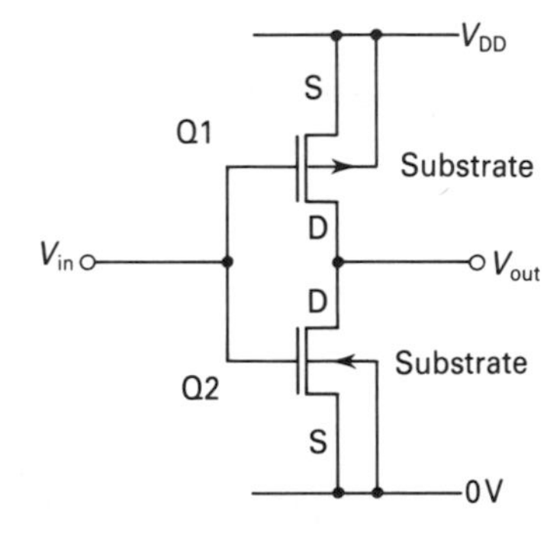

Figure 16.9 The CMOS inverter.

p-well. Ion implantation is used to adjust the threshold voltages so that V_T is positive for the n-channel device and negative for the p-channel.

16.3.1 Static characteristics

When the input is at ground potential the n-channel transistor (Q2) is turned OFF. The voltage on the gate of the p-channel transistor (Q1) with respect to its source is $-V_{DD}$ and a p-channel is formed, that is, Q1 is ON. No current flows but with a connection from the common drain connection to the positive rail the output is at V_{DD}. When the input is at V_{DD} then Q2 is turned ON with an n-type channel formed beneath the gate, while for Q2 the gate and source are both at the same potential (V_{DD}) and, therefore, Q2 is OFF. No current flows but with the output connected via the n-channel to ground, it is at 0 V. The CMOS inverter does not conduct any current when the input is at 0 V or V_{DD}, unlike the NMOS inverter which conducts heavily when the input is at V_{DD}. However, current does flow as the inverter changes from one state to the other. The dc transfer characteristic for the CMOS inverter is shown in Figure 16.10.

The variation of the drain current with input voltage is also shown. If the devices are matched then the peak current will occur when $V_{in} = V_{out} = V_{DD}/2$.

An important feature of the CMOS inverter is that the resistance ratio (R_L/R_D) is unity.

16.3.2 Transient analysis

When a capacitor is attached to the output of a CMOS inverter, as shown in Figure 16.11, the $I-V$ paths for both discharge and charge are the same. The capacitor discharges through the n-channel device when V_{DD} is applied to the input and the $I-V$ path is $P_1P_2P_3$. When 0 V is applied to the input the capacitor charges through the p-channel device towards V_{DD}, and the $I-V$ path is the same because the voltage across the p-channel device decreases from $V_{DSp\text{-}channel} = V_{DD}$ to $V_{DSp\text{-}channel} = 0$ V. The

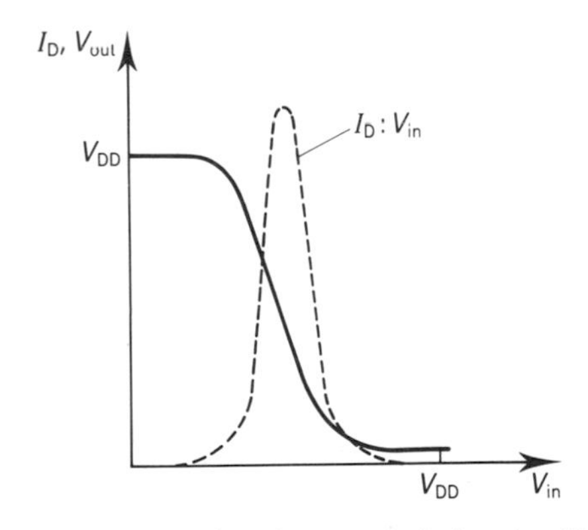

Figure 16.10 Static transfer characteristic for the CMOS inverter.

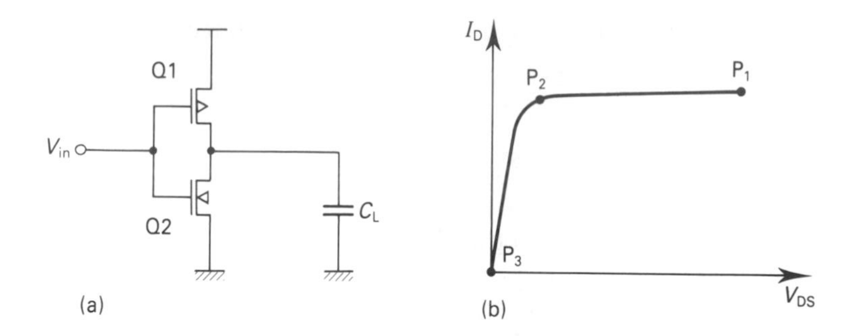

Figure 16.11 (a) CMOS inverter with a load capacitor and (b) the *I–V* charging and discharging paths.

asumptions made for the NMOS inverter still apply and the ratio of OFF to ON times is:

$$\frac{t_{\mathrm{OFF}}}{t_{\mathrm{ON}}} = \frac{r_{\mathrm{DSP}}}{r_{\mathrm{DSN}}}$$

where

$$r_{\mathrm{DSP}} = \frac{1}{\beta_{\mathrm{P}}(V_{\mathrm{GSP}} - V_{\mathrm{TP}})}$$

and

$$r_{\mathrm{DSN}} = \frac{1}{\beta_{\mathrm{N}}(V_{\mathrm{GSN}} - V_{\mathrm{TN}})}$$

where V_{TP} is the thresnold voltage of the p-channel device, V_{TN} is the threshold voltage of the n-channel device and V_{GSP}, V_{GSN} are the gate voltages of the p-channel and n-channel devices respectively.

Let $V_{\mathrm{TP}} = -0.2\,V_{\mathrm{DD}}$ and $V_{\mathrm{TN}} = 0.2\,V_{\mathrm{DD}}$. Then with $V_{\mathrm{GSP}} = -V_{\mathrm{DD}}$ and $V_{\mathrm{GSN}} = V_{\mathrm{DD}}$, the ratio of OFF to ON times is:

$$\frac{t_{\mathrm{OFF}}}{t_{\mathrm{ON}}} = \frac{\beta_{\mathrm{P}}}{\beta_{\mathrm{N}}}$$

From Section 15.3 $\beta = \varepsilon_{\mathrm{o}}\varepsilon_{\mathrm{ox}} \cdot W/(t_{\mathrm{ox}} \cdot L)$ and

$$\frac{t_{\mathrm{OFF}}}{t_{\mathrm{ON}}} = \frac{\mu_{\mathrm{P}}}{\mu_{\mathrm{N}}}$$

Typically, $\mu_{\mathrm{N}} = 2 \cdot \mu_{\mathrm{P}}$ and

$$t_{\mathrm{OFF}} \approx 2 \cdot t_{\mathrm{ON}}$$

Equal rise and fall times can be obtained if β_{N} is made equal to β_{P}. This can be achieved by making the channel width of the p-channel device twice as wide as that of the n-channel device, that is:

$$W_P = 2 \cdot W_N$$

In practice, other considerations such as power dissipation, which is greater for wider devices, and increased silicon area are often more important than a symmetrical switching waveform, and it is usual for CMOS devices to employ minimum geometry dimensions ($W = L$) for both device types.

16.4 Transmission gate

The equivalent of the pass transistor used in NMOS logic is the transmission gate shown in Figure 16.12. An n-channel device (Q2) and a p-channel device (Q1) are placed in parallel. In-phase and anti-phase clock waveforms are required, with ϕ being applied to the gate of the n-channel device and the complement of ϕ (ϕ_{bar}) to the gate of the p-channel device.

Notice that, unlike the inverter where both gates are joined together so that when one device is ON the other is OFF, in the transmission gate both transistors are either ON or OFF.

The operation of the transmission gate is best understood by considering how each transistor charges or discharges a capacitor attached to the output.

16.4.1 n-channel pass transistor

Assuming that the capacitor is initially discharged ($V_{out} = 0$ V), then with ϕ = '0', that is, $V_{GS2} = 0$ V, $I_D = 0$ irrespective of the state of V_{in}. When ϕ = '1' and V_{in} = '1', then Q2 conducts and current flows to charge the load capacitor. As the output voltage approaches $V_{DD} - V_{TN}$ the device begins to turn OFF because there is insufficient voltage between the gate and the substrate to support a conducting channel. Thus with $V_{in} = V_{DD}$, $V_{out} = V_{DD} - V_{TN}$ and the output for a logic '1' is degraded.

With V_{in} = '0' and ϕ = '1' the capacitor discharges towards 0 V. The gate-to-substrate voltage will always exceed V_{TN} and the output voltage falls to 0 V, and the output logic '0' is not degraded.

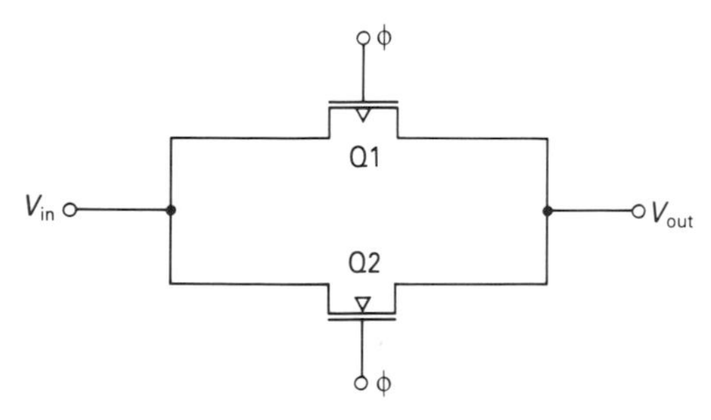

Figure 16.12 CMOS transmission gate.

16.4.2 p-channel pass transistor

With the gate of Q1 equal to V_{DD}, that is, ϕ_{bar} = '1', the transistor is OFF and if the capacitor is initially uncharged it will remain uncharged irrespective of the value of V_{in}.

When ϕ_{bar} = '0' and V_{in} = '1' the transistor conducts and current flows to charge the capacitor towards V_{DD}. With V_{GS2} at 0 V the output will rise to V_{DD}. If the capacitor is charged and V_{in} is reduced to 0 V then the current flows from the capacitor to ground through the transistor. As V_{out} falls, a point will be reached when the voltage between the gate and the substrate is not sufficient to support the channel. This occurs when V_{out} reaches V_{TP}, at which point the transistor ceases to conduct. Thus the logic '0' for the p-channel device is degraded.

The different operating conditions are summarized in Table 16.1. As can be seen from the table, when V_{in} = '1' the output from the n-channel device is $V_{DD} - V_{TN}$, for example $5 - 2 = 3$ V, but the output from the p-channel device is V_{DD}, that is, 5 V. Therefore, the output is 5 V. Similarly, when V_{in} = '0' the output from the n-device is 0 V, while the output from the p-device is V_{TP}, say 2 V, that is, $V_{out} = 0$ V. Thus the CMOS transmission gate is a perfect switch with no loss of signal for either input condition.

16.5 CMOS latch-up

A problem common to all CMOS circuits is latch-up. This is caused by the normal formation of parasitic bipolar transistors in the bulk or well.

A schematic of a CMOS inverter with the parasitic bipolar transistors is shown in Figure 16.13 together with the $I-V$ characteristics for a four-layer device. If minority carriers are injected into the base of one of the bipolar transistors then the normal transistor action will produce a collector current which is increased by the β of the transistor. The positive feedback which results from the way in which the transistors are connected multiplies this current and the four-layer diode switches into a low-impedance state and short-circuits the supply. This action can be initiated by a voltage transient, for example when power is applied to the device, but external radiation and electrostatic discharge may also induce carriers. Latch-up may cause permanent damage; it will certainly affect normal operation and it can only be removed by turning off the supply.

Susceptibility to latch-up is reduced by layout and process adjustments. Positive

Table 16.1 Transmission gate output (V_{out})

Device	V_{in} = '1'	V_{in} = '0'
n	$V_{DD} - V_{TN}$	'0'
p	'1'	V_{TP}

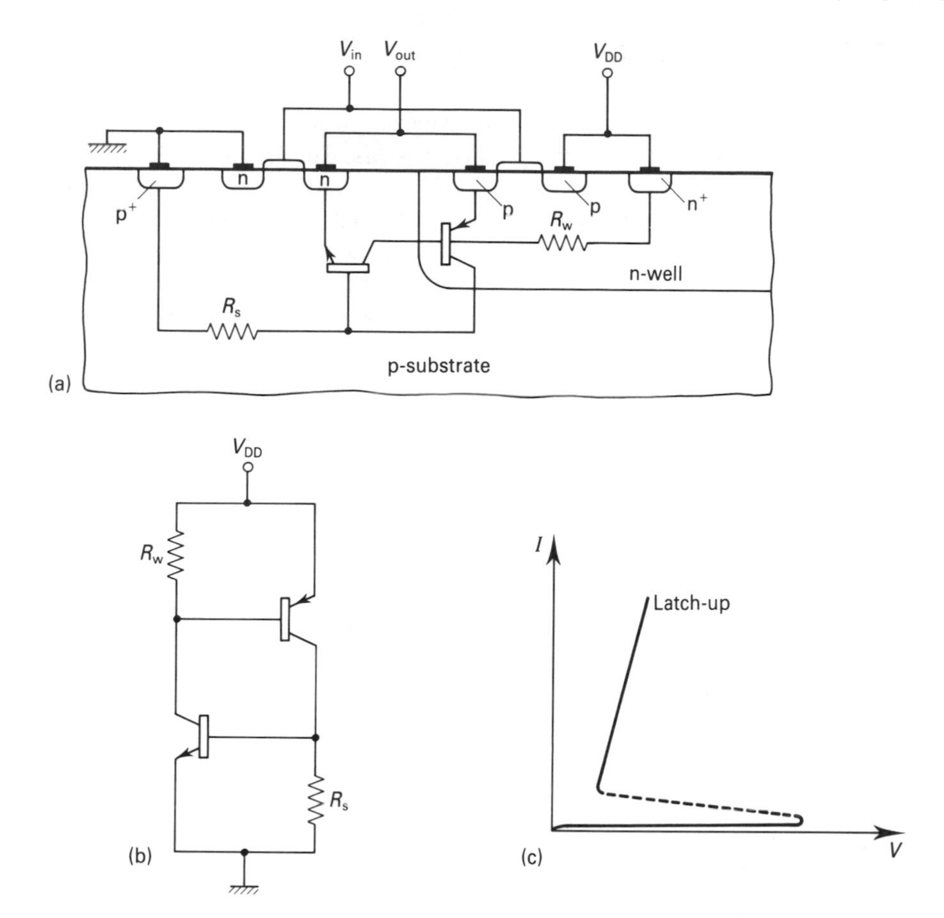

Figure 16.13 (a) Cross-section of a CMOS inverter with (b) a simplified equivalent circuit four-layer diode represented as two bipolar transistors and (c) the *I–V* characteristic.

feedback will not occur if the product of the gains of the two bipolar transistors is less than unity. This is often difficult to achieve for the vertical device in the well (pnp for an n-well process), but the gain of the lateral (npn) can be reduced by increasing lateral spacing and by the use of heavily doped guard rings around the well. Transistor action can be greatly reduced by ensuring that the series resistances R_s and R_w are as low as possible. These resistors shunt the emitter–base diodes and thus reduce the possibility of sufficient current being generated to forward bias the junctions. This is achieved by using a heavily doped substrate with a thin epitaxial layer. Alternatively, an overall ion implant is used as a first process step to produce a buried, heavily doped layer. Ion implantation can also be used to create a well region with a lightly doped surface and a heavily doped base. Finally, at system level, care should be taken to minimize the possibility of transients being generated when power is applied to the circuit.

16.6 Future developments

Many of the new circuits which are now being developed use CMOS but for some applications NMOS, often involving pass transistors, can provide a simpler solution particularly for non-critical, low-cost applications. An important advantage of CMOS is the low power consumption, but this is only achieved at low speeds, for example the counters required to divide down a 30 kHz clock frequency to provide minute, hour, day, month and even year counts. For high-frequency applications the difference in power consumption between NMOS and CMOS is less significant. Developments will continue in botsh technologies.

Problems

1. The process parameters for a silicon–gate NMOS process are:

gate oxide thickness:	80 nm
substrate doping:	$1 \times 10^{15}/\text{cm}^3$
work function ϕ_{ms}:	-0.6
surface charge (Q_{ss}/q):	$2 \times 10^{10}/\text{cm}^2$

$\varepsilon_{si} = 11.7$ $\quad \varepsilon_{ox} = 3.9$

$\mu_n = 580\ \text{cm}^{-2}\text{V}^{-1}\text{s}^{-1}$ $\quad \mu_p = 230\ \text{cm}^2\text{V}^{-1}\text{s}^{-1}$

(a) Determine the unimplanted threshold voltage for zero substrate bias.
[0.225 V.]

(b) Determine the type and number of impurities that must be implanted to produce threshold voltages of 1 V for the enhancement device and -4 V for the depletion device.
[$2.1 \times 10^{11}/\text{cm}^2$, $1.1 \times 10^{12}/\text{cm}^2$.]

(c) Determine the body-effect factor.
[0.4.]

(d) If a substrate voltage of 5 V is applied, what is the effective threshold voltage for the unimplanted condition.
[1.2 V.]

2. (a) For the inverter fabricated according to the process described in Problem 1 with $V_{TE} = 0.225$ V and $V_{TD} = -4$ V determine the W/L ratio for the load device if the saturation current is 100 μA.
[0.5.]

(b) If the aspect ratio is 4:1 and the minimum dimension is 5 μm, determine the channel dimensions (L and W) for the driver and load.
[5 μm, 10 μm.]

3. An NMOS manufacturing process has the following parameters:

t_{ox} (gate):	60 nm
ε_{ox}:	3.9
μ_n (channel):	$580\ \text{cm}^2\text{V}^{-1}\text{s}^{-1}$

(a) An inverter has a resistance ratio of 4:1 with the W/L ratio for the driver device of unity. If the channel resistance of the driver is given by $r_{DS} = 1/[\beta(V_{DD} - V_{TE})]$ and if $V_{TE} = 0.2\ V_{DD}$, estimate the maximum switching speed for the inverter with $V_{DD} = 5$ V when driving a load capacitance of 0.1 pF.
[120 MHz.]

(b) Show that if the resistance ratio is 4:1 and $V_{DD} = 5$ V, then the maximum frequency for the inverter is given by:

$$f_{max} = \frac{0.36\beta_D}{C_{load}}$$

(c) If the load capacitance is increased to 1 pF, determine the geometrical changes which must be made to the driver and the load devices to maintain the switching frequency at the same value as determined in part (a). [Hint: if the capacitance is increased the resistance must be decreased by increasing W.]

(d) Determine the power dissipation for the two devices of parts (a) and (c) given that $P_{diss} = V^2/(r_{DSdriver} - r_{DSload})$.
[0.6 mW, 6.6 mW.]

4. Use an analog simulator to investigate the dc and transient response of the simple nMOS inverter driving a load capacitance of 1 pF for load to driver resistance ratios of 2:1, 4:1 and 8:1. Observe the dc transfer curve and the output waveform.

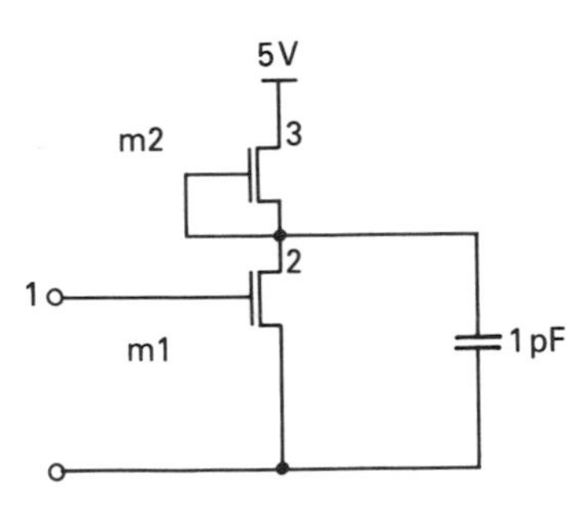

An input data file is as follows :

```
nmos inverter
m1 2 1 0 mehn l=5u w=5u
m2 3 2 2 mdep l=20u w=5u
cl 2 0 1p
vin 1 0 dc pulse(0 5V 0 1n 1n 1u 2u)
vdd 3 0 5V
.dc vin 0 5V 0.1V
.tran 10n 4u
.plot dc v(2)
.plot tran v(2) v(1)
.model mehn nmos(level=1 vto=1.0
kp=1.5e-5 gamma=0.5
+ cgso=4.5e-10 cgdo=4.5e-10
cj=1.0e-4 cjsw=1.0e-9)
.model mdep nmos(level=1 vto=-4.0
kp=1.5e-5 gamma=0.7
+ cgso=4.5e-10 cgdo=4.5e-10
cj=1.0e-4 cjsw=1.0e-9)
.end
```

The exact format will depend on the particular simulator being used. Note the difference between the threshold voltages for the enhancement and depletion devices in the 'model' statement. These values correspond to the $0.2\ V_{DD}$ and $-0.8\ V_{DD}$ values specified in Section 16.2.1.

5. For a CMOS inverter the following parameters apply;

gate oxide thickness:	50 nm
substrate doping (N_d):	$1 \times 10^{15}/cm^3$
p-well doping (N_a):	$1 \times 10^{16}/cm^3$
surface-state charge (Q_{ss}/q):	$2 \times 10^{10}/cm^2$
ϕ_{ms} (n-channel):	-0.6 V
ϕ_{ms} (p-channel):	0.85 V

$\varepsilon_{si} = 11.7$ $\qquad \varepsilon_{ox} = 3.9$
$\mu_n = 580\ cm^2 V^{-1} s^{-1}$ $\qquad \mu_p = 230\ cm^2 V^{-1} s^{-1}$

(a) Estimate the threshold voltages for each device.
[$V_{TN} = 0.135$ V, $V_{TP} = -0.214$ V.]

(b) If the devices are used as an inverter to drive a 1 pF capacitor estimate the maximum switching speed if W/L for each device is unity.
[26 MHz.]

(c) Suggest how the asymmetry in the switching may be improved and how the W/L ratio of one of the devices may be changed increase the switching speed to 45 MHz.

6. Use an analog simulator to investigate the dc and transient properties of the CMOS inverter driving a 1 pF capacitive load and investigate the effect of modifying the aspect ratio in order to improve the symmetry of the output waveform.

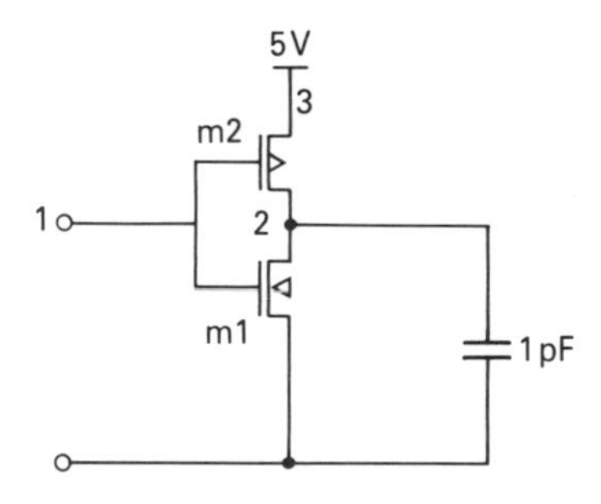

A suitable input file is:

```
cmos inverter
m1 2 1 0 0 mehn 1=5u w=5u
m2 2 1 3 3 mehp 1=5u w=5u
c1 2 0 1pf
vin 1 0 dc pulse(0 5V 0 1n 1n 100n 200n)
vdd 3 0 5v
```

```
.dc vin 0 5V 0.1V
.tran 2n 400n
.plot dc v(2)
.plot tran v(1) v(2)
.model mehn nmox(level=1 vto=1.0
+ kp=1.5e-5 gamma=0.5
+ cgso=4.5e-10 cgdo=4.5e-10
+ cj=1.0e-4 cjsw=1.0e-9)
.model mehp pmos(level=1 vto=-1.0
+ kp=0.75e-5 gamma=0.5
+ cgso=4.5e-10 cgdo=4.5e-10
+ cj=1.0e-14 cjsw=1.0e-9)
.end
```

CHAPTER 17

MOS circuits

17.1 Introduction

The simplest logic circuit is the inverter which was considered in Chapter 16. From this basic logic gate it is a relatively simple matter to design NOR gates and NAND gates using NMOS, but the equivalent CMOS gates are a little more complicated. Of equal importance to the configuration of the transistors for each logic gate is the geometrical layout of the various regions of each device.

Circuit designers have access to a number of computer assisted design (CAD) aids to assist in the layout of an integrated circuit, particularly for MOS circuits which, because of their simpler geometrical structure, have led to the development of programs which produce the layout automatically from a circuit diagram. A number of these CAD programs use a symbolic representation of the individual MOS transistors which closely resembles the geometrical layout, but also has the appearance of an electrical circuit symbol. These 'stick' symbols are available in a number of commercial CAD packages including PHASE software from Integrated Silicon Design Limited.

An example of a section of a completed CMOS circuit with bond pads and aluminium interconnections for signal tracks (fine lines) and power rails (broad tracks) is shown in Figure 17.1.

This section will consider some of the basic logic gates but more complex circuit elements involving flip-flops, registers, counters, memory elements, etc., are easily developed from a basic understanding of the simple gates.

17.2 Design rules

The source, drain and gate electrodes of the MOS transistor can be formed by a photographic mask composed of simple rectangles. With such simple shapes it is possible to establish design rules which can convert a circuit schematic directly into an integrated circuit layout. Many of these conversion programs use rules which are based on the work of Mead and Conway[1]. They defined an arbitrary unit of length called

Figure 17.1 Serial subsystem circuitry. [Copyright Motorola, reproduced with kind permission.]

Lambda (λ). For any particular technology there is a certain minimum dimension which can be accurately reproduced on the silicon wafer. This dimension becomes Lambda. All device dimensions are then based on multiples of Lambda. Some examples are shown in Figure 17.2, with wires, transistors and contacts.

In Figure 17.2(a) minimum-width tracks for diffusions and polysilicon are 2λ; for metal they are 3λ. The separation of diffused and polysilicon tracks is also 2λ, while the separation of a diffused track and a polysilicon track is 1λ; for metal tracks the separation is 3λ. The geometry of minimum-size transistors is shown in Figure 17.2(b) with $L = W = 2\lambda$ for both enhancement and depletion devices. For the depletion device ion implantation is used to adjust the threshold voltage, and the photographic mask for this process overlaps the active region by 2λ. Contacts are shown in Figure 17.2(c) with contact between diffusion and metal and a butt contact between polysilicon and a diffused region. This form of contact is required for the gate and source of a depletion-mode transistor where the gate is connected to the source.

Other structures can be developed from these simple rules to cover all possible layouts for both NMOS and CMOS. The rules greatly simplify the design process and are easily incorporated into CAD packages, but they do not necessarily result in the optimum layout. The advantage of such a simple set of rules is that the CAD software can be developed to produce the layouts for a number of different technologies and the

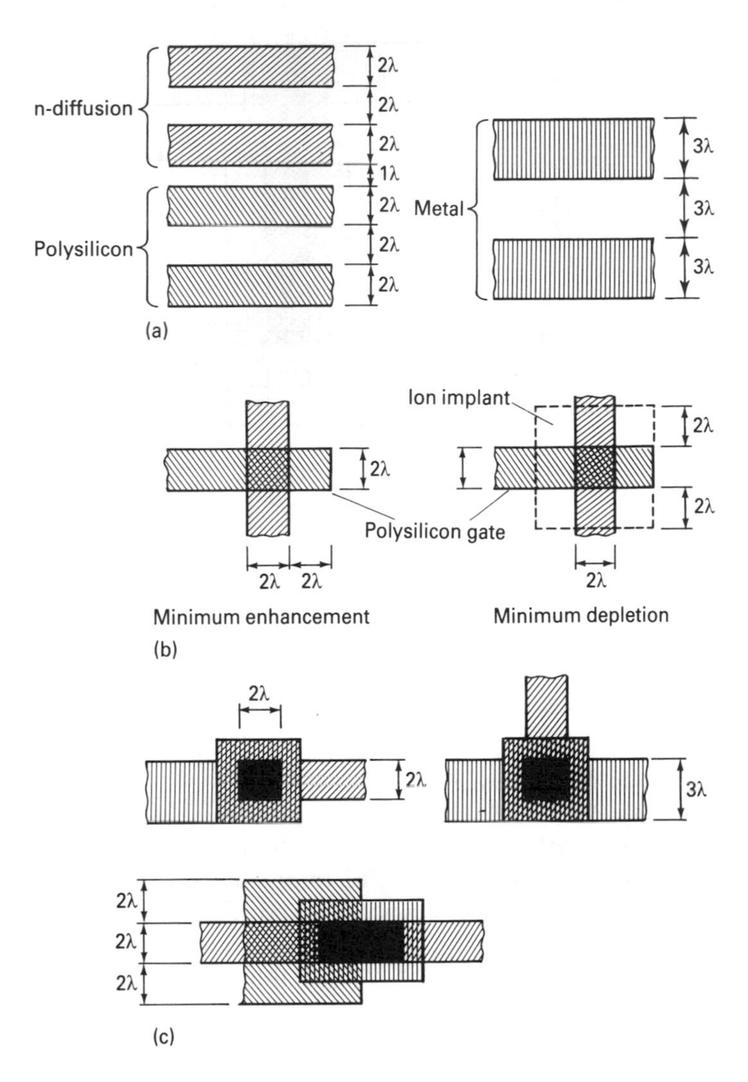

Figure 17.2 Lambda based design rules for (a) diffusion and polysilicon wires, (b) transistors and (c) contacts.

only important parameter is Lambda. This can be changed to suit the technology and as the technology improves and smaller dimensions are achieved, then it is a simple matter to change Lambda to produce a new mask set.

17.3 NMOS inverter

The electrical schematic and the layout for an NMOS inverter is shown in Figure 17.3. The layout is based on the design rules illustrated in Figure 17.2 and the inverter

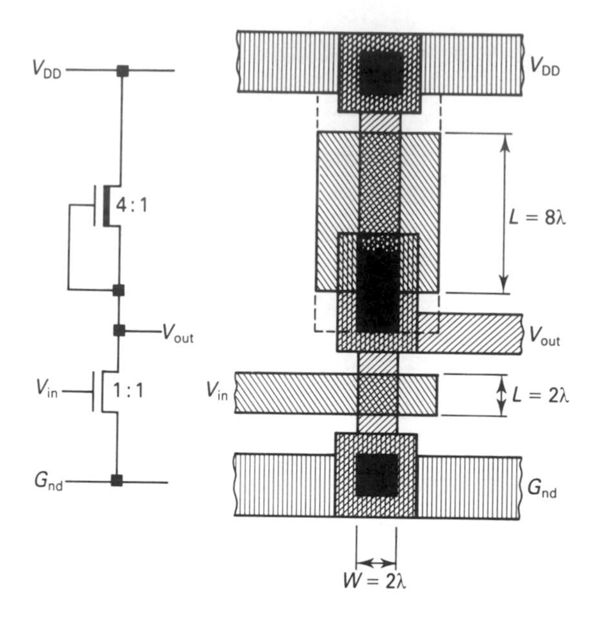

Figure 17.3 Circuit schematic and layout of an NMOS inverter.

described in Section 16.2.1 in which the impedance ratio is 4:1. The driver transistor is a minimum-dimension device with $L = W = 2\lambda$.

To achieve the 4:1 impedance ratio the load device has a width $W = 2\lambda$ and a length $L = 8\lambda$. If the inverter is driven by a pass transistor (Section 16.2.2), then the length of the load device is increased to 16λ to provide an impedance ratio of 8:1.

A much simplified masking sequence for producing an inverter is illustrated in Figure 17.4. A thick oxide is first grown on a p-type substrate and then an oxide window is opened, as shown in Figure 17.4(a). A thin oxide is grown in this window. Next, a photoresist mask is produced, as shown in Figure 17.4(b), for ion implantation for the depletion-mode transistor. The photoresist prevents the ions reaching the substrate except in the opening over the area where the load transistor is located. After the ions are implanted the photoresist is removed and polysilicon is deposited over the whole slice.

Photolithography is used to remove the unwanted polysilicon, leaving the two gate electrodes, as shown in Figure 17.4(c). The thin oxide in the regions D2, S2 and D1, S1 is removed, and with the polysilicon and the surrounding thick oxide acting as a mask, n-type impurities are introduced into the exposed silicon to form the source and drain regions for each transistor (D2, S2 and D1, S1). The oxide is regrown over the source and drain regions and further masks are used to define the contact windows and the metal interconnections, as shown in Figure 17.3. The above sequence is very much simplified and the actual manufacturing process has many more intermediate steps for cleaning, inspection and testing of the wafers, apart from additional photolithographic steps.

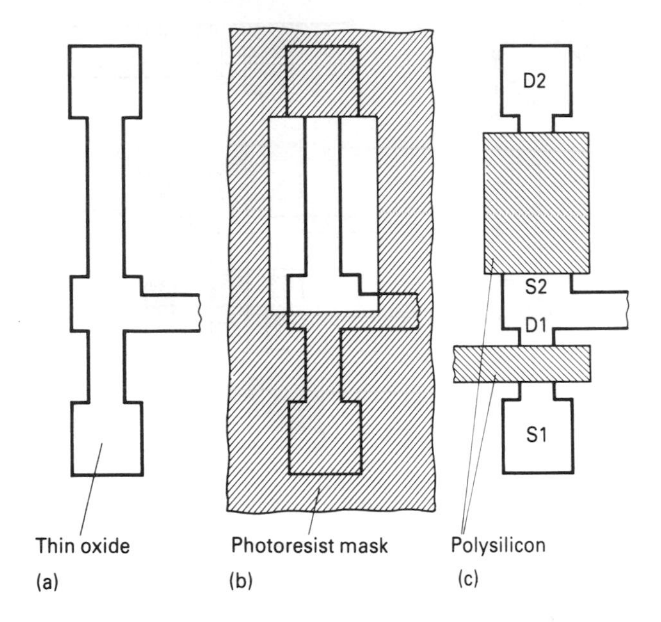

Figure 17.4 Mask sequence required to produce the NMOS inverter with (a) the thin oxide window, (b) the photoresist mask for ion implantation to form the depletion transistor and (c) the polysilicon for the gates.

17.4 CMOS inverter

The circuit schematic and layout for a CMOS inverter is shown in Figure 17.5. For a p-well process the starting material is n-type silicon and a p-type well is formed for the n-channel, that is, Q2 in the diagram.

The p-well is in the region below the demarcation line. In practice, a number of n-channel transistors are placed in the same well. For an n-well process an n-type region is created above the demarcation line to contain the p-channel transistors. Notice that the n- and p-type devices both have unity aspect ratios and occupy less silicon than the NMOS inverter, although because of the design rules, the overall size is not significantly different. The masking sequence, although more complex than that used for the NMOS, is similar, with thin oxide regions being formed which are then covered with polysilicon to form self-aligned gate electrodes. Ion implantation or diffusion of n- and p-type impurities is then used to form the source and drain regions. An important process is the use of ion implantation to adjust the threshold voltages of the n-channel and p-channel transistors. This takes place before the deposition of the polycrystalline silicon and involves a careful compromise between the impurity

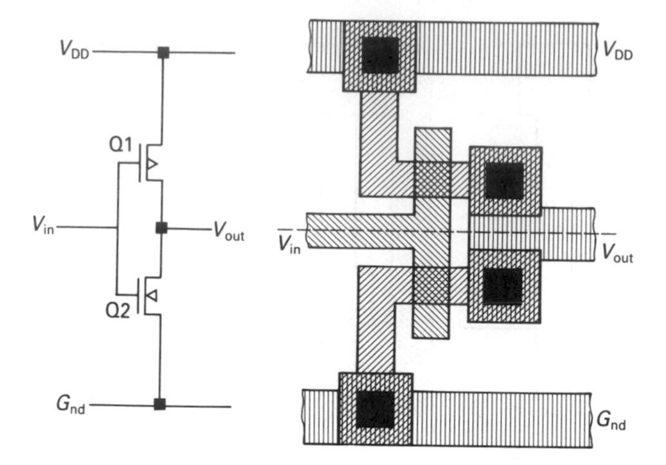

Figure 17.5 Circuit schematic and layout for CMOS inverter.

concentration in both the substrate and the well, and the ion implant dose, so that the single implant adjusts both thresholds.

17.5 Stick diagrams

The process of translating a circuit schematic to a layout is greatly simplified by the use of a symbolic form of the transistor, which, in addition to representing the circuit symbol, also closely resembles the circuit symbol with gate, source and drain terminations. The stick diagrams for NMOS and CMOS inverters are shown in Figure 17.6. The gate electrodes are always polysilicon, the supply and ground are metal and the active channel regions of the transistors are formed from thin oxide cuts. For the CMOS inverter a demarcation line is shown to separate the n- and p-channel devices. With CAD colour graphic terminals it is a simple matter to use colours to differentiate between the different parts of the circuit. Thus, for example, the thin oxide regions are green, the polysilicon gates are red, the metal interconnections are blue, the ion implant region is yellow and the contacts are black. Colour and the close similarity between the stick symbols and the more conventional circuit schematic symbols enable designers to develop stick diagrams for the most complex digital circuits. The computer program automatically converts the stick diagram into a mask layout and a database of x and y coordinates from which the photographic mask can be manufactured.

Other types of logic gate are easily produced, with the stick diagrams clearly showing the geometrical layout. Two input NAND gates are shown in Figure 17.7, with an NMOS gate in Figure 17.7(a) and a CMOS gate in Figure 17.7(b).

The W/L ratio for the driver transistors in the NMOS NAND gate must be adjusted to account for the fact that when both transistors are ON and the output

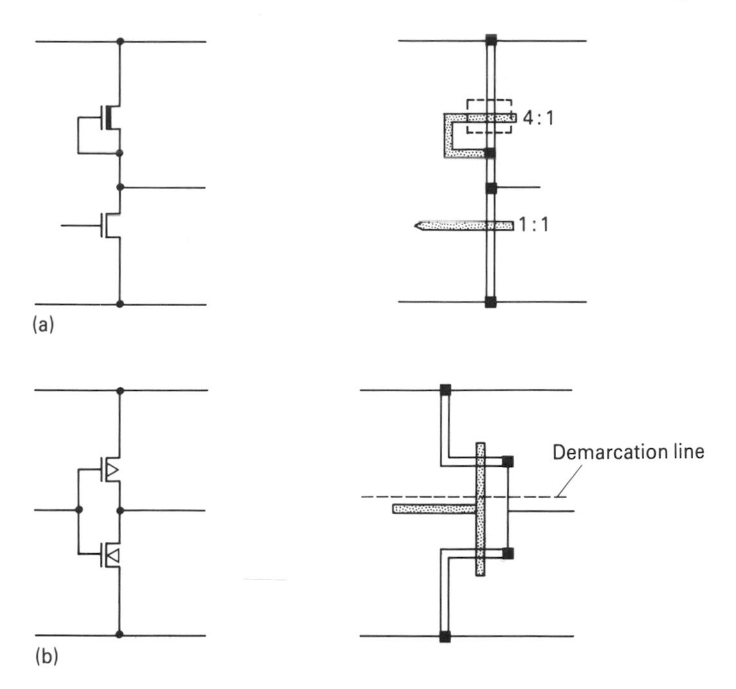

Figure 17.6 Circuit schematic and stick diagram for (a) an NMOS inverter and (b) a CMOS inverter.

capacitor is discharged to ground through them, the channels of the two transistors are in series. With no change to the W/L ratio this would result in an increase in the resistance and, therefore, an increase in the discharge time. This increase can be prevented by increasing the width of each driver transistor to reduce the channel resistance. For two inputs the width is doubled, thereby reducing the resistance by a factor of two. Thus the W/L ratio for the drivers is 2:1 with the load transistor W/L ratio remaining at 1:4. However, increasing the width of the drivers increases the area of the gate electrode and in consequence the gate capacitance. Thus the delay associated with preceding gates is increased. As a result NAND gates formed from NMOS transistors are only used where absolutely necessary.

The CMOS NAND gate is not affected by the number of inputs because switching does not depend on the impedance ratio of the transistors. However, it may be necessary to make some adjustments to the L/W ratio to improve the symmetry of the output waveform.

Two input NOR gates are shown in Figure 17.8. For the NMOS NOR gate any number of inputs can be added because only one input transistor needs to be switched ON to change the output. Thus the L/W ratio is not affected. The CMOS NOR gate is also unaffected by the number of inputs, although some adjustment may be required to optimize the waveform symmetry.

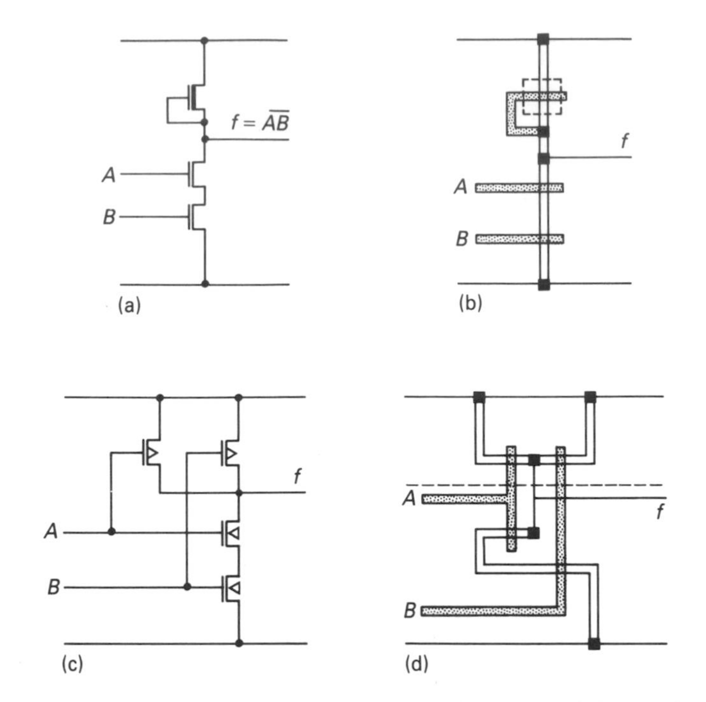

Figure 17.7 Schematic and stick diagrams of 2-input NAND with (a) NMOS transistors and (b) CMOS transistors.

17.6 Driving large capacitive loads

Large capacitive loads are associated with the output pads, where the electric signal is transferred to a printed circuit board, or with an internal 'bus' which connects together the inputs of a large number of gates. The capacitance associated with a printed circuit board or a bus may be several orders of magnitude greater than the on-chip capacitance of the interconnections. For example, the on-chip capacitance is usually the gate of a transistor plus the capacitance of the metal or diffused interconnection track from the output of one gate to the input of the next.

Capacitance may be expressed as:

$$C = \frac{Area \ \varepsilon_o \ \varepsilon_r}{t}$$

where ε_r is the dielectric constant and t is the distance between the effective plates of the capacitor.

For typical values of oxide thickness (or depletion width for a diffused track) the following capacitance values may be obtained:

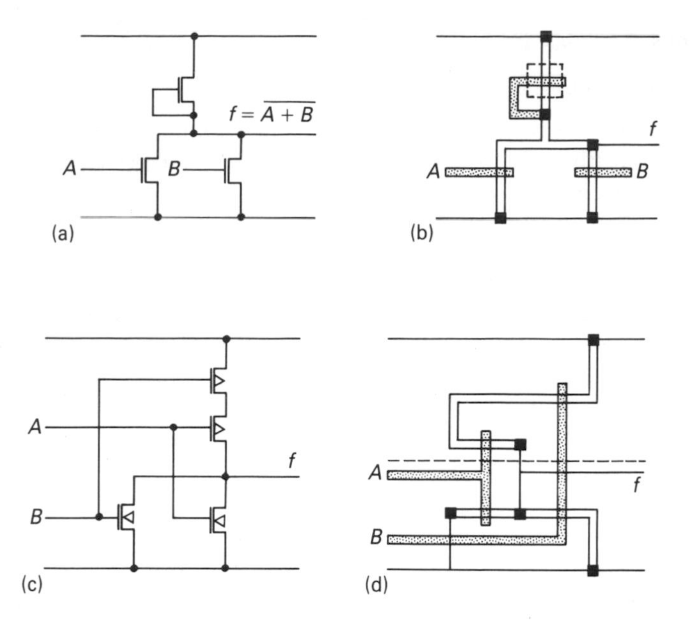

Figure 17.8 Schematic and stick diagram of 2-input NOR gate with (a) and (b) NMOS transistors and (c) and (d) CMOS transistors.

Gate to channel	$4 \times 10^{-4}\,\text{pF}/\mu\text{m}^2$
Metal to silicon	$0.4 \times 10^{-4}\,\text{pF}/\mu\text{m}^2$
Diffusion	$1 \times 10^{-4}\,\text{pF}/\mu\text{m}^2$

For a minimum-area $5\,\mu\text{m}$ transistor the gate area is $5\,\mu\text{m} \times 5\,\mu\text{m}$ and the gate capacitance is $\approx 0.01\,\text{pF}$. This value could be doubled with the inclusion of the interconnection track between the output of one gate and the input of the next. The off-chip capacitance could be 10^3–10^4 times greater, that is, 20–200 pF. Capacitors of this magnitude must be driven from logic gates with very low output resistance to prevent excessive delays occurring. Low resistance implies large W/L values, that is, the width must be very large compared with the length. The area of such an inverter is very large compared to the on-chip inverters and, as a result, it presents a significant load to the preceding stage. The remedy is to cascade several stages, each slightly larger than the preceding stage. Consider, for example, a load capacitance of 100 pF. If a single inverter is used, then the delay caused by discharging the capacitor through the driver transistor of an NMOS inverter is:

$$\tau \approx 2.2\, r_{\text{ds}}\, 100\,\text{pF}$$

or

$$\tau \approx 2.2\, r_{\text{ds}}\, 10\,000 C_{\text{g}}$$

where C_{g} is the capacitance of a minimum-area transistor with a typical value of

0.01 pF. Thus the delay is:

$$\tau \approx 10\,000\tau_d$$

where τ_d is the delay which occurs when driving a minimum-area transistor.

In addition, there is the delay caused by charging the capacitor through the pull-up transistor equal to 4τ. Thus the total delay is:

$$\tau_{total} = 50\,000\tau_d$$

Now consider four inverters in which the channel resistance is reduced by a factor of 10 in each inverter by increasing the width of the driver (and also changing the W/L ratio of the load in order to maintain the ratio of the channel resistances at 4:1). The inverter chain is shown in Figure 17.9.

The first inverter (1) is a minimum-area device driving the input gate of the second inverter (2), which has a channel length of 5 μm and a channel width of 50 μm. Thus the area of the gate of the second inverter is 500 μm^2, that is, $10C_g$. This capacitive load is shown as a discrete capacitor at the input to the second inverter. The third inverter has an aspect ratio of 1:100, which produces a load for the second inverter of $100C_g$. Finally, the last inverter has an aspect ratio of 1:1000. It drives the load capacitance of $10\,000C_g$ and presents a load to the third inverter of $1000C_g$.

The delay for the first inverter is:

$$\tau_1 \approx 2.2r_{ds}10C_g$$

or

$$\tau_1 \approx 10\tau_d$$

and

$$\tau_2 \approx 2.2\frac{r_{ds}}{10}100C_g$$

$$\tau_2 \approx 10\tau_d$$

Using the same approach for inverters 3 and 4 results in a total discharge time of $40\tau_d$. To this must be added the time required to charge the capacitor giving a total delay of:

$$\tau_{total} = 40\tau_d + 160\tau_d$$

$$= 200\tau_d$$

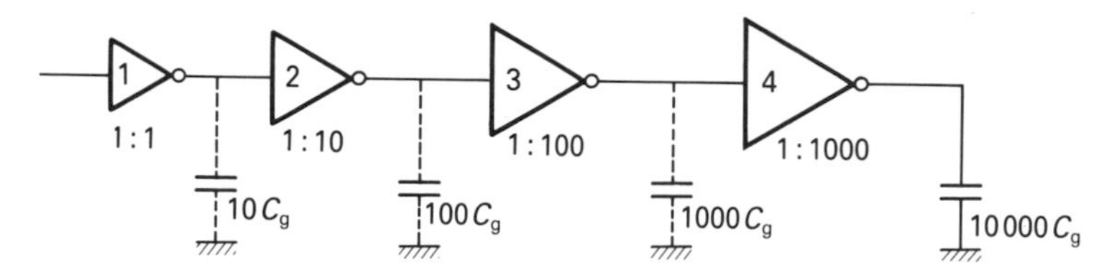

Figure 17.9 Representation of four inverters driving a load capacitor.

This value is considerably less than the $50\,000\tau_d$ of the single minimum-area inverter. The delay for N stages, each with an aspect ratio which increases by a factor f, is:

$$\tau_{total} = Nf\tau_{d'}$$

where $\tau_{d'}$ is the combined delay for both the discharge and the charging of the load capacitor. For example, for the NMOS inverter $\tau_{d'} = 5\tau_d$ and the delay for the four stages described above is:

$$\tau_{total} = (4)(10)(5\tau_d)$$
$$= 200\tau_d$$

If the ratio of the load capacitance to the transistor capacitance is y, then:

$$y = \frac{C_L}{C_g}$$

For the example above with $f = 10$ and $N = 4$ then:

$$y = f^N$$

or

$$\ln(y) = N\, ln(f)$$

and

$$N = \frac{\ln(y)}{\ln(f)}$$

Thus the total delay is:

$$Nf\tau_{d'} = \frac{(f)(5\tau_d)\ln(y)}{\ln(f)}$$
$$= \frac{(10)(5\tau_d)\ln(10\,000)}{\ln(10)}$$
$$= 200\tau_d$$

It can be shown that the delay is minimized by the correct selection of f. The delay for a number of different values of f for $y = 10\,000$, as shown in Table 17.1

Thus the minimum delay occurs for $f = 2.718$ for which the number of stages is:

$$N = \ln(y) = \ln(C_L/C_g)$$

For the example above this would mean y $= \ln(10\,000) = 9$. In practice, there would be a compromise on the value of the delay and the number of stages.

For CMOS inverters the rise and fall times are approximately equal and the delay per stage for a minimum-area transistor is $\tau_{d'} = 2\tau_d$. The inverters are cascaded to increase the drive current for large off-chip capacitive loads in the same manner that is used for NMOS inverters.

Table 17.1 Variation of delay with f

f	Delay $\{\ln(y)\,[f/\ln(f)]5\tau_d\}$
1.5	$170\tau_d$
2.7	$125\tau_d$
4.0	$170\tau_d$
10.0	$200\tau_d$

In addition to off-chip capacitance, large values of capacitance occur on-chip when one circuit is required to drive a large number of inputs. For example, a clock generator is usually required to drive a large number of gates in a synchronous logic design. This on-chip load relates to the fan-out capability of the circuit. The optimization of circuit designs for load-driving ability or fan-out is greatly simplified by the use of circuit simulators such as SPICE which are capable of accurately simulating devices and circuits with different values of load capacitance and also at different temperatures. Output plots from such simulations show the rise and fall times for different configurations and allow the designer to experiment, using the relationships described above, to produce the optimum design.

17.7 Circuit Design

A number of texts are available which deal with the design of integrated MOS logic circuits, for example, *Basic VLSI Design: Systems and Circuits*, by D. A. Pucknell and K. Eshraghian (2nd edn) (1988), Prentice Hall; *Introduction to NMOS and CMOS VLSI Systems Design*, by A. Mukherjee (1986), Prentice Hall; and *Introduction to VLSI Systems*, by C. Mead and L. Conway (1980), Addison Wesley. All these texts provide a brief introduction to the technology but the main emphasis is on design information on a wide range of circuit elements. It is not the intention of this text to try to repeat these design processes, and the interested reader who wishes to study circuit design is advised to seek the necessary information from texts on VLSI design such as those listed above.

Problems

1. On graph paper draw the layout for a two-input NAND gate for both NMOS and CMOS technologies.

2. Construct stick diagrams for NMOS and CMOS technologies for three-input NAND and NOR gates. Indicate the pull-up and pull-down ratios in each case.

3. Draw the transistor circuit diagram for a CMOS cascade of four inverters required to drive a large load capacitance where $C_L/C_g = 5000$.

4. Show that the total delay for the circuit in Problem 3 is $67\tau_d$ assuming equal rise and fall times.

5. Determine the number of stages required for the minimum delay and determine the minimum delay.
[9, $49\tau_d$.]

6. Use an analog simulator to design a CMOS pad driver comprising a number of inverters from a minimum dimension inverter to $1 : f^N$ length-to-width ratio for the final inverter, where N is the number of inverters and is given by:

$$N = \frac{\ln Y}{\ln f} \text{ where } Y = \frac{C_L}{C_G}$$

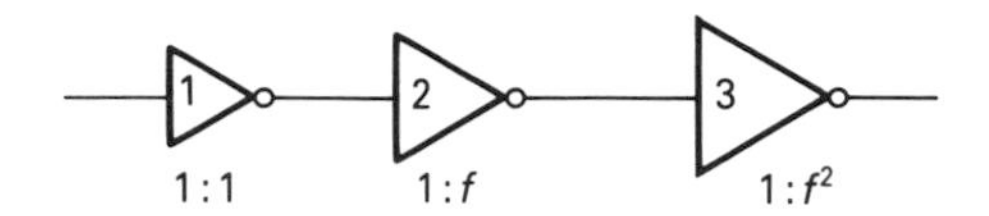

Assume that C_G is 0.01 pF and that C_L is 200 pF. Try first with four inverters and then determine the optimum number for the minimum time.

Suitable 'model' statements for a SPICE simulator are as follows:

```
.model mehn nmos(level=1 vto=1.0
+ kp=1.5e-5 gamma=0.5
+ cgso=4.5e-10 cgdo=4.5e-10
+ cj=1.0e-4 cjsw=1.0e-9)
.model mehp pmos(level=1 vto=-1.0
+ kp=0.75e-5 gamma=0.5
+ cgso=4.5e-10 cgdo=4.5e-10
+ cj=1.0e-4 cjsw=1.0e-9)
.end
```

References

[1] C. Mead and L. Conway (1980), *Introduction to VLSI Systems*, Addison-Wesley.

APPENDIX 1

Summary of important constants, equations and graphs

General constants

The following apply to silicon:

Atomic number	14
Atomic weight	28.1
Density (g/cm^3)	2.33
Dielectric constant, ε_{si}	11.7
Number of atoms/cm^3	5.0×10^{22}
Energy gap, E_{go}(eV)	1.12
n_i at 300 K (cm^{-3})	1.45×10^{10}
$\mu_n(cm^2V^{-1}s)$	1414
$\mu_p(cm^2V^{-1}s)$	471
$D_n(=\mu_n kT/q)$ (cm^2/s)	36.6
$D_p(=\mu_p kT/q)$ (cm^2/s)	12.2

Miscellaneous:

Boltzmann's constant, k (J/K)	1.38×10^{-23}
Electron charge, q (C)	1.6×10^{-19}
Free-space permittivity, ε_o (F/cm)	8.85×10^{-14}
Oxide dielectric constant, ε_{ox}	3.9
kT/q at 300 K (V)	0.0259

Chapter 1 Some electrical properties of silicon

Mass action law

$$np = n_i^2$$

Fermi potential, n-type

$$E_f - E_i = \phi_{fn} = \frac{kT}{q}\ln\left(\frac{N_d}{n_i}\right)$$

Fermi potential, p-type

$$E_i - E_f = \phi_{fp} = \frac{kT}{q}\ln\left(\frac{N_a}{n_i}\right)$$

Saturation velocity

$$v_s = 9.6 \times 10^6 \text{ cm/s}$$

Conductivity, n-type

$$\sigma_n = q\mu_n n_n$$

Conductivity, p-type

$$\sigma_p = q\mu_p p_p$$

Graph 1 mobility : impurity concentration
Graph 2 resistivity : impurity concentration

Chapter 2 Dielectric layers

Silicon consumed during oxidation

$$t_{si} = 0.45 t_{ox}$$

Graph 3 Oxide thickness : time for dry oxygen
Graph 4 Oxide thickness : time for steam
Graph 5 Oxide thickness : time at high pressure

Chapter 4 Diffusion

Graph 6 Diffusion coefficient : reciprocal of temperature
Graph 7 Solid solubility : temperature
Graph 8 $Y:X$ for $\exp - X^2$ and erfc X

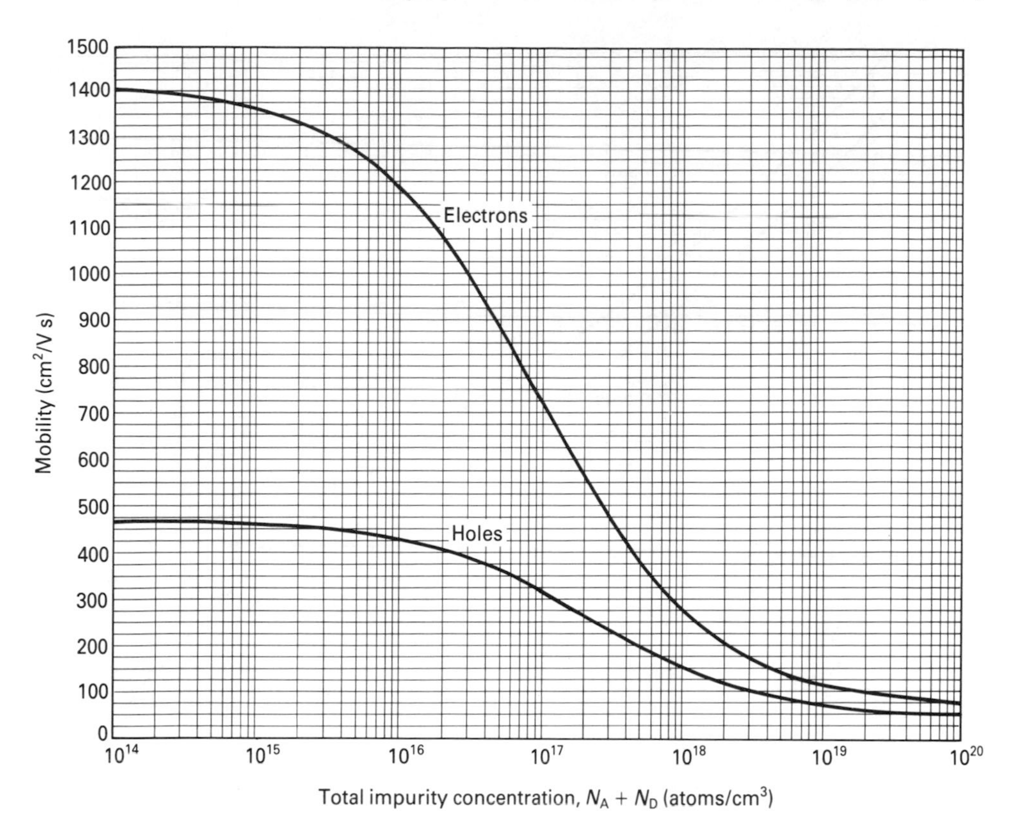

Graph 1 Variation of electron and hole mobility with impurity concentration.

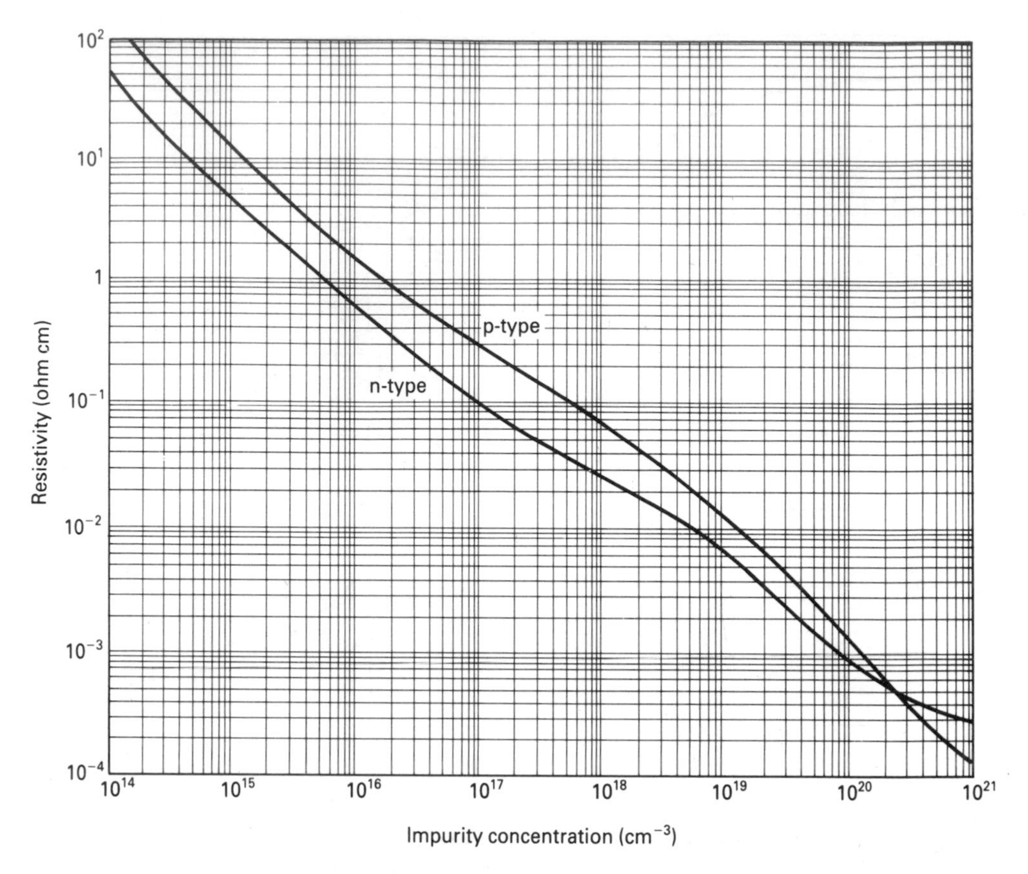

Graph 2 Variation of resistivity of silicon with impurity concentration.

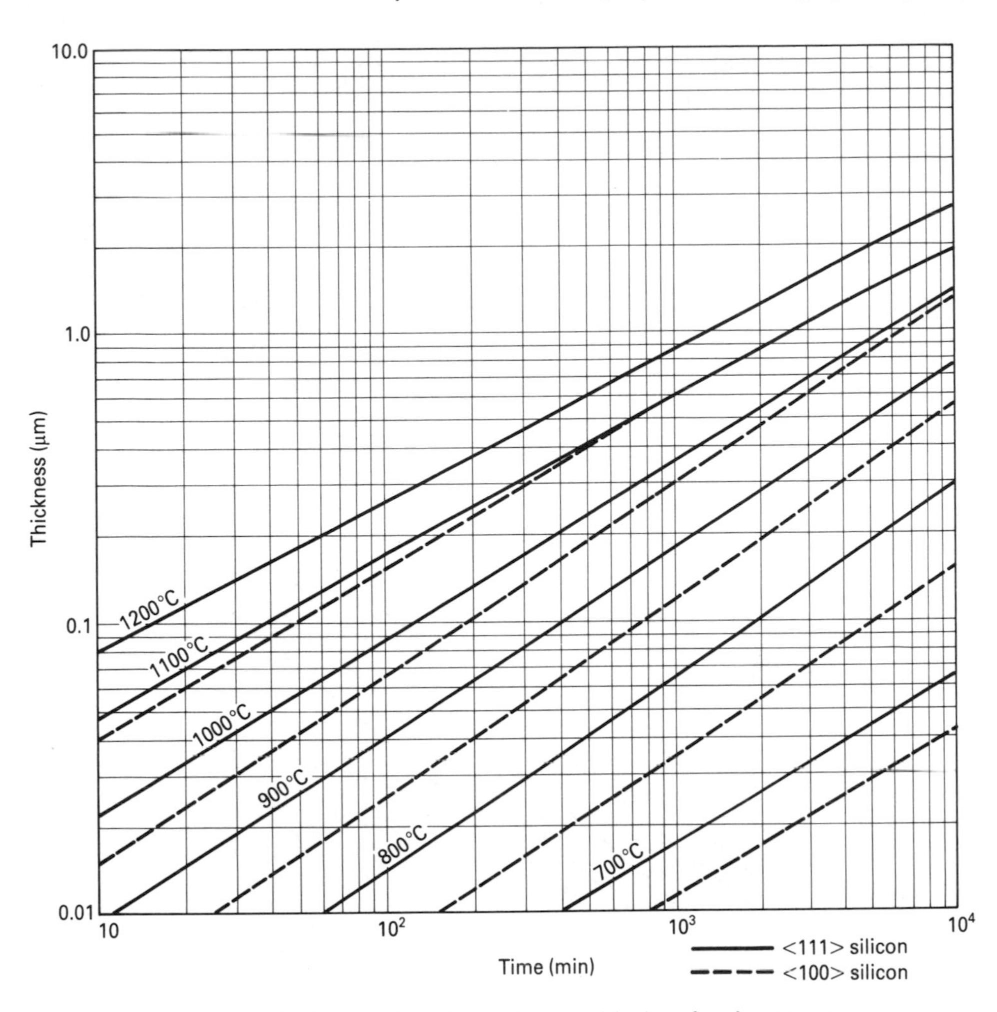

Graph 3 Variation of oxide thickness with time for dry oxygen.

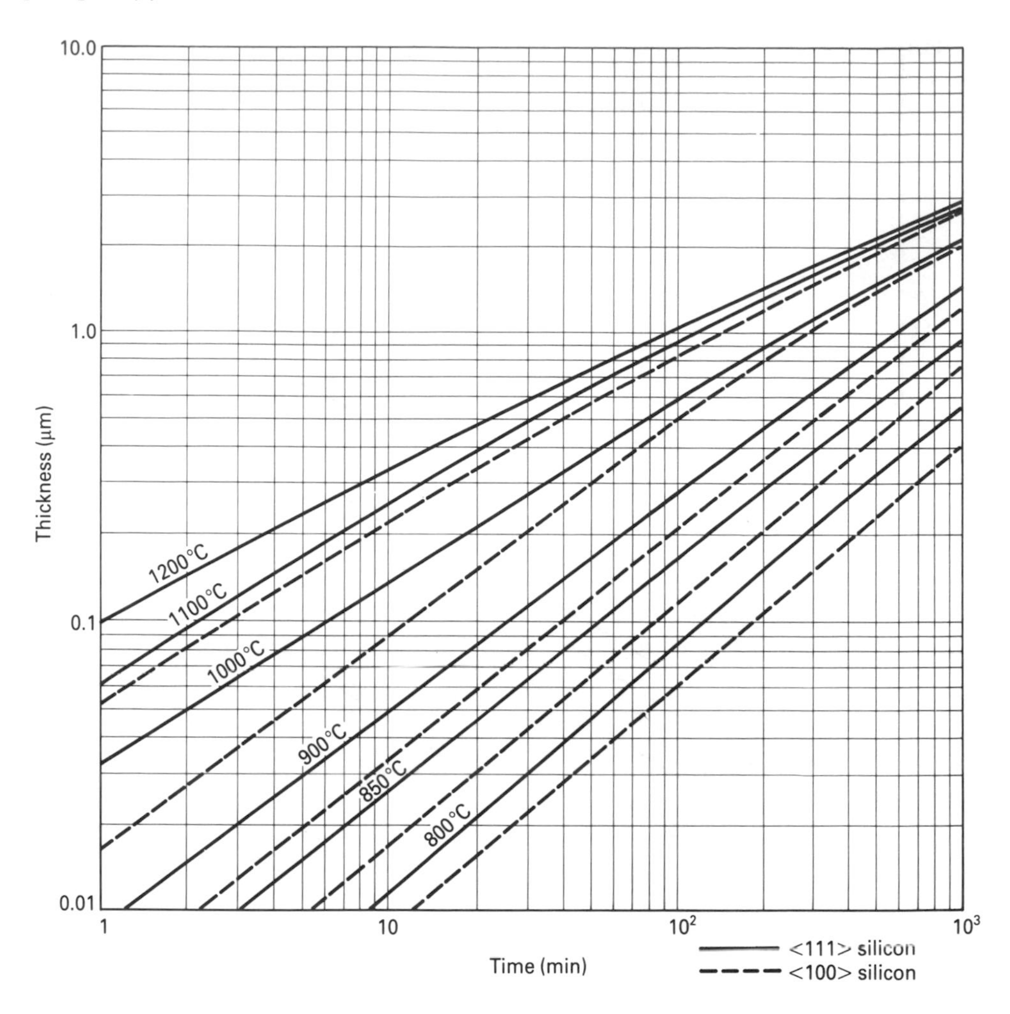

Graph 4 Variation of oxide thickness with time for steam.

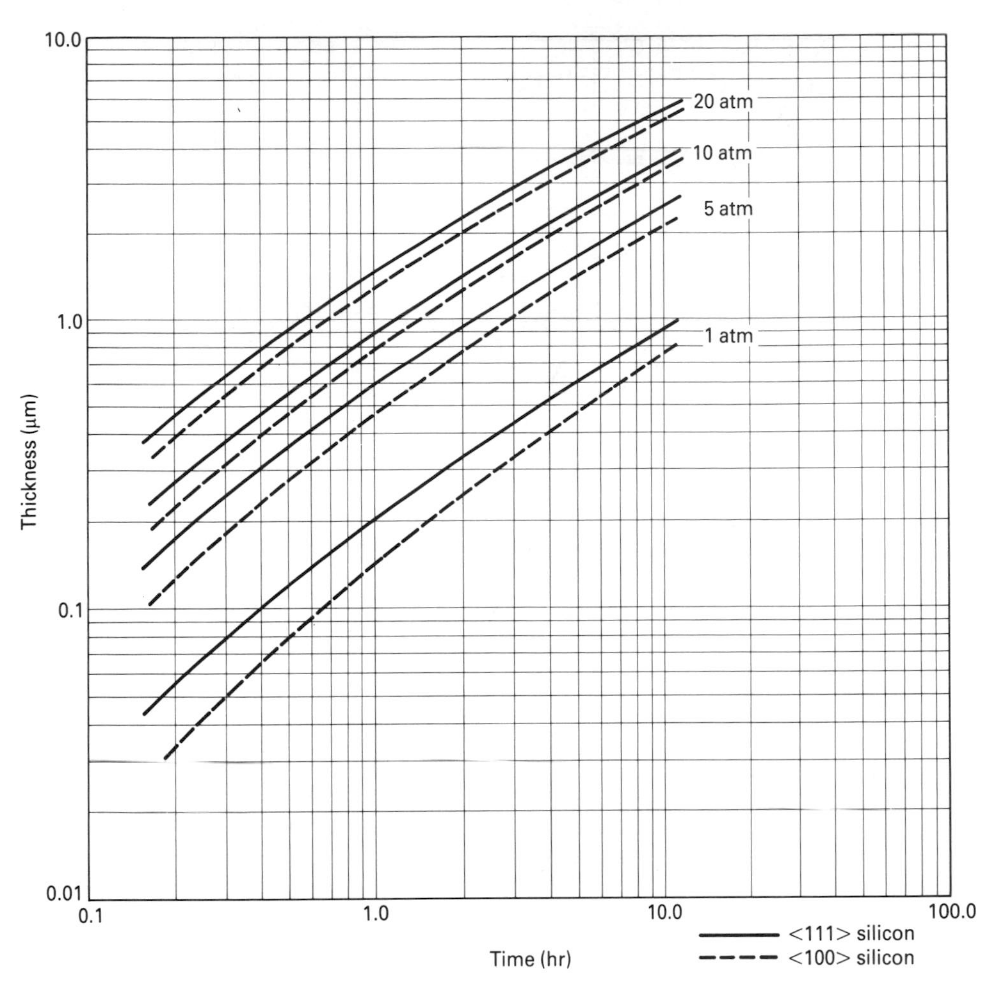

Graph 5 Variation of oxide thickness with time at 900 °C with high-pressure steam.

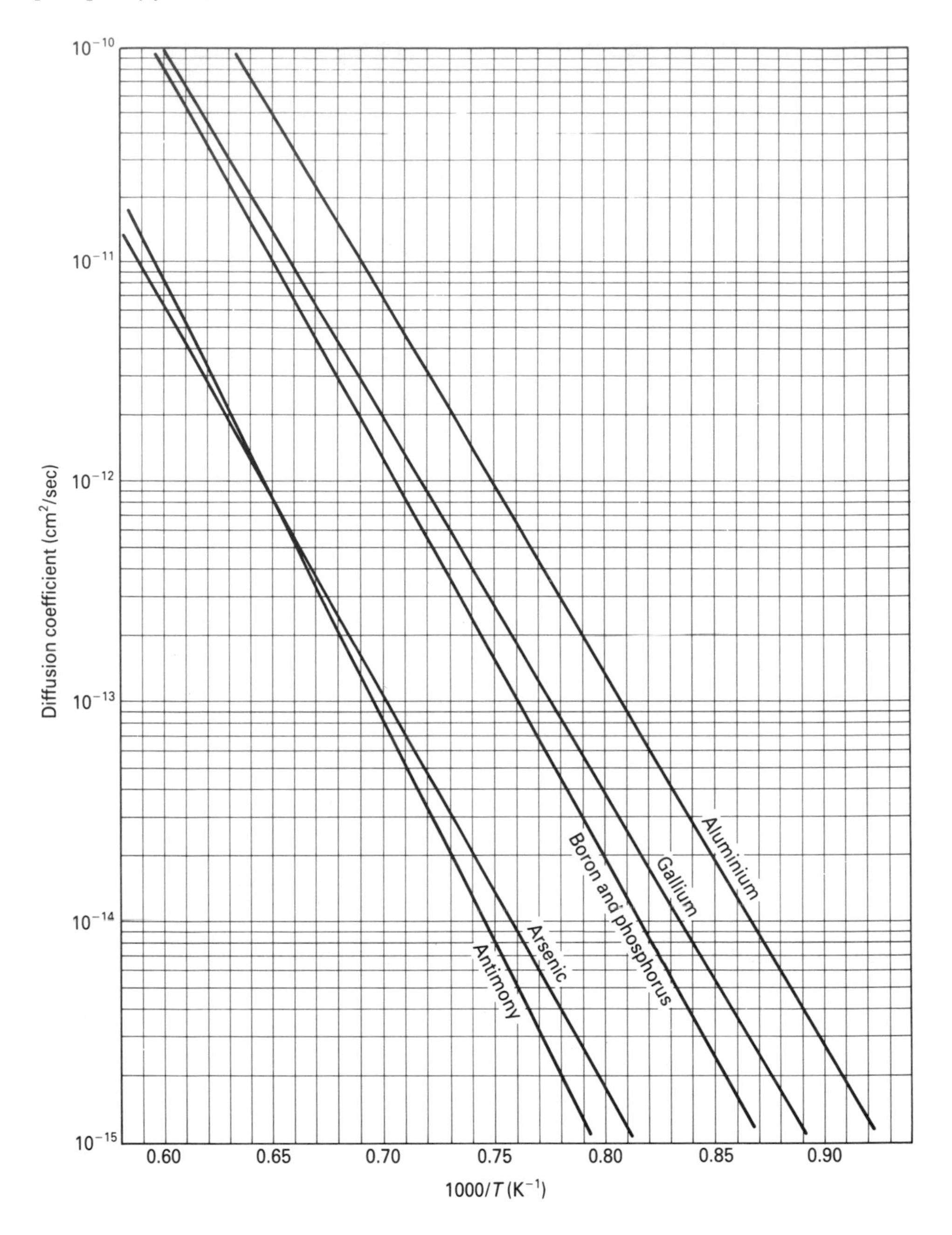

Graph 6 Variation of diffusion coefficient with reciprocal of temperature.

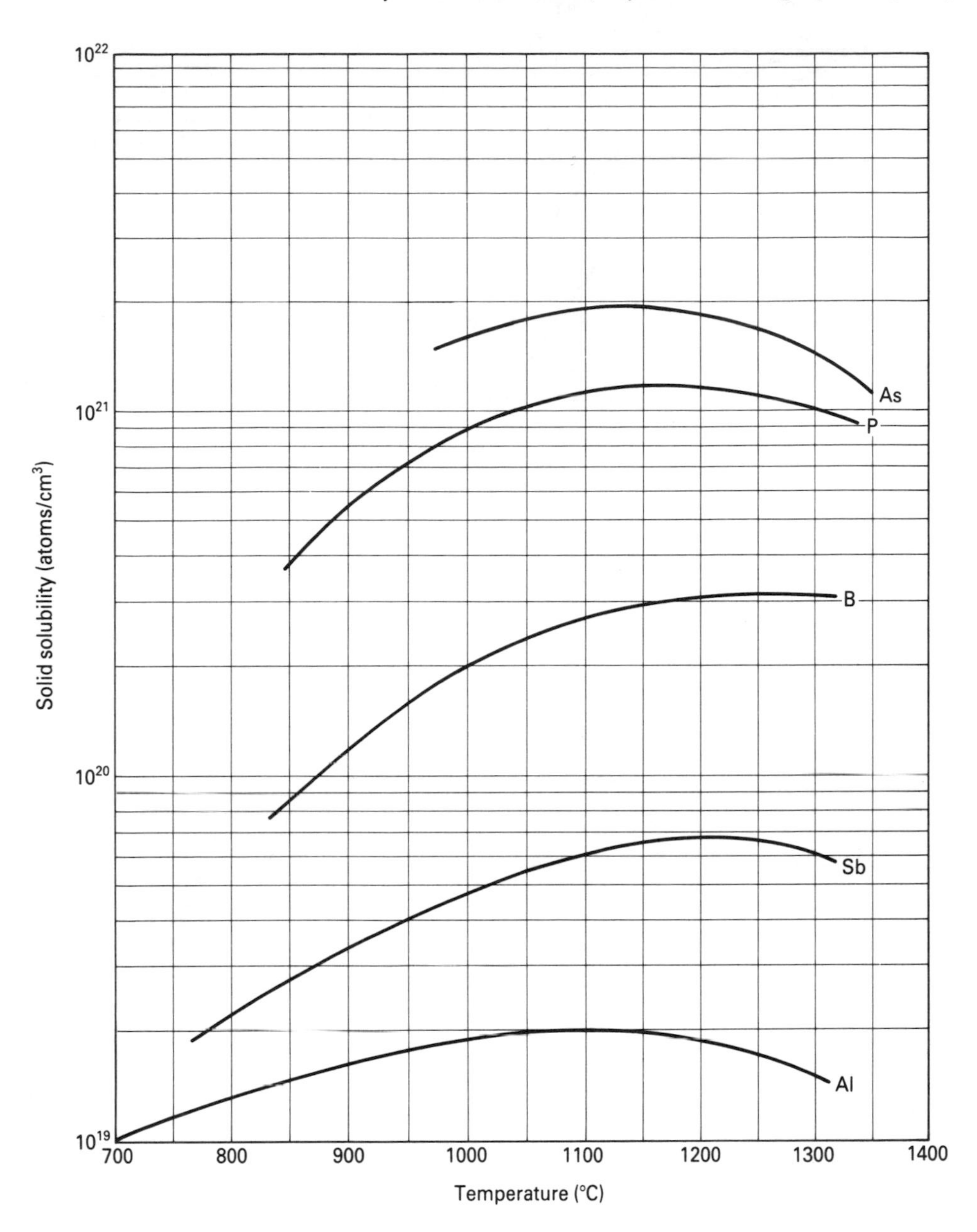

Graph 7 Variation of solid-solubility of impurities with temperature for silicon.

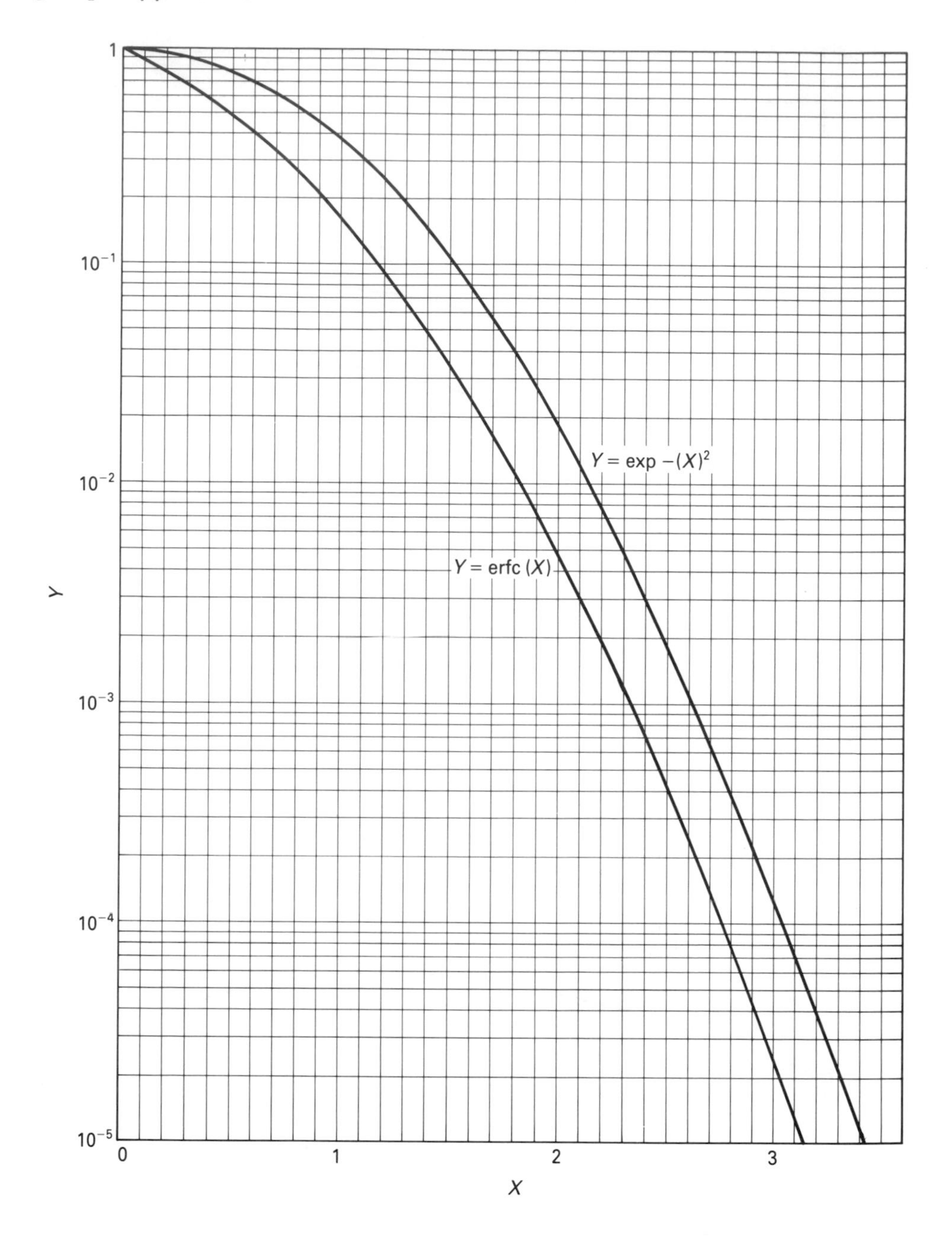

Graph 8 $Y = \exp - X^2$ and $Y = \text{erfc}\, X$.

Constant source profile

$$N_{x,t} = N_s \operatorname{erfc}\left(\frac{x}{2\sqrt{Dt}}\right)$$

Limited source profile

$$N_{x,t} = \frac{Q}{\sqrt{\pi Dt}} \exp -\left(\frac{x}{2\sqrt{Dt}}\right)^2$$

Pre-deposit

$$Q = N_s\left(\frac{D_1 t_1}{\pi}\right)^{1/2}$$

Multiple Dt steps

$$Dt_{\text{effective}} = D_1 t_1 + D_2 t_2 + \cdots$$

Chapter 5 Ion implantation

Implant dose

$$Q = \frac{It}{mq\ Area} \text{ atoms/cm}^2$$

Implant profile

$$N_x = \frac{Q}{\sqrt{2\pi \Delta R_p}} \exp -\frac{1}{2}\left(\frac{x - R_p}{\Delta R_p}\right)^2$$

where R_p is the projected range and ΔR_p is the standard deviation.

Chapter 9 Quality and reliability

Mean-time-to-failure

$$t_{av} = \sum_{r=1}^{N} \frac{t_r}{N}$$

Failure rate

$$f_r = K \exp -\frac{E}{kT}$$

where K is a constant and E is the activation energy.

Die per wafer

$$N_d = \frac{\pi(R - A^{1/2})^2}{A}$$

where R = radius of wafer and A = die area

Poisson's yield model

$$Y = \exp - (AD)$$

Murphy's yield model

$$Y = \left(\frac{1 - \exp - (AD)}{AD}\right)^2$$

Seed's yield model

$$Y = \exp - (\sqrt{AD})$$

where A = die area and D = defect density

Chapter 11 Bipolar transistor

Graph 9 Breakdown voltage: impurity concentration

Emitter injection efficiency

$$\gamma = \left[1 + \frac{D_{pE} N_B W_B}{D_{nB} N_E L_E}\right]^{-1}$$

or

$$\gamma = \left[1 + \frac{D_{pE} G_b}{D_{nB} G_e}\right]^{-1}$$

where G_b, G_e are the Gummel numbers for the emitter and base.

Base transport factor

$$\beta^* = 1 - \frac{W_B^2}{2L_B^2}$$

where $L_B = \sqrt{D_n \tau_n}$ for npn transistor

Emitter current

$$I_E \approx \frac{qAD_B n_i^2}{G_b} \exp \frac{V_{BE}}{V_T}$$

Band-gap narrowing

$$dE_g = 22.5\left(\frac{N_E}{10^{18}}\right)^{1/2} \text{ meV}$$

and

$$n_{iE}^2 = n_i \exp \frac{dE_g}{kT}$$

where n_{iE} is used in place of n_i in the equation for I_E above.

Base resistance

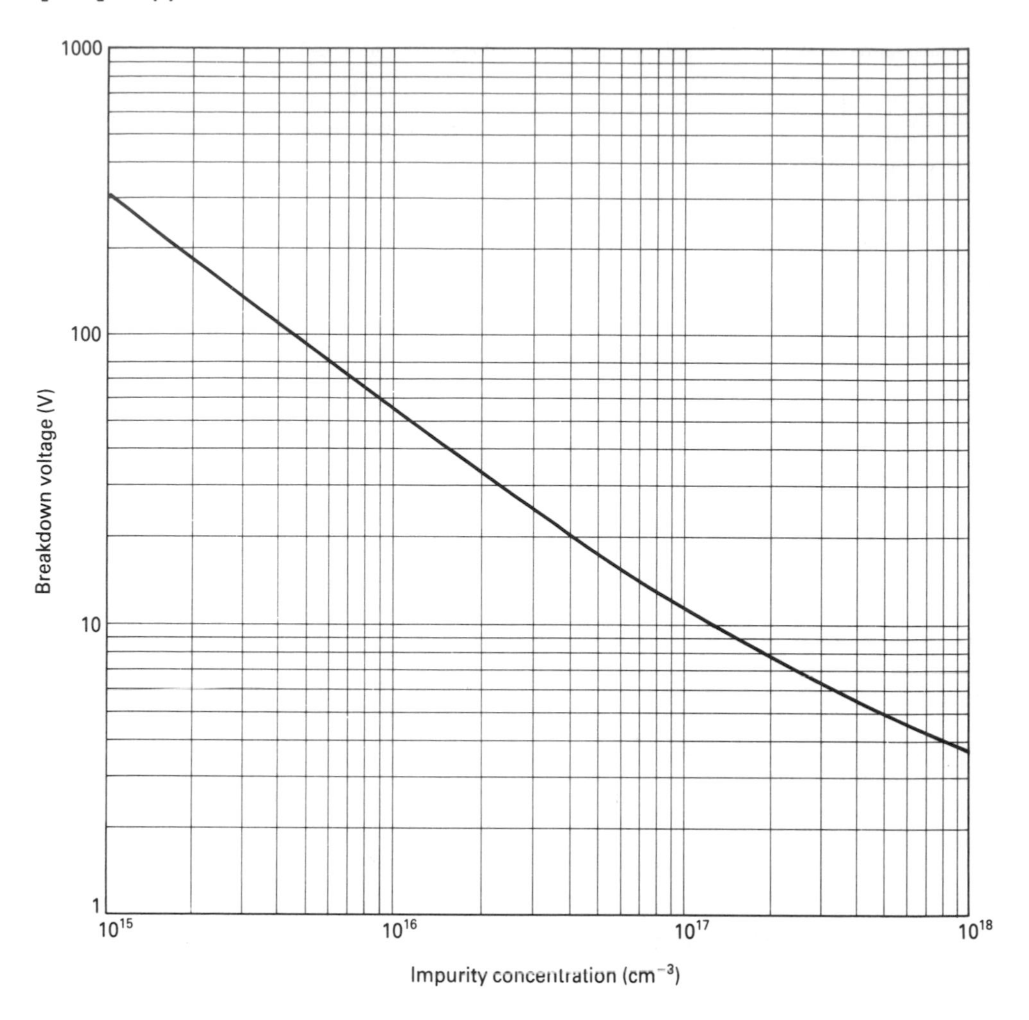

Graph 9 Variation of junction breakdown voltage with impurity concentration.

$$r_{bb} = \frac{\rho_B x_{BC}}{3d_B L_E} + \frac{\rho_B d_{EB}}{W_B L_E} + \frac{\rho_B d_E}{6W_B L_E}$$

Collector resistance

$$r_{cc} = \frac{\rho_C(x_{epi} - x_E)}{d_C L_C} + \frac{\rho_S d_{EC}}{(L_E + L_C)/2} + \frac{\rho_C(x_{epi} + x_{BC})}{d_E L_E}$$

Junction capacitance

$$C = A\left[\frac{\varepsilon_o \varepsilon_{si} q N}{2(V + \phi_o)}\right]^{1/2}$$

where ϕ_o is the built-in junction potential.

Depletion layer width

$$x = \left[\frac{2\varepsilon_o \varepsilon_{si}(V + \phi_o)}{qN}\right]^{1/2}$$

Cut-off frequency

$$f_t = \left\{2\pi\left[\frac{kTC_E}{qI_E} + \frac{W_B^2}{D_{nB}} + r_{cc}(C_{BC} + C_{CS}) + \frac{x_C}{v_s}\right]\right\}^{-1}$$

Chapter 12 Diodes, resistors and capacitors

Diode equation

$$I_F = I_S\left(\exp\frac{V}{V_T} - 1\right)$$

where

$$I_S = qA\left(\frac{D_p p_{no}}{L_p} + \frac{D_n n_{po}}{L_n}\right)$$

and

$$I_S \approx \frac{qAD_n n_{po}}{L_n} \text{ for an n}^+\text{p junction}$$

Generation current

$$I_g = qAgx_m$$

where g is the generation rate and x_m is the width of the depeletion layer.

Generation rate

$$g = \frac{n_i}{\tau}$$

where τ is the effective lifetime of minority carriers.

Schottky diode current

$$I_D = A^{**}T^2 \exp\left(\frac{-\phi_B}{V_T}\right)\left(\exp\frac{V}{V_T} - 1\right)$$

where A^{**} is Richardson's constant and ϕ_B is the metal–semiconductor barrier height.

$$A^{**} = 120\ \text{A/K}^{-2}$$

Chapter 14 The MOS capacitor

Inversion voltage (ideal)

$$V_{Ginv} = \frac{(2\varepsilon_o \varepsilon_{si} q N_a 2\phi_f)^{1/2}}{C_{ox}} + 2\phi_f \text{ p-type substrate}$$

$$V_{Ginv} = \frac{(2\varepsilon_o \varepsilon_{si} q N_d 2\phi_f)^{1/2}}{C_{ox}} - 2\phi_f \text{ n-type substrate}$$

Work function

$$\phi_{ms} = \phi_m - \phi_s = \phi_m - \chi + \frac{E_g}{2} - \phi_f \quad \text{for n-type}$$

$$\phi_{ms} = \phi_m - \phi_s = \phi_m - \chi + \frac{E_g}{2} + \phi_f \quad \text{for p-type}$$

where ϕ_m is the work function of the metal.

Inversion voltage:
Aluminium electrode

$$V_{inv} = -0.6 - \phi_f - \frac{Q_b}{C_{ox}} - \frac{Q_{ss}}{C_{ox}} \quad \text{p-type}$$

$$V_{inv} = -0.6 + \phi_f + \frac{Q_b}{C_{ox}} - \frac{Q_{ss}}{C_{ox}} \quad \text{n-type}$$

n-type polysilicon electrode

$$V_{inv} = +0.55 - \phi_f - \frac{Q_b}{C_{ox}} - \frac{Q_{ss}}{C_{ox}} \quad \text{p-type}$$

$$V_{inv} = -0.55 + \phi_f + \frac{Q_b}{C_{ox}} - \frac{Q_{ss}}{C_{ox}} \quad \text{n-type}$$

Substrate concentration

$$N_{sub} = \frac{4\phi_f}{q\varepsilon_o \varepsilon_{si}}\left[\frac{C_{s\,min}}{A}\right]^{1/2}$$

where

$$\phi_f = \frac{kT}{q}\ln\frac{N_{sub}}{n_i}$$

Flat-band capacitance

$$C_{fb} = \frac{A\varepsilon_o\varepsilon_{si}}{t_{ox} + (\varepsilon_{ox}/\varepsilon_{si})L_D}$$

where

$$L_D = \sqrt{kT\varepsilon_o\varepsilon_{si}/(q^2N_{sub})}$$

Oxide charge

$$Q_{ss} = \frac{C_{ox}}{A}(\phi_{ms} - V_{fb})$$

Chapter 15 The MOS transistor

Drain current (linear region)

$$I_D = K\frac{W}{L}\left[(V_{GS} - V_T)V_{DS} - \frac{V_{DS}^2}{2}\right]$$

where

$$K = \frac{\varepsilon_o\varepsilon_{ox}\mu}{t_{ox}}$$

Drain current (saturation)

$$I_D = \frac{KW}{2L'}(V_{GS} - V_T)^2$$

where

$$L' = L - \sqrt{2\varepsilon_o\varepsilon_{si}V_{DS}/(qN_{sub})}$$

Threshold voltage

$$V_T = \phi_{ms} - \frac{Q_{ss}}{C_{ox}} + 2\phi_f + \frac{\sqrt{2\varepsilon_o\varepsilon_{si}qN_a2\phi_f}}{C_{ox}}$$

n-channel

V_T with substrate bias

$$V_T = V_{T(0)} + \gamma\sqrt{V_{sub}}$$

where

$$\gamma = \frac{t_{ox}}{\varepsilon_o\varepsilon_{ox}}\sqrt{2\varepsilon_o\varepsilon_{si}qN_a} \quad \text{n-channel}$$

where $V_{T(0)}$ is the threshold voltage without substrate bias

Transconductance

$$g_m = \frac{\mu\varepsilon_o\varepsilon_{ox}}{t_{ox}}\frac{W}{L}(V_{GS} - V_T)$$

Maximum frequency

$$f_T = \frac{\mu}{2\pi L^2}(V_{GS} - V_T)$$

EPROM/EEPROM field

$$E_1 = \frac{V_G}{(t_{ox1} + t_{ox2})} + \frac{Q}{\varepsilon_o\varepsilon_{ox1}\left(1 + \frac{t_{ox1}}{t_{ox2}}\right)}$$

where the charge is

$$Q(t) = \int_0^t J_1(E_1)\,dt$$

and the shift in V_T is

$$\Delta V_T = Q\frac{t_{ox1}}{\varepsilon_o\varepsilon_{ox1}}$$

Chapter 16 The MOS inverter

Channel resistance

$$r_{DS} = \frac{L}{KW}\frac{1}{(V_{GS} - V_T)}$$

nMOS resistance ratio

$$r_{DS}|_{load} = 4r_{DS}|_{driver}$$

nMOS with pass transistor

$$r_{DS}|_{load} = 8r_{DS}|_{driver}$$

APPENDIX 2

Solution guide to end-of-chapter problems

Chapter 1 Some electrical properties of silicon

Q1 Use the relationship $\rho = 1/\sigma = 1/(q\mu_n N_d)$. Obtain a value for mobility from Graph 1 (Appendix 1) and substitute the number of arsenic atoms for N_d.

The number of electrons is equal to the number of arsenic atoms.

Use the mass action law (Appendix 1) to determine the number of holes.

Q2 N_d is equal to the number of phosphorus atoms; N_a is equal to the number of boron atoms.

$N_d > N_a$, therefore the number of electrons is $N_d - N_a$. Use the mass action law to determine the number of holes.

Resistivity is $1/(q\mu_n[N_d - N_a])$, where μ_n is determined for a concentration of $(N_d + N_a)$.

Q3 Number of holes corresponds to number of boron atoms. Use mass action law to determine number of electrons.

$N_a > N_d$, therefore material is p-type and number of holes is $N_a - N_d$. Use mass action law to find number of electrons.

Q4 $\phi_f = 0.026 \ln(N/n_i)$, solve for $N = 1 \times 10^{14}$ and $N = 1 \times 10^{18}$.

Chapter 2 Dielectric layers

Q2 Use Graphs 3 and 4, Appendix 1.

From Graph 3 determine thickness after 30 min – 140 nm. Determine the time it would have taken to produce 140 nm in steam at 1200 °C – 2 min. Add this to the 30 min to give an equivalent time of 32 min in steam at 1200 °C. The thickness of 580 nm. When part of the oxide is removed and the slice is reoxidized, then part of the slice will be clean silicon and the other part will be oxidized with 580 nm of oxide. For the bare silicon use the graph to determine the thickness after 30 min at 1000 °C – 30 nm. For the oxidized section determine an equivalent time required to produce 580 nm in dry oxygen at 1000 °C – 2500 min. The total time for the oxidized section is 2500 min + 30 min. The change in thickness is negligible.

Q4 As for Q2. After etching the oxide there will be a new region of exposed silicon which undergoes two oxidations, one at 1100 °C in dry oxygen for 20 min and another at 1100 °C for 50 min. The original silicon undergoes four oxidation steps.

Chapter 4 Diffusion

Q1 *n*-type Si with $N_B = 1 \times 10^{15}$.
Constant source with $N_S = 1 \times 10^{19}$ and erfc profile.

At 1100 °C find the diffusion coefficient from Graph 6 as $D = 3.8 \times 10^{-13}\,\text{cm}^2/\text{s}$. Then

$$1 \times 10^{15} = 1 \times 10^{19}\,\text{erfc}\,X$$

Use Graph 8 to determine X as $X = 2.76$. Then

$$2.76 = \frac{x_j}{2\sqrt{3.8 \times 10^{-13} \times 45 \times 60}}$$

and

$$x_j = 1.76 \times 10^{-4}\text{cm} \quad \text{or } 1.76\,\mu\text{m}$$

Q2 n-type Si with $N_B = 4 \times 10^{16}$ and a constant source (erfc profile) with $N_S = 4 \times 10^{18}$.
At 1200 °C the diffusion coefficient is $3.2 \times 10^{-12}\,\text{cm}^2/\text{s}$.

After 20 min

$$N_{x,t} = 4 \times 10^{14}\,\text{erfc}\frac{2 \times 10^{-4}}{2\sqrt{3.2 \times 10^{-12} \times 20 \times 60}}$$

Use Graph 8 to determine

$$Y = \text{erfc}\,X = 2.3 \times 10^{-2}$$

to give

$$N_{x,t} = 9.2 \times 10^{16}/\text{cm}^{-3}$$

The junction depth after 20 min + 30 min is:

$$4 \times 10^{16} = 4 \times 10^{18}\,\text{erfc}\,X$$

and from Graph 8, $X = 1.82$. Then

$$1.82 = \frac{x_j}{2\sqrt{3.2 \times 10^{-12} \times 50 \times 60}}$$

and

$$x_j = 3.6 \times 10^{-4} = 3.6\,\mu\text{m}$$

Q3 Determine the diffusion coefficients at 1200 °C, 1180 °C and 1100 °C from Graph 6 as 3.2×10^{-12}, 2×10^{-12} and $3.8 \times 10^{-13}\,\text{cm}^2/\text{s}$.

Determine the effective Dt product as

$$Dt_{\text{effective}} = (3.2 \times 10^{-12} \times 3 \times 60 \times 60) + (2 \times 10^{-12} \times 40 \times 60) + (3.8 \times 10^{-13} \times 20 \times 60)$$

$$Dt_{\text{effective}} = 3.98 \times 10^{-8}\,\text{cm}^2/\text{s}$$

For a Gaussian diffusion:

$$N_{x,t} = \frac{5 \times 10^{13}}{\sqrt{\pi \times 3.98 \times 10^{-8}}} \exp\left(-\frac{x}{2\sqrt{3.98 \times 10^{-8}}}\right)^2$$

The junction is formed when $N_{x,t} = N_B = 2 \times 10^{16}$ (0.35 ohm cm).

Solve to give a junction depth of $5.6\,\mu$m.

Q4 Determine the diffusion coefficients at 950 °C and 1100 °C as 8.5×10^{-15} and 3.8×10^{-13}.

For a two-step diffusion the first 15 min diffusion produces a thin surface layer for which the surface density Q is:

$$Q = 2N_s\left(\frac{Dt}{\pi}\right)^{1/2}$$

$$Q = 2 \times 5 \times 10^{19}\left(\frac{8.5 \times 10^{-15} \times 15 \times 60}{\pi}\right)^{1/2}$$

$$= 1.56 \times 10^{14}\,\text{cm}^2$$

The second diffusion is described by the Gaussian profile as:

$$N_{x,t} = \frac{1.56 \times 10^{14}}{\sqrt{\pi \times 3.8 \times 10^{-13} \times 30 \times 60}} \exp\left(-\frac{x}{2\sqrt{3.8 \times 10^{-13} \times 30 \times 60}}\right)^{1/2}$$

The junction is formed when $N_{x,t} = 5 \times 10^{16}$, that is,

$$0.015 = \exp -\left(\frac{x_j}{2\sqrt{6.84 \times 10^{-10}}}\right)^2$$

and

$$2.04 = \frac{x}{5.23 \times 10^{-5}} = 1.06\,\mu\text{m}$$

At the surface when $x = 0$ the surface concentration is:

$$N_{0,t} = \frac{Q}{\sqrt{\pi Dt}} = \mathbf{3.4 \times 10^{18}/cm^3}$$

Chapter 9 Quality and reliability

Q1 Failure rate is given by:

$$f_r = \frac{N}{\sum_{r=1}^{N} t_r} = \frac{3}{200 \times 400} = 3.75 \times 10^{-5}$$

and the constant K is given by:

$$K = \frac{3.75 \times 10^{-5}}{\exp - \left[\dfrac{1.6 \times 10^{-19} \times 0.68}{1.38 \times 10^{-23} \times (200 + 273)}\right]}$$

$$= 650$$

and the failure rate at 70 °C is:

$$f_r = 650 \exp - \left(\frac{1.6 \times 10^{-19} \times 0.68}{1.38 \times 10^{-23} \times (70 + 273)}\right)$$

$$= 6.77 \times 10^{-8}$$

or

$f_r = 0.68$ devices per 10^9 hr

Q2

	#1	#2
Size	100 mm	150 mm
Defects	4	2.5
Die area	30 mm^2	25.5 mm^2
Die/wafer	207	602
Yield	33.4%	45%
Good die/wafer	69	270
Cost/wafer	$90	$150
Cost/die	$1.3	$0.55

Q3 Sketch the control charts.

Chapter 11 The bipolar transistor

Q1 Refer to Figure 11.2 for guidance, but note that the band diagram is required for a pnp transistor.

Q2 Use the relationship:

$$|E_f - E_i| = \frac{kT}{q} \ln\left(\frac{N}{n_i}\right)$$

Q3 For the npn transistor:

emitter $N_d = 2 \times 10^{19}$, $\mu_p = 50$ (from Graph 1)

base $N_a = 5 \times 10^{17}$, $\mu_n = 370$

$L_e = 1\ \mu m$

$W_B = 0.6\ \mu m$

$L_B = 20\ \mu m$

The emitter injection efficiency is:

$$\gamma = \frac{1}{1 + \dfrac{D_{pE} G_b}{D_{nB} G_e}}$$

where

$$D_{pE} = 0.026 \times \mu_p = 0.026 \times 50 = 1.3\ \text{cm}^2/\text{s}$$

and

$$D_{nB} = 9.6\ \text{cm}^2/\text{s}$$

The Gummel numbers are:

$$G_b = \frac{(5 \times 10^{17} - 1 \times 10^{15}) \times 0.6 \times 10^{-4}}{2}$$

$$= 1.497 \times 10^{13}/\text{cm}^2$$

and

$$G_e = 2 \times 10^{19} \times 1 \times 10^{-4} = 2 \times 10^{15}/\text{cm}^2$$

and

$$\gamma = \frac{1}{1 + \dfrac{1.3 \times 1.497 \times 10^{13}}{9.6 \times 2 \times 10^{15}}} = 0.998\,98$$

The base transport factor is:

$$\beta^* = 1 - \frac{W_B^2}{2L_B^2} = 1 - \frac{(0.6 \times 10^{-4})^2}{2 \times (20 \times 10^{-4})^2}$$

$$= 0.999\,55$$

and

$$\alpha_F = \gamma\beta^* = 0.998\,98 \times 0.999\,55 = 0.998$$

and

$$\beta_F = \frac{\alpha_F}{1 - \alpha_F} = \frac{0.998}{1 - 0.998} = 679$$

Q4 Use the equation for the critical current above which the Kirk effect reduces the current gain, is illustrated in Figure 11.10:

$$3 \times 10^{-3} = 25 \times 10^{-8} \times 1.6 \times 10^{-19} \times 1 \times 10^{7} \times \left[N_c + \frac{2 \times 8.85 \times 10^{-14} \times 11.7 \times 5}{1.6 \times 10^{-19} \times (2.5 \times 10^{-4})^2} \right]$$

Solve for N_c.

Q5 Assume a uniform base impurity distribution with $N_a = 2 \times 10^{17}$ atoms/cm^3; obtain the mobility for holes from Graph 1 as 250 cm^{2}V^{-1}s.

The resistivity of the base is

$$\rho = 1/(1.6 \times 10^{-19} \times 250 \times 2 \times 10^{17})$$
$$= 0.125 \text{ ohm cm}$$

Determine the three separate components of the base resistance as:

$$r_a = \frac{0.125 \times 1 \times 10^{-4}}{3 \times 4 \times 10^{-4} \times 10 \times 10^{-4}} = 10.4$$

$$r_b = \frac{0.125 \times 2 \times 10^{-4}}{0.4 \times 10^{-4} \times 10 \times 10^{-4}} = 625$$

$$r_c = \frac{0.125 \times 5 \times 10^{-4}}{6 \times 0.4 \times 10^{-4} \times 10 \times 10^{-4}} = 260$$

to give a total of 895 ohm.

Q6 The built-in junction potential is:

$$\phi_o = 0.026 \ln \frac{(5 \times 10^{19})(2 \times 10^{17})}{(1.45 \times 10^{10})^2} = 0.998 \text{ V}$$

Use the equation for junction capacitance:

$$C_E = \frac{area\sqrt{1.6 \times 10^{-19} \times 8.85 \times 10^{-14} \times 11.7 \times 2 \times 10^{17}}}{\sqrt{2(0.998 - 0.6)}}$$

For the bottom of the emitter well the area is $50 \times 10^{-8}\ \mu$m^2, while for the sides of the emitter the area is $18 \times 10^{-8}\ \mu$m.

Q7 Use the appropriate part of the equation for f_t:

$$f_t = 2\pi \left[\frac{0.026 \times (0.1 \times 10^{-12} + 0.04 \times 10^{-12})}{10 \times 10^{-6}} + \frac{(0.4 \times 10^{-4})^2}{2 \times 20} \right] = 15 \text{ GHz}$$

Chapter 12 Diodes, resistors and capacitors

Q1 Use the diode equation and determine values of I_F for a range of voltages from 0 V to 0.7 V and plot $I_F : V$. That is, use:

$$I_F = I_S\left(\exp\frac{V}{V_T} - 1\right)$$

Use a plotting routine or spread sheet to plot the graph. If it is possible to change the y-axis to a logarithmic scale, note that the graph of $\log(I_F) : V$ is a straight line.

Q2 Notice that the equation for the Schottky diode is the same as that for the pn junction diode, but where I_S is:

$$I_S = A^{**}T^2 \exp\left(\frac{-\phi_B}{V_T}\right)$$

and that with $\phi_B = 0.84$ and at 300 K, $I_S = 1 \times 10^{-7}$ A. This is considerably larger than the typical value of I_S for a junction diode.

NB: if an analog circuit simulator does not contain a model for a Schottky diode then it is quite acceptable to use the model for a pn junction diode, but to increase the value of I_S.

Q3 The value of a simple rectangular integrated resistor is:

$$R = \rho_s \frac{L}{W}$$

where ρ_s is the sheet resistance and has values of either 200 ohm/square or 600 ohm/square. Determine the length L when the width W is 5 μm for each value of resistor. The area is $W \times L\ \mu$m^2. Determine the junction capacitance from:

$$C = Area\sqrt{\frac{q\varepsilon_o\varepsilon_{si} N}{2(V + \phi_o)}}$$

where

$$\phi_o = 0.026 \ln \frac{N}{n_i} = 0.33 \text{ V}$$

The frequency is given by:

$$f_c = \frac{1}{3RC}$$

Q4 The nominal resistor is given by:

$$R = \rho_s \frac{L}{W}$$

With $W = 5\,\mu m$ then L is $50\,\mu m$ for 10 kohm and $125\,\mu m$ for 25 kohm. Now assume that both L and W are increased by $0.1\,\mu m$ through over-etching. The new values of resistance are 0.543 kohm and 23.828 kohm.

The change in absolute value is aproximately 5%, but the ratio for both the nominal resistors and the over-etched resistors is the same.

Q5 For the resistors to have the same length then the width must vary. If the minimum width is $2\,\mu m$ then this must apply to the larger value–the 25 kohm resistor. The nominal width for the 10 kohm resistor which has the same length as the 25 kohm resistor ($125\,\mu m$) is $5\,\mu m$.

Over-etching results in values of 9.811 kohm and 23.828 kohm. The percentage change in absolute value is now different for each resistor, but in addition the ratio also changes.

The main point to note is that the effect of over- or under-etching is minimized by increasing the minimum dimensions, that is, the width. If the ratio is important then the width must be the same for both resistors.

Chapter 13 Bipolar circuit elements

Q1 With one of the inputs low Q1 is forward biased and

$$I_{R1} = \frac{5 - 0.2 + 0.7}{2.7\,k} = 1.52\,mA$$

Q2 The output of a gate connects to a number of inputs of following gates. When the output is low so that the inputs of the following gates are forward biased then the output transistor Q4 must sink the input currents of each of the gates attached to it. If the output transistor can sink 20 mA then the number of gates that can be attached to it is 20 mA/1.52 mA = 13.

Q3 Inputs high

$$P_{diss} = (I_{R1} + I_{R2})V_{CC}$$

$$I_{R1} = \frac{5 - 1.4 - 0.7}{2.7\,k} = \frac{2.9}{2.7\,k} = 1.07\,mA$$

$$I_{R2} = \frac{5 - 0.9}{750} = 5.46\,mA$$

$$P = (1.07\,mA + 5.46\,mA)5 = 32\,mW$$

One input low

$$I_{R2} = 1.52\,mA$$

$$P = (1.52\,mA)(5) = 7.6\,mA$$

and

$$P_{average} = \frac{32\,mW + 7.6\,mW}{2} = 20\,mW$$

Q4 If the area of Q2 is 1/10 of the area of Q1 then $I_{C2} = I_{C1}/10$, and $I_{Ref} = 10I_{C1} = 0.1$ mA. Then

$$R_{ref} = \frac{10 - 0.7}{0.1\,mA} = 93\,k\Omega$$

Q5

$$I_{C2} = \frac{0.026}{R_E}\ln\left(\frac{I_{ref}}{I_{C2}}\right)$$

If $I_{ref} = 1$ mA then
$R_{ref} = (10 - 0.7)/1\,mA = 9.3\,k\Omega$.
R_E is:

$$R_E = \frac{0.026}{10 \times 10^{-6}}\ln\left(\frac{1 \times 10^{-3}}{10 \times 10^{-6}}\right) = 11.97\,k\Omega$$

Q6 The equation for the current is:

$$I = \frac{0.026}{10 \times 10^{3}}\ln\frac{7.15 \times 10^{-4}}{I'}$$

where the value of I on the right-hand side of the equation is shown as I'. The equation is transcendental. It can be solved by trial and error by letting I' have some value and solving for I. The value of I is unlikely to be the same as I' for the first guess. For the second guess substitute the value of I for I' and repeat. The table below shows the progress.

I'	I
20 μA	9.2 μA
9.27 μA	11.29 μA
11.29 μA	10.78 μA
10.78 μA	10.9 μA
10.9 μA	10.87 μA
10.87 μA	10.88 μA

Thus the current is approximately 11 μA.

Q7 Repeat the process for a supply voltage of 10 V.

Q8 Determine I_{ref} as (30 – 0.7)/30 k and calculate I_{C4} using the same procedure as shown in Q6. Assume that the current divides equally between Q1 and Q2. The gain is given by:

$$A_v = g_m R_{load}$$

$$g_m = \frac{Ic}{0.026}$$

Solving for I_C gives 300 μA/2, for which g_m is 5769 μmho, and the gain is ≈170.

Q9 The gain is $A_v = R_{o2}/R_{o4}$. The current through Q2 and Q4 is 5 μA and the output resistance for each transistor is:

$$R_{o2} = \frac{\textit{Early voltage}}{I_C} = \frac{30}{5 \times 10^{-6}} = 6\,\text{M}\Omega$$

$$R_{o4} = \frac{50}{5 \times 10^{-6}} = 10\,\text{M}\Omega$$

and the g_m is 5 μA/0.026 = 192 μmho. The gain is:

$$A_v = 192 \times 10^{-6}\left(\frac{6 \times 10^6 \times 10 \times 10^6}{6 \times 10^6 + 10 \times 10^6}\right) = 720$$

Q10 The gain of the single-ended common-emitter stage is:

$$A_v = g_m R_{o1}//R_{o2}$$

The reference current through Q1 is given by:

$$I_{C2} = \frac{(20 - 0.7)}{39\,\text{k}} = 0.495\,\text{mA}$$

and

$$I_{C1} = \frac{I_{C2}}{5} \quad \text{because of the ratio of the peripheries}$$

and

$$g_m = \frac{I_{C2}}{0.026}$$

The output resistance of each transistor is

$$R_{out} = \frac{\textit{Early voltage}}{I_C} = \frac{50}{0.099 \times 10^{-3}}$$

$$= 252\,\text{k}\Omega$$

and the gain is

$$A_v = g_m 252\,\text{k}//252\,\text{k}$$

Q11 The gain for the input differential stage can be approximted by:

$$A_v = 20\log\left[\frac{1}{0.026}\left(\frac{100 \times 60}{100 + 60}\right)\right] = 63\,\text{dB}$$

and the current in the Widlar current mirror is:

$$I_{C5} = \frac{0.026}{4\,\text{k}}\ln\left(\frac{(30\,\text{V} - 0.7\,\text{V})/4\,\text{k}\Omega}{I_{C5}}\right)$$

Solve for I_{C5} to obtain a value of approximately 23 μA, which should be compared with the value observed in the output from the simulator.

Q13 The compensating capacitor placed between the collector of Q11 and the base of Q7 is reflected to the input of Q7 by Miller's theorem as $C \times A_v$, where A_v is the gain of the common-emitter amplifier Q10. The output resistance of the differential input stage together with this effective capacitance act as a low-pass filter with a corner frequency $1/2\pi R_{out} C_{Miller}$. Observe the open-loop frequency response of the complex amplifier and compare the corner frequency from the simulation with the value predicted by using the Miller capacitance.

Chapter 14 The MOS capacitor

Q1 For an ideal surface:

$$V_{inv} = \frac{\sqrt{2\varepsilon_o \varepsilon_{si} q N_a 2\phi_f}}{C_{ox}} + 2\phi_f$$

where

$$C_{ox} = \frac{\varepsilon_o \varepsilon_{ox}}{t_{ox}} = \frac{8.85 \times 10^{-14} \times 3.9}{80 \times 10^{-7}}$$

$$= 4.31 \times 10^{-8}\ \mathrm{F/cm^2}$$

and

$$\phi_f = 0.026 \ln \frac{N_a}{n_i} = 0.026 \ln \frac{5 \times 10^{16}}{1.45 \times 10^{10}}$$

$$= 0.391\ \mathrm{V}$$

Then

$$V_{inv} = \frac{\sqrt{2 \times 8.85 \times 10^{-14} \times 11.7 \times 1.6 \times 10^{-19} \times 5 \times 10^{16} \times 2 \times 0.391}}{4.31 \times 10^{-8}} + 2 \times 0.391$$

$$V_{inv} = 2.64 + 0.782 = 3.4\ \mathrm{V}$$

Q2 If V_{gate} is equal to $2V_{inv}$ then the additional charge to balance the equation gives:

$$2V_{inv} = \frac{\sqrt{2\varepsilon_o \varepsilon_{si} q N_a 2\phi_f}}{C_{ox}} + 2\phi_f + \frac{Q_{inv}}{C_{ox}}$$

where Q_{inv} is the accumulated charge in the inversion layer. Q_{inv}/C_{ox} must be 3.4 V. Then Q_{inv} is $C_{ox}3.34/q$.

Q3 The inversion voltage is given by:

$$V_{inv} = -0.6 + \phi_f + \frac{Q_B}{C_{ox}} + \frac{Q_{SS}}{C_{ox}}$$

Proceed as shown in Q1 above to determine Q_B, C_{ox} and ϕ_f for the two different oxide thicknesses. NB: this sort of calculation is necessary for MOS circuits to ensure that channels are only formed between source and drain electrodes and not beneath the field oxides.

Q4 Proceed as for Q3, but with different values of Q_{SS}. NB: this calculation identifies the difference between ⟨111⟩ and ⟨100⟩ silicon.

Much more accurate values of threshold voltage can be obtained with a process simulator such as SUPREM.

Chapter 15 The MOS transistor

Q1 The threshold voltage is given by:

$$V_T = -\phi_{MS} + 2\phi_f + \frac{Q_B}{C_{ox}} - \frac{Q_{SS}}{C_{ox}}$$

Solve as for V_{inv} in the preceding chapter on the MOS capacitor. The final equation should have values equal to:

$$V_T = -0.6 + 0.66 + 0.784 - 0.186 = 0.66\ \mathrm{V}$$

For part (b) the equation is:

$$V_T = V_{T(0)} + \frac{Q_I}{C_{ox}}$$

where Q_I is the implanted charge, and $V_{T(0)}$ is the unimplanted threshold voltage for zero substrate bias. Determine Q_I and then calculate the number of ions as $N_I = Q_I/q$.

Q2 From the equation for the drain current I_D determine β as:

$$\beta = \frac{\varepsilon_o \varepsilon_{ox}}{t_{ox}} \mu \frac{W}{L} = 2.5 \times 10^{-5}$$

Solve for I_D as:

$$I_D = \frac{2.5 \times 10^{-5}}{2}(5 - 0.66) = 54\ \mu\mathrm{A}$$

Q3 The g_m is given by:

$$g_m = \frac{\delta I_D}{\delta V_{GS}} = \frac{\delta}{\delta V_{GS}} \left\{ K \frac{W}{L} \left[\frac{(V_{GS} - V_T)^2}{2} \right] \right\}$$

$$g_m = K \frac{W}{L} (V_{GS} - V_T)$$

Solve to give $g_m = 460\ \mu\mathrm{mho}$.

The maximum frequency of operation is given by:

$$f_{max} = \frac{g_m}{2\pi C_g}$$

where

$$C_g = \mathit{area} \times C_{ox} = W \times L \times C_{ox}$$

$$\approx 3.88 \times 10^{-14}\ \mathrm{F}$$

and f_{max} is approximately 2 GHz.

Chapter 16 The MOS inverter

Q1 For parts (a) and (b) follow the solution shown for Q1 in Chapter 15, that is, for an n-channel device the threshold voltage is:

$$V_T = -0.6 + 2\phi_f + \frac{Q_B}{C_{ox}} - \frac{Q_{ss}}{C_{ox}}$$

and for the depletion device

$$-4 = V_{T(0)} - \frac{Q_I}{C_{ox}}$$

For part (c) the body effect factor is given by:

$$\gamma = \frac{1}{C_{ox}}\sqrt{2\varepsilon_o\varepsilon_{si}qN_a}$$

Solve to obtain a value of $\gamma \approx 0.42$.

For part (d) the effective voltage with a substrate bias of 5 V is:

$$V_T^* = 0.225 + \gamma(V_{sub})^{1/2}$$

Q2 The current in the load device is given by:

$$I_D = \frac{\varepsilon_o\varepsilon_{si}}{2t_{ox}}\mu_n\frac{W}{L}(V_{GS} - V_T)^2$$

Solve for $I_D = 100\,\mu$A and for $V_{GS} = 0$ V (gate shorted to source for depletion load).

For part (b)

$$\left.\frac{W}{L}\right|_{load} = 0.5, \text{ that is, } W = \frac{L}{2}$$

For the driver

$$\left.\frac{W}{L}\right|_{driver} = 4\left.\frac{W}{L}\right|_{load}$$

Q3 Calculate β and the channel resistance of the driver as:

$$\beta = \frac{8.85 \times 10^{-14} \times 3.9}{60 \times 10^{-7}} \times 580 \times \frac{W}{L}$$

$$= 3.34 \times 10^{-5}\frac{W}{L}$$

and the channel resistance is:

$$r_{DS}|_{driver} = \frac{1}{3.34 \times 10^{-5}(5 - 0.2 \times 5)}$$

$$\approx 7.4\,\text{k}\Omega \text{ if } W/L = 1$$

The load resistance is $4 \times r_{DS}|_{driver} = 29.6\,\text{k}\Omega$. The maximum switching speed is given by:

$$f_{max} = \frac{1}{t_{ON} + t_{OFF}}$$

where $t_{ON} \approx 2.2r_{DS}|_{driver}C_{load}$ and $t_{OFF} \approx 2.2r_{DS}|_{load}C_{load}$. Solve for f_{max}.

For part (b) $t_{OFF} = t_{ON}$ and thus $f_{max} = 1/(5t_{on})$, but $t_{ON} = 2.2r_{DS}C_{load}$ and $r_{DS} = 1/(\beta_D 0.8V_{DD})$. Thus solve for f_{max}.

For part (c) the load capacitance is increased by a factor of 10. Therefore, in order to maintain the frequency the channel resistance must be decreased by a factor of 10. The largest time constant is due to the load device. Thus if $r_{DS}|_{load}$ is reduced by 10 to 3 kΩ, then the channel resistance for the driver in order to maintain the 4:1 ratio must be 750 Ω. Thus

$$t_{ON} = 2.2 \times 750 \times 1 \times 10^{-12} = 1.65\,\text{ns}$$

and

$$t_{OFF} = 4t_{ON} = 6.6\,\text{ns}$$

and

$$f_{max} = \frac{1}{1.65\,\text{ns} + 6.6\,\text{ns}} = 121\text{ MHz}$$

Thus if W/L for the driver was originally unity, then for the modified device with a channel resistance of 750 Ω, the width (W_D) must be $10 \times L_D$. For the load device the width (W_L) is $2.5 \times L_L$.

For part (d) the original power dissipation is:

$$P_{diss} = \frac{5^2}{7.4\,\text{k} + 29.6\,\text{k}} = 0.67\,\text{mW}$$

and for the modified circuit the dissipation is:

$$P_{diss} = \frac{5^2}{750 + 3000} = 6.6\,\text{mW}$$

Q4 The threshold voltages for the n-channel and p-channel devices are:

$$V_T = \phi_{ms} + 2\phi_f + \frac{Q_B}{C_{ox}} - \frac{Q_{ss}}{C_{ox}} \quad \text{n-channel}$$

$$V_T = \phi_{ms} + 2\phi_f + \frac{Q_B}{C_{ox}} - \frac{Q_{ss}}{C_{ox}} \quad \text{p-channel}$$

Use the data provided to obtain:

$V_T = -0.6 + 0.58 + 0.201 - 0.046$ n-channel

$V_T = 0.85 - 0.70 - 0.318 - 0.046$ p-channel

For part (b) the channel resistance is:

$$r_{DS} = \frac{1}{\beta(V_{DD} - V_{TE})} \approx \frac{1}{\beta V_{DD}} \quad \text{if } V_{TE} \ll V_{DD}$$

Solve to give $r_{DS}|_{\text{n-channel}} = 5\,\text{k}\Omega$ and $r_{DS}|_{\text{p-channel}} = 12.6\,\text{k}\Omega$. Then if $C_L = 1$ pF

$$t_1 + t_2 = 2.2 \times 1 \times 10^{-12} \times (5\,\text{k} + 12.6\,\text{k})$$
$$= 38.7\,\text{ns}$$

and the maximum frequency is 1/38.7 ns or approximately 26 MHz.

For part (c) the asymmetry in the waveform is caused by the different values of channel resistance for the two devices because of the different mobilities. The symmetry can be improved by changing the W/L ratio for the p-channel device so that it equals 580/230.

Chapter 17 MOS circuits

Q4 The number of stages is given by:

$$N = 4 = \frac{\ln Y}{\ln f} = \frac{\ln 5000}{\ln f}$$

and

$$f = 8.41$$

Then the total delay is

$$t_{total} = 4 \times 8.41 \times 2 \times \tau_d = 67\tau_d$$

Q5 For minimum delay $f = 2.718$ and the number of stages is:

$$N = \frac{\ln 5000}{\ln 2.718} = 8.52$$

and the total delay is

$$t_{total} = 9 \times 2.718 \times 2 \times \tau_d = 49\tau_d$$

Index